CELL AGING

CELL BIOLOGY RESEARCH PROGRESS

Additional books in this series can be found on Nova's website
under the Series tab.

Additional E-books in this series can be found on Nova's website
under the E-book tab.

CELL BIOLOGY RESEARCH PROGRESS

CELL AGING

JACK W. PERLOFT

AND

ALEXANDER H. WONG

EDITORS

Nova Science Publishers, Inc.

New York

Library of Congress Cataloging-in-Publication Data

Cell aging / editor, Jack W. Perloft and Alexander H. Wong.
    p. cm.
 Includes index.
 ISBN 978-1-61324-369-5 (hardcover)
 1. Cells--Aging. I. Perloft, Jack W. II. Wong, Alexander H.
 QP86.C427 2011
 571.6--dc22
                    2011011461

Published by Nova Science Publishers, Inc. †New York

CONTENTS

PREFACE

This book studies the cell aging of individual animals as observed by light and electron microscopic radioautography. Also discussed in this compilation is sarcopenia, which is the age-related loss of skeletal muscle mass, as well as recent advances in basic molecular mechanisms that underlie aging with findings that are shedding light on the pathogenesis of Alzheimer's disease. The authors also study Akt/protein kinase B (PKB), which plays a prominent role in the regulation of cellular homeostasis including cell survival, cell growth and gene expression; and understanding the process and mechanisms involved in erythrocyte senescence.

Chapter 1 – The term "Cell Aging" initially means how the cells change due to their aging. It contains 2 meanings, one how a cell changes when it is isolated from original animals or plants such as in vitro cells in cell culture, while the other means how all the cells of an animal or a plant changes in vivo due to the aging of the individual animal or plant.

The author has been studying the latter changes from the viewpoint of the cell nutrients, the precursors for the macromolecular synthesissuch as DNA, RNA, proteins, glucides and lipids, which are incorporated and synthesized intovarious cells of individual animals. Therefore, this article deals with only the cell aging of animal cells in vivo, how the metabolism, i.e., incorporations and syntheses of respective nutrientprecursors in various kinds of cells change due to the aging of individual experimental animals such as mice and rats by means of microscopic radioautography to localize the RI-labeled precursors. The incorporations and syntheses of various precursors for macromolecules such as DNA, RNA, proteins, glucides, lipids and others in various kinds of cells of various organ in respective organ systems such as skeletal, muscular, circulatory, digestive, respiratory, urinary, reproductive, endocrine, nervous and sensory systems are reviewed referring many original papers already published from the authors' laboratory during these 50 years since the late 20C to the early 21C.

Chapter 2 – The world's elderly population is expanding rapidly, and we are now faced with the significant challenge of maintaining or improving physical activity, independence, and quality of life in the elderly. Sarcopenia, the age-related loss of skeletal muscle mass, is characterized by a deterioration of muscle quantity and quality leading to a gradual slowing of movement, a decline in strength and power, increased risk of fall-related injury, and often, frailty. Since sarcopenia is largely attributed to various molecular mediators affecting fiber size, mitochondrial homeostatis, and apoptosis, the mechanisms responsible for these deleterious changes present numerous therapeutic targets for drug discovery. Muscle loss has been linked with several proteolytic systems, including the calpain, ubiquitin-proteasome and lysosome-autophagy systems. In addition, reseachers have indicated defects of Akt-mTOR (mammalian target of rapamycin) and RhoA-SRF (Serum response factor) signaling in sarcopenic muscle. In this chapter, the authors summarize the current understanding of the mechanisms underlying the progressive loss of muscle mass and provide an update on therapeutic strategies (resistance training, myostatin inhibition, amino acids supplementation, calorie restriction, etc) for counteracting sarcopenia. Resistance training combined with amino acid-containing supplements would be the best way to prevent age-related muscle wasting and weakness. Myostatin inhibition seems to be the most interesting strategy for attenuating sarcopenia as well as muscular dystrophy in the near future. In contrast, muscle loss with age may not be influenced positively by treatment with a proteasome inhibitor or antioxidant.

Chapter 3 – Because of the increased lifespan of our population, problems linked to age-associated disabilities are becoming more important and are absorbing a growing fraction of the costs associated with health care. In particular, the disability for cognitive loss and dementia combined is currently the second most expensive among medical conditions affecting the aging population and is expected to become the most expensive in the next decade. In fact, aging is accompanied by cognitive loss and reduced high-order brain functions in a large segment of the population. Aging is also the most important risk factor for sporadic Alzheimer's disease, which is one of the most common forms of dementia in the world. Despite this important role in disease pathogenesis and morbidity, we have yet to fully dissect the aging process of the brain at a molecular level. However, the powerful tools of molecular biology applied to model organisms have recently revealed that the natural process of aging is driven by specific signaling pathways, which in turn regulate molecular events that are essential for the general physiology of the cell. Careful dissection of these events is now providing new information as well as possible ways of intervention for age-associated diseases. In this chapter the authors seek to integrate recent advances in understanding basic molecular mechanisms that underlie aging with findings that are shedding light on the pathogenesis and development of age-associated cognitive loss and Alzheimer's disease neuropathology.

Chapter 4 – The increased costs associated with an ever-growing aged population worldwide are expected to pose a significant burden on health care resources. From a biological standpoint, aging is an accelerated deteriorative process in tissue structure and function that is associated with higher morbidity and mortality. The Akt / protein kinase B (PKB) is a family of serine/threonine protein kinases, which play prominent roles in regulation of cellular homeostasis including cell survival, cell growth, gene expression, apoptosis, protein synthesis, and energy metabolism. It has been demonstrated that diminished Akt activity is associated with dysregulation of cellular metabolism and cell death while Akt over-activation has been linked to inappropriate cell growth and proliferation. It is likely that age-related changes in tissue structure (such as atrophy and hypertrophy) and function are related to alterations in Akt expression and Akt-dependent signaling. Although the regulation of Akt function has been well characterized *in vitro*, much less is known regarding the function of Akt*in vivo*. In this article the authors examine how Akt and Akt-dependent signaling may be regulated in aged cells/tissues, and how age-related alterations in Akt signaling may play a role in influencing cellular metabolism and the response of cells/tissues to environmental factors. Finally, the authors attempt, where possible, to highlight how such changes may have clinical relevance while also commenting on potential new directions of inquiry.

Chapter 5 – Multipotential mesenchymal stem cells (MSC) are present as a rare subpopulation within any type of stroma in the body of higher organisms. They are deliberately considered of being capable to terminally differentiating into a broad variety of tissue cell types. When required, MSC are activated accordingly and deployed for tissue remodeling and regeneration after injury. In this vein, MSC are not active only during early development and growth, yet also in repair and regeneration throughout adulthood up to highly advanced ages. Besides replenishing mesenchymal tissues, MSC also modulate hematopoiesis as well as immune response.

During aging, tissue function declines, as does hematopoietic activity, or immune function. Consequences are bone loss, fat redistribution, cartilage defects, autoimmune diseases, and increased incidence of sarcomas. While numerous studies have unveiled the multipotentiality of these cells and demonstrated their capacity to regenerate and repair tissues, only recently the concept of stem cells to encounter aging has been acknowledged as an increasingly important field to appropriately tackle these unanswered questions.

Scientists are just at the beginning to conceive the complexity and importance of mechanisms that undermine and attenuate MSC function as these cells age. Notably, the authors and others have found that during aging, overall MSC number and their proliferative potential are hardly ever affected. Yet in aging donors, the authors could reveal distinct MSC subpopulations which express marked levels of vascular cell-adhesion molecule 1 (VCAM1/CD106) or leptin receptor (LEPR/CD295). MSC show such changes as a response to extrinsic factors such as decreased oxygen supply, extracellular matrix aging, e.g. collagen crosslinking, or when getting in close contact with infiltrating lymphocytes which establish a chronic proinflammatory milieu (inflamm-aging), as well as in respect of age-associated intrinsic changes with regard to altered DNA methylation, and the shifted expression of miRNAs.

Herein, the authors review the current perception regarding possible causes of MSC aging, and consequences thereof.

Chapter 6 – Mature red blood cells (RBCs) lack protein synthesis and are unable to restore inactivated enzymes or damaged cytoskeleton and membrane proteins. An oxidation breakdown of band 3 is probably part of the mechanism leading to the generation of a senescent cell antigen (SCA). It serves as a specific signal for the clearance of these cells by inducing the binding of autologous IgG and phagocytosis.

Whole blood samples from volunteer donors were processed. Senescent (Se) RBCs and Young (Y) RBCs were obtained by self-formed Percoll gradients. The separation of both populations was demonstrated by statistically significant changes in hematological parameters and creatine concentration. The antioxidant response in RBCs of different ages was studied. Activities of glucose-6-phosphate dehydrogenase (G6PD), soluble NADH-cytochrome b5 reductase (b5Rs) and membrane-bound b5R (b5Rm) were determined spectrophotometrically. The G6PD and b5Rm activities in SeRBCs were significantly lower than that observed in YRBCs. The decline in those activities would indicate a decrease in the antioxidant response associated to RBC aging.

Membrane proteins modifications in RBCs of different ages were assessed. Membrane proteins were analyzed by SDS-PAGE, band 3 by immunoblotting, and protein oxidation by measuring the carbonyl groups. Densitometric analysis showed no differences between mean percentage values obtained from the major bands of SeRBCs and YRBCs membrane proteins. On the contrary an increase in band 3 and its degradation products were found by immunoblotting in SeRBCs. A higher protein oxidation level was also encountered in this population. These results provide experimental evidence about protein modifications occurring during the RBC lifespan.

Then, considering that the accumulation of autologous IgG on RBCs membrane provides a direct mechanism for the removal of SeRBCs, the IgG content of intact RBCs was measured using an enzyme linked anti-immunoglobulin test. In addition, the presence of bound IgG was observed by confocal microscopy. It was shown that the amount of IgG bound to SeRBCs was significantly higher than that observed for YRBCs.

The interaction between different RBCs populations (SeRBCs, YRBCs and desialiniysed RBCs) and peripheral blood monocytes was further investigated through a functional assay. The increase observed in the percentage of erythrophagocytosis with SeRBCs confirmed the involvement of autologous IgG in the selective removal of erythrocytes. Also, the percentage of monocytes with phagocytosed desialiniysed RBCs was higher than that obtained with YRBCs. This finding suggests that a decrease in sialic acid content of RBCs may be involved in the physiological erythrophagocytosis.

In addition, cells of different ages in whole blood were characterized using light scatter, binding of autologous IgG and externalization of phosphatidylserine measurements. Dot-plot analysis based on the forward scatter versus side scatter parameters showed two RBC populations of different sizes and density. RBCs were further incubated with FITC-conjugated mouse anti-human IgG o PE-annexin-V. Binding of IgG to RBCs was analyzed by mean fluorescence intensity. The percentage of IgG positive cells was significantly higher in SeRBCs. The fraction of annexin-V positive RBCs was also larger in SeRBCs. These results indicate that flow cytometry permits differentiating erythrocyte populations of different ages. This methodology could be an alternative tool to study erythrocyte aging.

Taken together, these findings will contribute to a better understanding of the process and mechanisms involved in the erythrocyte senescence.

In: Cell Aging
Editors: Jack W. Perloft and Alexander H. Wong

ISBN: 978-1-61324-369-5
2012 Nova Science Publishers, Inc.

Chapter 1

CELL AGING OF INDIVIDUAL ANIMALS AS OBSERVED BY LIGHT AND ELECTRON MICROSCOPIC RADIOAUTOGRAPHY

*Tetsuji Nagata**

Professor Emeritus, Department of Anatomy and Cell Biology, Shinshu University
School of Medicine, Matsumoto 390-8621, Japan, and
Professor of Anatomy, Shinshu Institute of Alternative Medicine and Welfare,
Nagano 380-0816, Japan

ABSTRACT

The term "Cell Aging" initially means how the cells change due to their aging. It contains 2 meanings, one how a cell changes when it is isolated from original animals or plants such as in vitro cells in cell culture, while the other means how all the cells of an animal or a plant changes in vivo due to the aging of the individual animal or plant.

I have been studying the latter changes from the viewpoint of the cell nutrients, the precursors for the macromolecular synthesissuch as DNA, RNA, proteins, glucides and lipids, which are incorporated and synthesized intovarious cells of individual animals. Therefore, this article deals with only the cell aging of animal cells in vivo, how the metabolism, i.e., incorporations and syntheses of respective nutrientprecursors in various kinds of cells change due to the aging of individual experimental animals such as mice and rats by means of microscopic radioautography to localize the RI-labeled precursors. The incorporations and syntheses of various precursors for macromolecules such as DNA, RNA, proteins, glucides, lipids and others in various kinds of cells of various organ in respective organ systems such as skeletal, muscular, circulatory, digestive, respiratory, urinary, reproductive, endocrine, nervous and sensory systems are reviewed referring many original papers already published from our laboratory during these 50 years since the late 20C to the early 21C.

---

* E-mail: nagatas@po.cnet.ne.jp

1. INTRODUCTION

The term "Cell Aging" initially means how the cells change due to their aging. It contains 2 meanings, one how a cell changes when it is isolated from in vivo original animals or plants such as in vitro cells in cell culture, while the other means how all the cells of an animal or a plant change in vivo due to the aging of the individual animal or plant. I had first studied the meaning of cell aging many years ago (more than 50 years) how a cell changed when it was isolated from original experimental animals such as mice and rats by cell culture (Nagata 1956, 1957a,b), and then moved to the study on the latter cell aging, i.e., how all the cells of an experimental animal change in vivo due to the aging of the individual prenatal and postnatal animal (Nagata 1959, 1962, Nagata and Momoze 1959, Nagata et al. 1960a,b).

Recently, I have been studying the aging changes from the viewpoint of the cell nutrients which were incorporated and synthesized into various cells in individual animals during their aging (Nagata 2010c). Therefore, this article deals with only the cell aging of animal cells in vivo, how the metabolism, i. e., incorporations and syntheses of respective nutrients, the macromolecular precursors, in various kinds of cells change due to the aging of individual experimental animals such as mice and rats by means of microscopic radioautography. The incorporations and syntheses of various nutrients such as DNA, RNA, proteins, glucides, lipids and others in various kinds of cells of various organ in respective organ systems such as skeletal, muscular, circulatory, digestive, respiratory, urinary, reproductive, endocrine, nervous and sensory systems should be reviewed referring many original papers already published from our laboratory.

1.1. Radioautography

In order to observe the localizations of the incorporations and syntheses of various nutrients synthesizing macromolecules in the bodies such as DNA, RNA, proteins, glucides and lipids in various kinds of cells of various organ in respective organ systems such as skeletal, muscular, circulatory, digestive, respiratory, urinary, reproductive, endocrine, nervous and sensory systems, we employed the specific techniques developed in our laboratory during these 50 years (Nagata 2002). The technique is designated as radioautography using RI-labeled compounds. To demonstrate the localizations of macromolecular synthesis by using such RI-labeled precursors as ^{3}H-thymidine for DNA, ^{3}H-uridine for RNA, ^{3}H-leucine for protein, ^{3}H-glucosamine or ^{35}SO$_4$ for glucides and ^{3}H-glycerol for lipids are divided into macroscopic radioautography and microscopic radioautography. The techniques employ both the physical techniques using RI-labeled compounds and the histochemical techniques treating tissue sections by coating sections containing RI-labeled precursors with photographic emulsions and processing for exposure and development. Such techniaques can demonstrate both the soluble compounds diffusible in the cells and tissues and the insoluble compounds bound to the macromolecules (Nagata 1972b). As the results, specimens prepared for EM RAG are very thick and should be observed with high voltage electron microscopes in order to obtain better transmittance and resolution (Nagata 2001a,b). Such radioautographic techniques in details should be referred to other literature (Nagata 2002). On the other hand, the systematic results obtained by radioautography should be designated as radioauographology, or science of radioautography (Nagata 1998b, 1999e, 2000e). This article deals with the results dealing with the radioautographic changes of individual cell by aging that should be included in radioautographology.

1.2. Macromolecular Synthesis

The human body as well as the bodies of any experimental animals such as mice and rats consist of various macromolecules. They are classified into nucleic acids (both DNA and RNA), proteins, glucides and lipids, according to their chemical structures. These macromolecules can be demonstrated by specific histochemical staining for respective molecules such as Feulgen reaction (Feulgen and Rossenbeck 1924) which stains all the DNA contained in the cells. Each compounds of macromolecules such as DNA, RNA, proteins, glucides, lipids can be demonstrated by respective specific histochemical stainings (Pearse 1991) and such reactions can be quantified by microscpectrophotometry using specific wave-lengths demonstrating the total amount of respective compounds (Nagata 1972a). To the contrary, radioautography can only demonstrate the newly synthesized macromolecules such as synthetic DNA or RNA or proteins depending upon the RI-labeled precursors incorporated specifically into these macromolecules such as ^{3}H-thymidine into DNA or ^{3}H-uridine into RNA or ^{3}H-amino acid into protein (Nagata 2002).

Concerning to the newly synthesized macromolecules, the results of recent studies in our laboratory by the present author and co-workers should be reviewed in this article according to the classification of macromolecules as follows.

2. The DNA Synthesis

The DNA (deoxyribonucleic acid) contained in cells can be demonstrated either by morphological histochemical techniques staining tissue sections such as Feulgen reaction or by biochemical techniques homogenizing tissues and cells. To the contrary, the synthetic DNA or newly synthesized DNA but not all the DNA can be detected as macromolecular synthesis together with other macromolecules such as RNA or proteins in various organs of experimental animals by either morphological or biochemical procedures employing RI-labeled precursors. We have studied the sites of macromolecular synthesis in almost all the organs of mice during their aging from prenatal to postnatal development to senescence by means of microscopic radioautography, one of the morphological methods (Nagata 1992, 1994b,c,d, 1996a,b,c,d, 1997a). The results should be here described according to the order of organ systems in anatomy or histology.

2.1. The DNA Synthesis in the Skeletal System

The skeletal system of men and experimental animals consists of bones, joints and ligaments. We studied the DNA synthetic activities of the bones and joints of experimental animals (Kobayashi and Nagata 1994, Nagata 1998c).

2.1.1. The DNA Synthesis in the Bone

We studied the ossifications of salamander skeletons from hatching to senescence (Nagata 1998c). The fore-limbs (Figure 1A) and hind-limbs (Figure 1B) of salamanders were composed of skeletons consisting of bones and cartilages which were covered with skeletal muscles, connective tissues and epidermis consisting of stratified squamous epithelial cells in the outermost layer. The bones of juvenile salamanders at 4 weeks consisted of the hyaline cartilage (Figures. 1A, 1B). The hyaline cartilage consisted of spherical or polygonal cartilage cells or chondrocytes at the center. They were surrounded by rich interstitial ground substance which stained deep blue with toluidine blue staining. The spherical cartilage cells at the center of the bone changed their shapes to flattened shape under the perichondrium or free joint surfaces. Some of the nuclei of the chondrocytes were covered with silver grains when labeled with ^{3}H-thymidine (Figures. 1A, 1B). Mitotic figures were frequently seen in spherical cartilage cells in young animals. Examination of radioautograms at the young stages such as 4 weeks after hatching showed that many spherical cartilage cells and flattened cartilage cells were predominantly labeled. At 6 weeks after hatching, the size of bones enlarged and the number of cartilage cells increased. At this stage, however, the number of labeled cells in the cartilage cells in both fore-limbs (Figure 1C) and hind-limbs (Figure 1D) decreased as compared with the previous stage (Figures. 1A, 1B). The size of bones in juvenile animals at 8, 9, 10, and 11 weeks enlarged gradually (Figure 1E, 1F). Radioautograms at these stages showed that the number of the labeled cells remarkably reduced as compared with those of 4 and 6 weeks. In the adult salamanders at 8 months up to 12 months, the bones showed complete mature structure and examination of radioautograms revealed that the number of labeled cells reached almost zero. No difference was found on the morphology and labeling between the fore-limbs and hind-limbs at any stage. The labeling indices of respective cell types changed with aging as expressed by mean in each group. The labeling index of the cartilage cells was lower than the epithelial cells. The peak of the labeling index of the cartilage cells in both fore-limbs and hind-limbs was found about 15-18% at 4 weeks after hatching (Figure 2). The labeling index of the cartilage cells in both limbs at 6 weeks rapidly decreased to about 4-6%, then increased at 8 weeks to about 7-8% and finally decreased to 2-3% gradually from 8 weeks to 9 weeks with aging and fell down to 0-1% at 10 weeks. The labeling index of cartilage cells from 10 weeks to 12 months kept very low around 0-1% (Figure 2). Thus, the cartilages and bones of fore-limbs and hind-limbs of salamanders are demonstrated to complete the development by 10 weeks after hatching.

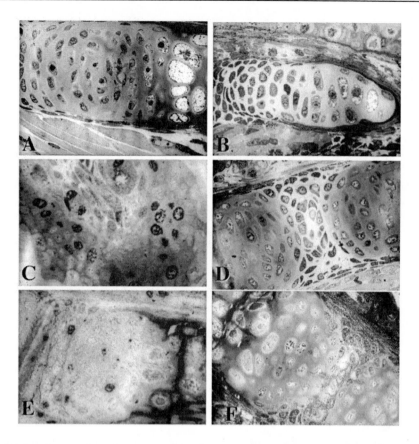

Figure1A. Light microscopic radioautogram of the bone of a fore-limb of a salamander at 4 weeks after hatching. Many cartilage cells (arrows) are labeled with silver grains due to ^{3}H-thymidine. Magnification.x 1200.

Figure1B. Light microscopic radioautogram of the bone of a hind-limb of a salamander at 4 weeks after hatching. Many cartilage cells (arrows) are labeled with silver grains due to to ^{3}H-thymidine. Magnification.x 1200.

Figure 1C. Light microscopic radioautogram of the bone of a fore-limb of a salamander at 6 weeks after hatching. Only a few cartilage cells (arrow) are labeled. The numbers of labeled cells are fewer than the bone of a younger salamander at 4 weeks after hatching (Figure 1A). Magnification x 1200.

Figure 1D. Light microscopic radioautogram of the bone of a hind-limb of a salamander at 6 weeks after hatching. Only a few cartilage cells (arrow) are labeled. Magnification x 1200.

Figure 1E. Light microscopic radioautogram of the bone of a fore-limb of a salamander at 8 weeks after hatching. Only a few cartilage cells (arrow) are labeled. Magnification x 1200.

Figure 1F. Light microscopic radioautogram of the bone of a hind-limb of a salamander at 8 weeks after hatching. Only a few cartilage cells (arrow) are labeled. Magnification x 1200.

Figure 1. Light microscopic radioautograms of the bones of either fore-limbs or hind-limbs of salamanders at various ages from 4 weeks to weeks after hatching, injected with 3H-thymidine, fixed and processed for radioautography. Some of the cartilage cells (arrows) are labeled with silver grains due to ^{3}H-thymidine incorporation demonstrating DNA synthesis. From Nagata, T.: Bulletin Shinshu Inst. Alternat. Med. Vol. 2, p. 53, 2006, Nagano, Japan

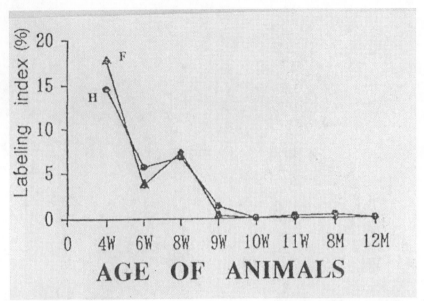

From Nagata, T.: Bulletin Shinshu Inst. Alternat. Med. Vol. 2, p. 54, 2006, Nagano, Japan.

Figure 2.Transitional curves of the labeling indices of the cartilage cells in the bones of the fore-limbs and the hind-limbs of salamanders labeled with ^{3}H-thymidine at various ages from 4 weeks to 60 months (5 years) after hatching. Mean ± S.D.

2.1.2. The DNA Synthesis in the Joint

The joints of an experimental animal such as mouse or human beings are consisted of either 2 or 3 bones and the synovial membranes covering the ends of the bones. The synovial membranes are composed of the collagenous fibers interspersed with the synovial cells which are fibroblasts and lining cells. We studied macromolecular synthesis, both DNA and RNA syntheses of the synovial cells of the joints surgically obtained from 15 elderly human patients of both sexes aged from 50 to 70, suffering from rheumatoid arthritis (Kobayashi and Nagata 1994). Both the normal and rheumatoid cells were cultured and labeled in vitro with ^{3}H-thymidine or ^{3}H-uridine and radioautographed. DNA synthetic cells labeled with silver grains were observed by LMRAG in both normal and rheumatoid cells. As the results, some labeled synovial cells with ^{3}H-thymidine were found. However, no significant difference was observed between the labeling indices of normal and rheumatoid cells labeled with ^{3}H-thymidine. From the results, it was concluded that the synovial cells synthesized DNA in both normal and rheumatoid conditions. However, the quantities of these macromolecules synthesized in these synovial cells varied in respective individuals and no significant difference was found between the labeling indices and grain counts in both normal and rheumatoid cells (Kobayashi and Nagata 1994).

2.2. The DNA Synthesis in the Muscular System

The muscular system consists of various skeletal muscles amounting to around 200 in number in men and less in experimental animals such as rats and mice. We studied the aging changes of DNA synthesis in the intercostal muscles of aging ddY mice from prenatal day 13 through postnatal 24 months by ^{3}H-thymidine RAG (Hayashi et al. 1993). Many nuclei were labeled in myotubes at embryonic day 13-17 (Figure 3C), then the number of labeled nuclei decreased to embryonic day 18-19 (Figure 3D), and less after birth. The labeling indices revealed chronological changes, showing a peak at embryonic day 13 and decreasing gradually to 0% at 3 months after birth (Figure 4). We classified the graduation of the embryonic muscle development into 5 stages. Among them, the labeling index (LI) at stage I was the highest, while the LI at stage II was significantly lower than stage I, the LI at stage IV was significantly higher than stage II, and the LI at stage V was significantly lower than stage IV (Figure 4). These changes accorded well with the primary and secondary myotube formation during the embryonic muscle development. We also studied the DNA synthesis of rat thigh muscles during the muscle regeneration after injury (Sakai et al. 1977). When the skeletal muscles, i. e., the diaphragma, the rectus abdominis muscles and the gastrocnemius femoris muscle of adult Wistar rats were mechanically injured and labeled with ^{3}H-thymidine, satellite cells were labeled during their regeneration. The satellite cells in the muscles of dystrophy chickens and normal control chickens were also labeled with ^{3}H-thymidine, demonstrating DNA synthesis (Oguchi and Nagata 1980, 1981), which was later described in details in the review (Nagata 2002). Briefly, 2 groups of chickens, 4 dystrophy chickens and normal control chickens of both sexes aged 1 day and 21 days after hatching were used. All the animals received 4 times every 6 hrs intraperitoneal injections of ^{3}H-thymidine successively and sacrificed. The superficial pectoral muscles were taken out, fixed, embedded in Epon and processed for LM and EMRAG. The results demonstrated that many nuclei of the satellite cells in all the experimental groups were labeled but none of the nuclei in the muscle fibers were labeled. The labeling indices of normal chickens at 1 day and 21 days were 4.59 and 3.86%, respectively. These results showed that the LI decreased after hatching.

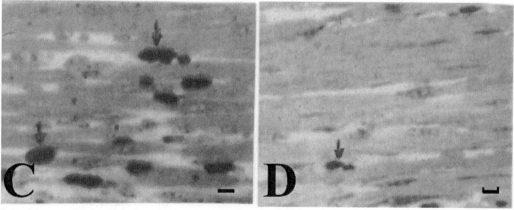

From Nagata, T.: Special Cytochemistry in Cell Biology, In, Internat. Rev. Cytol. Vol. 211, No. 1, p. 62, 2001, Academic Press, San Diego, USA, London, UK.

Figure 3. Light microscopic radioautograms of the skeletal muscle cells in the myotubes labeled with ^{3}H-thymidine at embryonic day 13-17 (Figure 3C), then the number of labeled nuclei decreased to embryonic day 18-19 (Figure 3D), and less after birth. x260.

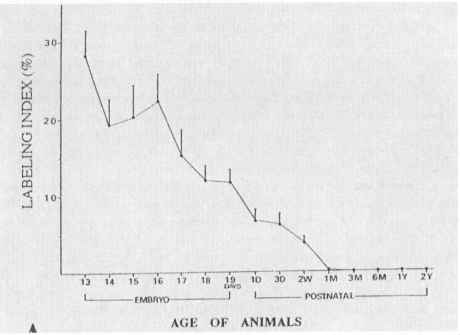

From Nagata, T.: Radioautographology, General and Special. In, Progr.Histochem.Cytochem.Vol. 37, No. 2, p. 114, 2002, Urban & Fischer, Jena, Germany.

Figure 4. The labeling indices of the muscular cells labeled with ^{3}H-thymidine revealed chronological changes, showing a peak at embryonic day 13 and decreasing gradually to 0% at 3 months after birth.

2.3. The DNA Synthesis in the Circulatory System

The circulatory system or cardiovascular system consists of the heart, the arteries, the veins, the capillaries, the blood, the lymphatic organs and the spleen. Among these cardiovascular organs, we studied the heart, the artery, some blood cells and the spleen.

2.3.1. The DNA Synthesis in the Heart

The nucleic acid synthesis, both DNA and RNA, in cultured fibroblasts from the hearts of chick embryos was studied by LMRAG (Nagata and Nawa 1966a,b). The fibroblasts of chick hearts in culture proliferated extensively and produced many binucleate cells. We compared the nucleic acid synthesis in mononucleate cells and binucleate cells in the heart fibroblasts. The incorporation of ^{3}H-thymidine into each nucleus of a binucleate cell was a little less than that of a mononucleate cell, but the total of the two nuclei of a binucleate cell almost twice as that of a mononucleate cell. The incorporation of ^{3}H-uridine in the two nuclei of a binucleate cell was almost twice as that of a mononucleate cell, while the incorporation of ^{3}H-uridine in the cytoplasm of a binucleate cell was not so much as twice as a mononucleate cell. From these results, it was concluded that the nucleic acid synthesis both DNA and RNA increased in binucleate cells than mononucleate cells of chick embryo heart fibroblasts (Nagata and Nawa 1966a,b).

2.3.2. The DNA Synthesis in the Artery

The blood vessels, both arteries and veins consist of 3 layers, from inside to outside, the tunica intima, the media and the adventitia. Those layers are formed with connective tissues and the smooth muscles. We studied the localization of anti-hypertensive drugs in the supramesenteric arteries of the spontaneous hypertensive rats (Suzuki et al. 1994). Two kinds of anti-hypertensive drugs, labeled with RI, ^{3}H-benidipine hydrochloride (Kyowa Hakko Kogyo Co., Shizuoka, Japan) and ^{3}H-nitrendipine (New England Nuclear, Boston, MA, USA) were used. Both intravenous administrations into rats and in vitro incubation for 10 to 30 min were employed. For light and electron microscopic radioautography, both the conventional wet-mounting radioautograms after chemical fixation for insoluble compounds and the dry-mounting radioautograms after cryo-fixation and freeze-substitution for soluble compounds were prepared. The silver grains due to the anti-hypertensive drugs were localized over the plasma membranes and the cytoplasm of the fibrocytes in the intima and the smooth muscle cells in the media, suggesting the pharmacological active sites. However, the localization of synthetic DNA was not studied.

2.3.3. The DNA Synthesis in the Blood Cells

The mature blood cells circulating in the blood vessels of mammals are classified into 3 types, the erythrocytes, the leukocytes and the blood platelets. Those mature cells are formed either in the lymphatic tissues in the lymphatic organs or the myeloid tissues in the bone marrow, where various immature cells, lymphoblasts, erythroblasts, myeloblasts, meylocytes, and megakaryocytes can be observed. Among these blood cells, we studied macromolecular synthesis and cytochemical localization in leukocytes, megakaryocytes and blood platelets. As for the granulocytes, normal rabbit granulocytes were shown by EMRAG and X-ray microanalysis to incorporate ^{35}SO$_4$ into the Golgi apparatus and to the granules demonstrating glucosaminoglycan synthesis (Murata et al. 1978, 1979).

On the other hand, the DNA, RNA and mucosubstance synthesis of mast cells from Wistar strain rats were studied by ^{3}H-thymidine, ^{3}H-uridine and ^{35}SO$_4$ radioautography, demonstrating incorporation changes of those normal mast cells from abnormal mastocytoma cells (Murata et al. 1977a). Mast cells are widely found distributing in the loose connective tissues of most mammals, as well as in the serous exudate in the peritoneal cavity as one of the free cells. We studied the fine structure and nucleic acid and mucosubstance syntheses of normal mast cells and Dunn and Potter's mastocytoma cells in mice and rats by electron microscopic radioautography (Murata et al. 1977b, 1979). As the results, some of the normal mast cells and mastocytoma cells incorporated ^{3}H-thymidine, ^{3}H-uridine and ^{35}SO$_4$, demonstrating DNA, RNA and mucosubstance syntheses. The incorporation of ^{3}H-thymidine was observed in the nuclei and mitochondria. The labeling index of ^{3}H-thymidine incorporation in the nuclei and mitochondria of normal mast cells was very low (0.37%) while that of mastocytomas cells was high (2-5%). These results suggest that the macromolecular synthesis such as nucleic acids (DNA, RNA) and mucosubstances were higher in tumor cells than normal cells.

2.3.4. The DNA Synthesis in the Spleen

The spleen is one of the blood cell forming organs and is composed of the lymphatic tissues. We studied ^{3}H-thymidine incorporation into the splenic cells of aging mice from newborn to adult and senescence in connection with the lysosomal acid phosphatase activity (Olea 1991, Olea and Nagata 1991, 1992a). The acid phosphatase activity as demonstrated by means of cerium substrate method was observed in the splenic tissues at various ages from postnatal day 1, week 1 and 2, month 1, 2 and 10. Electron dense deposits were localized in the lysosomes of macrophages, reticular cells and littoral cells in all the aging groups. The intensity of the reaction products as visually observed increased from day 1 to week 1, reaching the peak at 1 week, and decreased from week 2 to month 10 due to aging. The incorporation of ^{3}H-thymidine, on the other hand, demonstrating DNA synthesis, was mainly observed in the hematopoietic cells in the spleens from postnatal day 1 to month 10 animals (Olea and Nagata 1991, 1992a). The labeling index was the maximum at day 1 and decreased to week 1, 2, 4, 8 and 40. A correlation between DNA synthesis and AcPase activity was examined by comparing two cell populations in the cell cycle, the S-phase cells which were labeled with ^{3}H-thymidine and the non-S-phase cells or the interphase cells which were not labeled. It was demonstrated that the former showed an increase and decrease of much more AcPase activity with the aging while the latter less activity and no change. On the other hand, the number of labeled cells and the grain counts in the hematopoietic cells in the spleens labeled with ^{3}H-uridine, demonstrating RNA synthesis, from postnatal day 1 increased to 1 and 2 weeks, reaching the maximum, and decreased to 4, 8 and 40 weeks, different from the DNA synthesis (Olea and Nagata 1992b). These results demonstrated that AcPase activity, DNA and RNA synthetic activity changed due to aging.

2.4. The DNA Synthesis in the Digestive System

The digestive system consists of the digestive tract and the digestive glands. The digestive tract can be divided into several portions, the oral cavity, the pharynx, the esophagus, the stomach, the small and large intestines and the anus, while the digestive glands consist of the large glands such as the salivary glands, the liver and the pancreas and the small glands affiliated to the digestive tracts in the gastrointestinal walls such as the gastric glands, intestinal glands of Lieberkuehn or duodenal glands of Brunner. We have published many papers from our laboratory dealing with the macromolecular synthesis of respective digestive organs from the oral cavity to the gastrointestinal tracts and the digestive glands (Nagata 1992, 1993a,b, 1994a,b,c, 1995a,b, 1999a,b,c, 2002, Nagata et al. 1979, 1982a, 2000a, Chen et al. 1995). The outline of the results concerning to the synthetic DNA in the digestive organs should be here described in the order of systematic anatomy and special histology as follows.

2.4.1. The DNA Synthesis in the Oral Cavity

The oral cavity consists of the lip, tongue, tooth, and the salivary gland. The DNA synthesis of mucosal epithelia of the 2 upper and lower lips and the tongues as well as the 3 large salivary glands and many small glands of aging mice from fetal day 19 to postnatal 2 years were studied by LM and EMRAG labeled with ^{3}H-thymidine. The glucide and glycoprotein syntheses by ^{3}H-glucosamine and radiosulfate incorporations of the submandibular and sublingual glands of aging mice were also studied.

We first studied the DNA synthesis of the submandibular glands in 10 groups of aging mice at various ages from embryo to postnatal 2 years (Chen et al. 1995, Nagata et al. 2000a).

The submandibular gland of male mouse embryonic day 19 consisted with the glandular acini and duct system (Figure 5A). The duct system was composed of juxtaacinar cells (JA), intercalated duct cells (ICD) and striated duct cells (SD). Many labeled developing acinar cells (AC), JA and ICD cells were observed. At postnatal day 1 to 3 (Figure 5B), there was more JA cells and secretory granules than those of former stage. JA cells were cuboidal cells, characterized by small darkly stained granules in the supranuclear cytoplasm and by basophilic mitochondria mostly at the basal half of the cells. JA cells were present at the acinar-intercalated duct junction of the mouse submandibular gland. Many labeled AC, JA, ICD and SD cells were also observed by electron microscopy (Figure 5C). At postnatal 2 weeks to 3 months, developing immature acinar cells gradually matured to acinar cells, and JA cells increased and granular convoluted duct cells (GCT) appeared. At postnatal 6 months

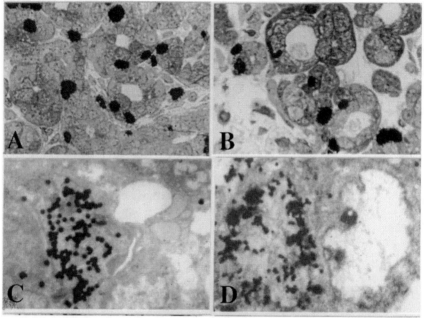

From Nagata, T. Radioautographology, General and Special.In, Progr.Histochem.Cytochem.Vol. 37, No. 2, p. 118, 2002, Urban & Fischer, Jena, Germany.

Figure 5A. LMRAG of the submandibular gland of male mouse embryonic day 19 labeled with ^{3}H-thymidine consisted with the glandular acini and duct system. The duct system was composed of juxtaacinar cells (JA), intercalated duct cells (ICD) and striated duct cells (ICD). Many labeled developing acinar cells (AC), JA and ICD cells were observed. x500.

Figure 5B. LMRAG of the submandibular gland at postnatal day 3, labeled with ^{3}H-thymidine. There were more JA cells and secretory granules than those of former stage (Figure 5A). x500.

Figure 5C. EMRAG of an ICD cell of a mouse at postnatal day 3, labeled with ^{3}H-thymidine observed by electron microscopy. Many silver grains are observed over the nucleus of an ICD. x10,000.

Figure 5D. EMRAG of the esophageal epithelial cells of a newborn mouse at postnatal day 1, labeled with ^{3}H-thymidine. Many silver grains are observed over one of the nuclei at left. x10,000.

Figure 5. (Continued).

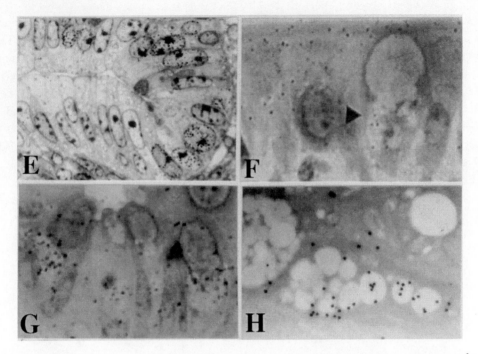

Figure 5E.LMRAG of the colonic epithelial cells of a mouse embryo at fetal day 19, labeled with ^{3}H-thymidine. Many silver grains are observed over the nuclei of several epithelial cells in the bottom of the crypt. x800.

Figure 5F. LMRAG of the ileum epithelial cells labeled with ^{3}H-glucosamine of an old mouse at postnatal 6 months. Many silver gains are localized over the Golgi region of the 3 goblet cells as well as over the cytoplasm of several absorptive columnar epithelial cells. x 1,000.

Figure 5G. LMRAG of the colonic epithelial cells of a mouse at postnatal month 1, labeled with ^{35}SO$_4$ in vitro and radioautographed. x1,000.

Figure 5H. EMRAG of a goblet cell in the deeper crypt of the colonic epithelial cells of an adult mouse after injection of ^{35}SO$_4$ and radioautographed.Many silver grains are observed over the Golgi region and mucous droplets of the goblet cell, demonstrating the incorporation of radiosulfate into sulfomucins. x4,800.

Figure 5.LM and EM RAG of the digestive organs.

to 2 years, the GCT cells were very well developed and were composed of the taller cells packed with many granules and became highly convoluted, and only a few labeled cells were found. The aging changes of frequency of 5 main individual cell types in submandibular glands of male mouse from embryonic day 19 to postnatal 2 years of age were counted. On embryonic day 19 of age, the gland consisted of developing acinar cells (49%), intercalated duct cells (37%), juxta-acinar (JA) cells (3%), striated duct (SD) cells (11%). At birth, JA cells increased rapidly to 32%, thereafter decreased gradually. At 1 month of age, JA cells disappeared and granular convoluted tubule (GCT) cells appeared and increased rapidly in number with age. They reached a maximum at 6 months. Then they decreased gradually from 6-21 months. The quantity proportion of acini was relatively stable during these periods. The frequency of ICD cells (Figure 5C) was the highest (37%) at 1 day after birth. Thereafter it gradually decreased month by month and reached 2.6% at 21 months, while the ratio of SD cells persisted in 7%-12% from embryonic day 19 to postnatal 2 weeks and it disappeared at 3 months after birth. The proliferative activity of the cell population is expressed by the labeling index which is defined as the percentage of labeled nuclei with ^{3}H-thymidine in a given cell population. The labeling index of the entire gland cells increased from 13.6% at embryonic 19 to 18.3% at neonate, when it reached the first peak (Figure 6A, B). Then it declined to 2.2% at 1 week of age. A second small peak (2.9%) occurred at 2 weeks. Thereafter, the labeling index decreased progressively to less than 1% at 4 weeks of age and then remained low. The analysis of the labeling indices of respective cell types revealed that the first peak at neonate was due to the increased labeling indices of AC, ICD and JA cells, and the second peak at 2 weeks was due to the increase of ICD and SD cells. Thereafter, the labeling index of ICD cells decreased steadily but remained higher than those of any other cell types. Since the labeling index of ICD cells was more than the other cell types and persisted for a long time, it was suggested that ICD cells concerned with the generation of other cell types (Nagata 2002).

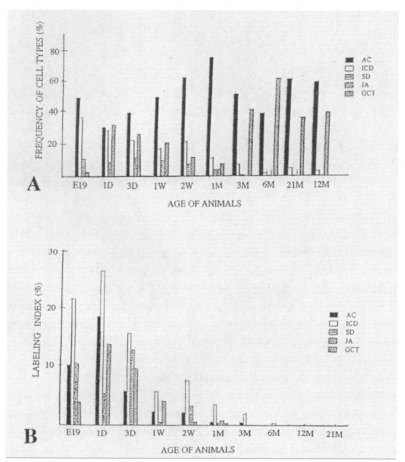

From Nagata, T. Radioautographology, General and Special.In, Progr.Histochem.Cytochem. Vol. 37, No. 2, p. 118, 2002, Urban & Fischer, Jena, Germany.

Figure 6.Histogram showing the frequencies (A) and labeling indices (B) of the five individual cell types in the submandibular glands of male ddY mice at respective ages.

2.4.2. The DNA Synthesis in the Esophagus

The esophagus is the characteristic digestive tract including all the layers, the mucous membrane covered with the stratified squamous epithelia, the submucosa, the muscular layer and the serosa or adventitia. We studied the DNA synthesis of the esophagus of aging mice labeled with ^{3}H-thymidine by LM and EMRAG (Duan et al. 1992, 1993). The labeled cells were mainly found in the basal layer of the esophageal epithelium (Figure 5D). By electron microscopy the nuclei and nucleoli of labeled cells were larger than those of unlabeled cells, but contained fewer cell organelles (Duan et al. 1993). The labeling indices in respective aging groups showed a peak at postnatal day 1 and decreased with aging keeping a constant level around a few % from 6 months to 2 years after birth.

2.4.3. The DNA Synthesis in the Stomach

The stomach consists of the mucosa covered with the surface epithelia of the columnar epithelia, including the gastric glands, the submucosa, the muscular layer and the serosa. As for the turnover of fundic glandular cells shown by ^{3}H-thymidine radioautography, it was extensively investigated with LMRAG by Leblond and co-workers (Leblond 1943, 1981, Leblond et al. 1958). They demonstrated that the DNA synthesis in the stomach increased at perinatal stages and decreased due to aging and senescence. However, the activity never reached zero but low activity continued until senescence. We studied the macromolecular synthesis including DNA, RNA, protein and glycoproteins in the gastric mucosa of both human and animal tissues by LM and EMRAG. As for the DNA synthesis, we obtained the same results as Leblond et al (1958). Therefore, the minute details will be here omitted.

2.4.4. The DNA Synthesis in the Intestines

The intestines of mammals are divided into 2 portions, small and large intestines, which can be further divided into several portions, the duodenum, the jejunum, the ileum, the caecum, the colon and the rectum. The intestinal tracts in any portions consist of the mucosa covered with columnar epithelial cells including absorptive and secretory cells, the submucosa, the smooth muscular layer and the serosa. We studied the macromolecular synthesis by LM and EMRAG mainly in the epithelial cells (Nagata 2002). The DNA synthesis of small and large intestines of mice were studied by ^{3}H-thymidine RAG. (Figure 5E). The cells labeled with ^{3}H-thymidine were localized in the crypts of both small and large intestines, a region defined as the proliferative zone. In the colon of aging mice from fetal to postnatal 2 years, the labeled cells in the columnar epithelia were frequently found in the perinatal groups from embryo to postnatal day 1. However, the labeling indices became constant from the suckling period until senescence (Morita et al. 1994). On the other hand, we examined the labeling indices of respective cell types in each layer of mouse colon such as columnar epithelial cells, lamina propria, lamina muscularis mucosae, tunica submucosa, inner circular muscle layer, outer longitudinal muscle layer, outer connective tissue and serous membrane of the colon and found that most labeling indices decreased after birth to 2 months except the epithelial cells which kept constant value to senescence (Jin and Nagata 1995a,b, Jin 1996) (Figure 7). Similar results were also obtained from the cecal tissues of mouse by LM and EMRAG. We also studied immunostaining for PCNA/cyclin and compared to the results obtained from RAG (Morita et al. 1994). We fixed the colonic tissues of litter mice of six aging groups from the embryonic day 19, to newborn postnatal day 1, 5,

21, adult 2 months and senescent 12 months in methacarn and immunostained the colonic epithelium for cyclin proliferating nuclear antigen (PCNA/cyclin), which appeared from G1 to S phase of the cell cycle, with the monoclonal antibody and the avidin-biotin peroxidase complex technique. The immunostaining positive cells were localized in the crypts of colons similarly to the labeled cells with ^{3}H-thymidine by radioautography, a region defined as the proliferative zone. The positive cells in the columnar epithelia were frequently found in the perinatal groups from embryo to postnatal day 1, and became constant from postnatal day 5 until senescence. Comparing the results by immunostaining with the labeling index by radioautography, it was found that the former was higher in each aging group than the latter. The reason for the difference should be due to that PCNA/cyclin positive cells included not only S-phase cells but also the late G1 cells.

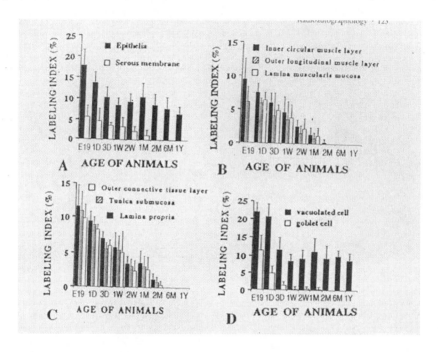

From Nagata, T. Radioautographology, General and Special.In, Progr.Histochem.Cytochem.Vol. 37, No. 2, p. 123, 2002, Urban & Fischer, Jena, Germany.

Figure 7.Histogram showing aging changes of average labeling indices in respective tissue layers and cells of mouse colons at various ages from embryo to postnatal year 1, labeled with ^{3}H-thyidine.

2.4.5. The DNA Synthesis in the Liver

The liver is the largest gland in the human and the mammalian body and consists of several types of cells (Nagata 2010c). The hepatocyte is the main component of the liver, composing the liver parenchyma which form the hepatic lobules, surrounded by other types of cells such as the connective tissue cells, sinusoidal endothelial cells, satellite cells of Kupffer, Ito's fat-storing cells and bile epithelial cells. In the livers of perinatal animals, the liver tissues include hematopoietic cells such as erythroblasts, myeloblasts and magakaryocytes.

We first studied the liver tissues at various ages from embryo to postnatal 2 years (Nagata 1993a,b, 1994a,b,c, 1995a,b,c,d, 1996a,b,c,d,e, 1997a,b,c, 1998a,b,c, 1999a,b,c, 2002, Nagata et al. 1977a, Nagata and Nawa 1966b). The results obtained from the tissues of 3 groups of animals injected separately with 3 kinds of RI-labeled precursors, ^{3}H-thymidine, ^{3}H-uridine and ^{3}H-leucine were already reported as several original articles and reviews (Ma 1988, Ma and Nagata 1988a,b, 1990a,b, 2000, Ma et al. 1991, Nagata 1996a,b, 1997a, 1998a, 1999c, 2001c, 2002, 2003, 2006a,b, 2007a,b, 2009a,b,c,d) or as a monograph in the series of Prog. Histochem.Ccytochem (Nagata 2002, 2004, 2010c). Therefore, the results from the livers in aging mice are briefly summarized in this article.

2.4.5.1. The DNA synthesis in hepatocyte nuclei

We studied macromolecular synthesis by LM and EMRAG mainly in hepatocytes of rats and mice (Nagata 1993b, 1994a,b,c,d, 1995a,b,c, 1996a, 1997a, 1999c, 2002, 2003, 2006b, 2007a, 2009a,d,h,i, 2010c,h). As for the nucleic acid synthesis, we first studied the difference between the mononucleate and binucleate hepatocytes of adult rats, injected with ^{3}H-thymidine and radioautographed (Nagata 1962, 1994d). The results showed that the frequency of labeled cells was greater in the mononucleate cells (Figure 8A) than in the binucleate cells.

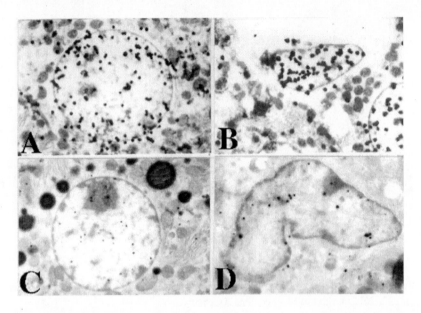

From Nagata, T. Radioautographology, General and Special.In, Progr.Histochem.Cytochem.Vol. 37, No. 2, p. 130, 2002, Urban & Fischer, Jena, Germany.

Figure 8A. EMRAG of a hepatocyte of the liver of a 14 day old mouse labeled with ^{3}H-thymidine. Many silver grains were observed over the nucleus and mitochondria.

Figure 8B. EM RAG of a sinusoidal endothelial cells of the liver of a 14 day old mouse labeled with ^{3}H-thymidine. Many silver grains were observed over the nucleus and mitochondria.

Figure 8C. EM RAG of a hepatocyte of the liver of a 14 day old mouse labeled with ^{3}H-uridine. Many silver grains were observed over the nucleus and mitochondria.

Figure 8D. EM RAG of an Ito's fat-storing cell of the liver of a newborn 14 day old mouse labeled with ^{3}H-uridine. Many silver grains were observed over the nucleus and mitochondria.

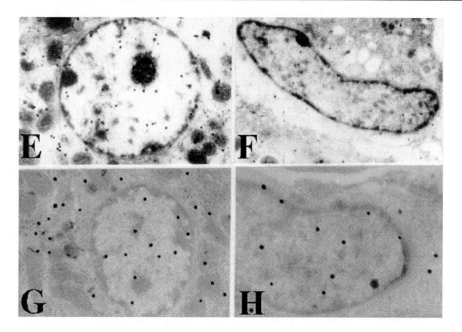

Figure 8E. EM RAG of a hepatocyte of the liver of a 1 month old mouse labeled with ^{3}H-leucine. Many silver grains were observed over the nucleus and mitochondria.

Figure 8F. EM RAG of a sinusoidal endothelial cells of the liver of a newborn 14 day old mouse labeled with ^{3}H-leucine. Many silver grains were observed over the nucleus and mitochondria.

Figure 8G. EMRAG of a hepatocyte of the liver of an adult 2 month old mouse labeled with ^{3}H-proline. Many silver grains were observed over the nucleus and mitochondria.

Figure 8H. EMRAG of a Kupffer cell of the liver of a newborn 1 day old mouse labeled with ^{3}H-proline. Many silver grains were observed over the nucleus and mitochondria.

Figure 8. EM RAG of the liver.

The labeled binucleate cells were classified into two types, i. e., a hepatocyte whose one of the two nuclei was labeled and a hepatocyte whose two nuclei were labeled. The former was more frequently observed than the latter. Grain counts revealed that the amount of DNA synthesized in the binucleate cell whose one nucleus was labeled was the same as the mononucleate cell, while the total amount of DNA synthesized in the binucleate cell whose two nuclei were labeled was almost twice as that of the mononucleate cell. These results suggest that the two nuclei of binucleate hepatocytes synthesize DNA independently from each other.

On the other hand, LM and EMRAG of prenatal and postnatal normal mice at various ages labeled with 3H-thymidine revealed that many silver grains were localized over the nuclei of various cell types consisting the liver, i. e., hepatocytes (Figure 8A), sinusoidal endothelial cells (Figure 8B), Kupffer's cells, Ito's fat-storing cells, bile ductal epithelia cells, fibroblasts and hematopoietic cells (Ma 1988, Ma and Nagata 1988a,b, 1990, Nagata 1995a). In hematopoietic cells in the livers of perinatal animals, silver grains were observed over the nuclei of erythroblasts, myeloblasts, lymphoblasts and megekaryocytes. However, most hematopoietic cells disappeared on postnatal day 14. At fetal day 19, the liver tissues were chiefly consisted of hepatocytes and haematopoietic cells and no lobular orientation was observed. At postnatal day 1 and 3, lobular formation started and finally the hepatic lobules were formed at day 9 after birth. During the perinatal period, almost all kinds of cells were labeled with 3H-thymidine. Percentage of labeled hepatocytes was the highest at fetal day 19, and rapidly decreased after birth to day 3. From day 9 to 14, percentage of labeled hepatocytes (labeling index) decreased gradually and finally to the lowest at 24 months (Figure 9A). When the labeling indices of hepatocytes in 3 hepatic acinar zones were analyzed, the indices decreased in zone 2 (intermediate zone) and zone 3 (peripheral zone) on days 3 and 9 after birth, whereas they increased in zone 1 (central) on day 9, and then they

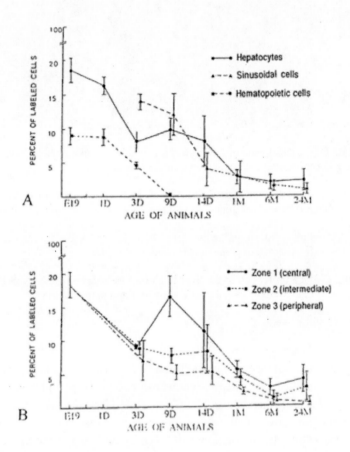

From Nagata, T.: Special Cytochemistry in Cell Biology, In, Internat. Rev. Cytol. Vol. 211, No. 1, p. 92, 2001, Academic Press, San Diego, USA, London, UK.

Figure 9A.Labeling indices of hepatocytes, sinusoidal endothelial cells and hematopoietic cells, respectively.

Figure 9B.Labeling indices of hepatocytes, sinusoidal endothelial cells and hematopoietic cells, respectively.

Figure 9. Transitional curves of the labeling indices in the livers of aging mice after injection of ^{3}H-thymine. Mean ± Standard Deviation.

altogether decreased from day 14 to 24 months (Figure 9B). When the size and number of cell organelles in both labeled and unlabeled hepatocytes were estimated quantitatively by image analysis with an image analyzer, Digigramer G/A (Mutoh Kogyo Co. Ltd., Tokyo, Japan) on EMRAG, the area size of the cytoplasm nucleus, mitochondria, endoplasmic reticulum, and the number of mitochondria in the unlabeled hepatocytes were more than the labeled cells (Ma and Nagata 1988a,b, Nagata 1995). These data demonstrate that the cell organelles of the hepatocytes which synthesized DNA were not well developed as compared to those not synthesizing DNA during the postnatal development. In some of unlabeled hepatocytes, several silver grains were occasionally observed localizing over mitochondria and peroxisomes as was formerly reported (Nagata et al. 1967a,b, 1982a,b). The mitochondrial DNA synthesis was first observed in cultured hepatocytes of chickens and mice in vitro (Nagata et al. 1967a,b). The percentages of labeled cells in other cell types in the liver of aging mice such as sinusoidal endothelial cells, Kupffer's cells, Ito's fat-storing cells, bile ductal epithelia cells and fibroblasts showed also decreases from perinatal period to postnatal 24 months.

2.4.5.2. Mitochondrial DNA synthesis in hepatocytes

When we observed DNA synthesis in the nuclei of mononucleate and binucleate hepatocytes, we also observed DNA synthesis in hepatocyte mitochondria (Ma 1988, Ma and Nagata 1988a,b, 1990a,b, Nagata and Ma 2005a). The results of visual grain counts on the number of mitochondria labeled with silver grains obtained from 10 mononucleate hepatocytes of each animal labeled with ^{3}H-thymidine demonstrating DNA synthesis in 7 aging groups at perinatal stages, prenatal embryo day 19, postnatal day 3, 9 and 14, month 1, 6, 12 and 24, were counted. The number of total mitochondria per cell increased from perinatal stage (35-50/cell) increased to postnatal month 6 (95-105/cell), reaching the maximum, decreased to month 24 (85-90/cell), while the number of labeled mitochondria per cell increased from perinatal stage to postnatal day 14, reaching the maximum, decreased to month 6, then increased again to month 12, reaching the second peak and decreased again to month 24. Thus, the labeling indices in respective aging stages were calculated from the number of labeled mitochondria which showed an increase from perinatal stage to postnatal day 14, reaching the maximum and decreased to month 24. The results showed that the numbers of labeled mitochondria with ^{3}H-thymidine showing DNA synthesis increased from prenatal embryo day 19 (3.8/cell) to postnatal day 14 (6.2/cell), reaching the maximum, and then decreased to month 6 (3.7/cell) and again increased to year 1 (6.0/cell), while the labeling indices increased from prenatal day 19 (11.8%) to postnatal day 14 (16.9%), reaching the maximum, then decreased to month 6 (4.1%) and year 1 (6.4%) and year 2 (2.3%). The increase of the total number of mitochondria in mononucleate hepatocytes was stochastically significant (P<0.01), while the changes of number of labeled mitochondria and labeling index in mononucleate hepatocytes were not significant (P<0.01).

As for the binucleate hepatocytes, on the other hand, because the appearances of binucleate hepatocytes showing silver grains in their nuclei demonstrating DNA synthesis were not so many in the adult and senescent stages from postnatal month 1 to 24, only binucleate cells at perinatal stages when reasonable numbers of labeled hepatocytes were found in respective groups were analyzed. The number of mitochondria in binucleate hepatocytes at postnatal day 1 to 14 kept around 80 (77-84/cell) which did not show such remarkable changes, neither increase nor decrease, as shown in mononucleate cells. Thus, the number of mitochondria per binucleate cell, the number of labeled mitochondria per binucleate cell and the labeling index of binucleate cell in 4 groups from postnatal day 1 to 14 were counted. The number of mitochondria and the number of labeled mitochondria were more in binucleate cells than mononucleate cells (Nagata 2007a,b,c,d, Nagata and Ma 2005a,b, Nagata et al. 1977a).

2.4.6. The DNA Synthesis inthe Pancreas

The pancreas is a large gland, next to the liver in men and animals, among the digestive glands connected to the intestines. It consists of exocrine and endocrine portions and takes the shape of a compound acinous gland. The exocrine portion is composed of ductal epithelial cells, centro-acinar cells, acinar cells and connective tissue cells, while the endocrine portion, the islet of Langerhans, is composed of 3 types of endocrine cells, A, B, C cells and connective tissue cells. Intracellular transport of secretory proteins in the pancreatic exocrine

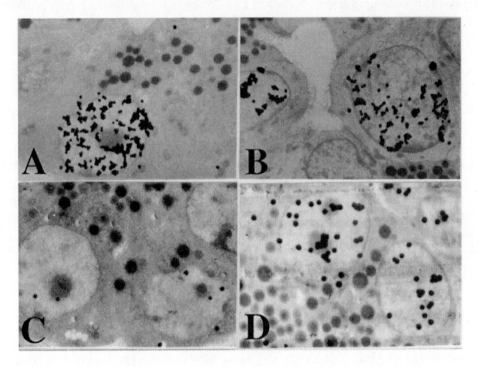

From Nagata, T. Radioautographology, General and Special.In, Progr.Histochem.Cytochem.Vol. 37, No. 2, p. 140, 2002, Urban & Fischer, Jena, Germany.

Figure 10A. EM RAG of 2 pancreatic acinar cells of a 14 day old mouse labeled with ^{3}H-thymidine, showing DNA synthesis. x10,000.

Figure 10B. EM RAG of 2 centro-acinar cells of a 14 day old mouse labeled with ^{3}H-thymidine, showing DNA synthesis. x10,000.

Figure 10C. EM RAG of 3 pancreatic acinar cells of a 1 day old mouse labeled with ^{3}H-uridine, showing RNA synthesis. x10,000.

Figure 10D. EM RAG of 3 pancreatic acinar cells of a 14 day old mouse labeled with ^{3}H-uridine, showing RNA synthesis. x10,000.

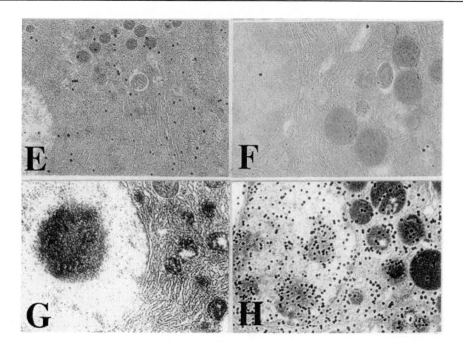

Figure 10E. EM RAG of a pancreatic acinar cell of a 30 day old mouse labeled with ^{3}H-leucine, showing protein synthesis. x10,000.

Figure 10F. EM RAG of a pancreatic acinar cell of a 12 month old mouse labeled with ^{3}H-leucine, showing protein synthesis. x10,000.

Figure 10G. EM RAG of a pancreatic acinar cell of a 1 day old mouse labeled with ^{3}H-glucosamine, showing glucide synthesis. x10,000.

Figure 10H. EM RAG of apancreatic acinar cell of a 14 day old mouse labeled with ^{3}H-glucosamine, showing glucide synthesis. x10,000.

Figure 10. EM RAG of the pancreas.

cells were formerly studied by Jamieson and Palade (1967) by EMRAG. We studied the macromolecular synthesis of the aging mouse pancreas at various ages. We first studied the DNA synthesis of mouse pancreas by LM and EMRAG using ^{3}H-thymidine (Nagata et al. 1986a,b). Light and electron microscopic radioautograms of the pancreas revealed that the nuclei of pancreatic acinar cells (Figure 10A), centro-acinar cells (Figure 10B), ductal epithelial cells, and endocrine cells were labeled with ^{3}H-thymidine. The labeling indices of these cells in 5 groups of litter mate mice, fetal day 15, postnatal day 1, 20, 60 (2 months) and 730 (2 years) were analyzed. The labeling indices of these cells reached the maxima at day 1 after birth and decreased gradually to 2 years (Figure 11). The maximum in the acinar cells proceeded to the ductal and centro-acinar cells, suggesting that the acinar cells completed their development earlier than the ductal and centro-acinar cells (Nagata et al. 1986a,b).

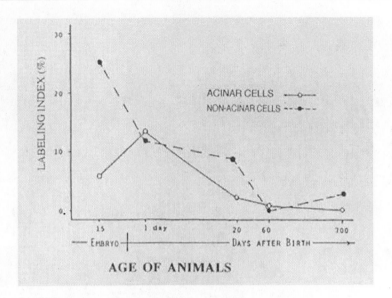

From Nagata, T. Radioautographology, General and Special.In, Progr.Histochem.Cytochem.Vol. 37,
    No. 2, p. 142, 2002, Urban & Fischer, Jena, Germany.

Figure 11.Transitional curves of the labeling indices of respective cell types of the pancreas of aging
mice labeled with[3]H-thymidine, showing DNA synthesis.Mean ± Standard Deviation.

2.5. The DNA Synthesis inthe Respiratory System

The respiratory system consists of 2 parts, the air-conducting portion and the respiratory
portion. We studied the macromolecular synthesis in the pulmonary tissues as well as the
tracheal tissues at various ages from embryo to postnatal 2 years (Sun et al. 1994, 1995a,b,
1997a,b).

2.5.1. The DNA Synthesis inthe Trachea
The tracheas of mammals are composed of ciliated pseudostratified columnar epithelia,
connective tissues, smooth muscles and hyalin cartilages. The changes of DNA synthesis of
tracheal cells in aging mice were studied by LM and EMRAG (Sun et al. 1997a, Nagata
2000d). The tracheae of 8 groups of mice from fetal day 18 to 2 years after birth were
examined. The results demonstrated that the DNA syntheses and morphology of tracheal cells
in the mouse tracheae changed due to aging. The radioautograms revealed that the DNA
synthesis in the nuclei of ciliated cells was observed only in the fetal animals (Figure 12A).
However, the DNA synthesis in nonciliated cells and basal cells was observed in both
prenatal and postnatal animals (Figure 12B). The labeling indices of respective cell types
were analyzed (Sun et al. 1997a). As the results, the labeling indices of the epithelial cells
showed their maxima on fetal day 18, then fell down from postnatal day 3 to 2 years (Figure
13A). The ciliated cell could not synthesize DNA and proliferate in the postnatal stage. They
are supposed to be derived by the division and transformation of the basal cells. On the other
hand, the DNA synthesis of chondrocytes was the highest on embryonic day 18, and rapidly

declined on postnatal day 3 (Figure 12C). The chondrocytes lost the ability of synthesizing DNA at 2 months after birth (Figure 13B). The labeling indices of other cells (including fibroblasts, smooth muscle and glandular cells) were the highest on fetal day 18 and fell down markedly on the third day after birth and decreased progressively due to aging (Figure 13C).

2.5.2. The DNA Synthesis inthe Lung

We studied the pulmonary tissues at various ages from embryo to postnatal 2 years of mice (Sun et al. 1995a,b). The pulmonary tissues obtained from ddY strain mice at various ages from embryo day 19 to adult postnatal day 30 and to year 2 consisted of several types of cells, i. e., the type I epithelial cells or the small alveolar epithelial cells, type II epithelial cells or large alveolar epithelial cells, interstitial cells and endothelial cells, which incorporated macromolecular precursors respectively (Figures. 12E,F, G, H).

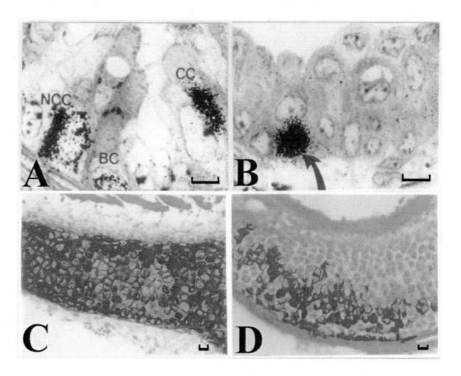

From Nagata, T.: Special Cytochemistry in Cell Biology, In, Internat. Rev. Cytol. Vol. 211, No. 1, p. 102, 2001, Academic Press, San Diego, USA, London, UK.

Figure 12A. LM RAG of the tracheal epithelial cells of a prenatal day 18 mouse embryo labeled with ^{3}H-thymidine, showing DNA synthesis. x1,125.

Figure 12B. LM RAG of the tracheal epithelial cells of a postnatal 1 month old mouse labeled with ^{3}H-thymidine, showing DNA synthesis. x1,125.

Figure 12C. LM RAG of the tracheal cartilage cells of a prenatal day 19 mouse embryo labeled with ^{35}SO$_4$ showing mucosubstance synthesis. x400.

Figure 12D. LM RAG of the tracheal cartilage cells of a postnatal day 3 mouse labeled with ^{35}SO$_4$ showing mucosubstance synthesis. x400.

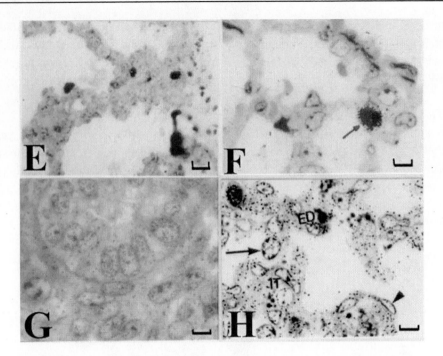

Figure 12E. LM RAG of the lung of a 1 day old mouse labeled with ^{3}H-thymidine, showing DNA synthesis. x750.

Figure 12F. LM RAG of the lung of a 1 month old mouse labeled with ^{3}H-thymidine, showing DNA synthesis. x750.

Figure 12G. LM RAG of the lung of a prenatal day 16 mouse embryo labeled with ^{3}H-uridine, showing RNA synthesis. x750.

Figure 12H. LM RAG of the lung of a 7 day old mouse labeled with ^{3}H-leucine, showing protein synthesis. x750.

Figure 12. LM RAG of the respiratory organs.

The pulmonary tissues obtained from ddY strain mice at embryonic to early postnatal stages consisted of undifferentiated cells (Figure 12E). However, they differentiated into several types of cells due to aging, the type I epithelial cells or the small alveolar epithelial cells (Figure 12E), the type II epithelial cells or the large alveolar epithelial cells (Figure 12F), the interstitial cells (Figure 12G), the endothelial cells (Figure 12H) and alveolar phagocytes or dust cells as we had formerly observed (Sun et al. 1995a,b). At embryonic day 16 and 18, the fetal lung tissues appeared as glandular organizations consisting of many alveoli bordering undifferentiated cuboidal cells and no squamous epithelial cells were seen (Figures. 12E, 12G). Mitotic figures were frequently observed in cuboidal epithelial cells. After birth, the structure of the alveoli was characterized by further development of the alveolar-capillary networks from postnatal day 1 to 3 and 7 (Figure 12H). During the development, the cellular composition of the alveolar epithelium resembled that of the adult lung, with a mixed population of the type I and type II epithelial cells. Up to 1 and 2 weeks after birth, the lung tissues showed complete alveolar structure and single capillary system

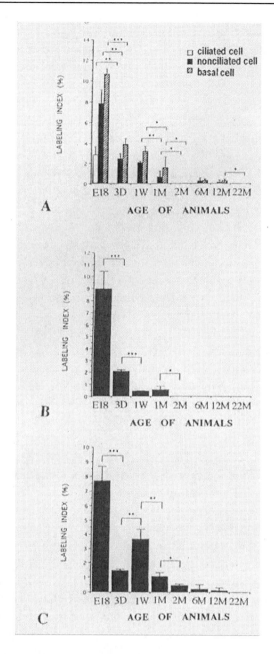

From Nagata, T. Radioautgraphology, General and Special.In, Progr.Histochem.Cytochem.Vol. 37, No. 2, p. 148, 2002, Urban & Fischer, Jena, Germany.

Figure 13A.Epithelial cells.

Figure 13B.Chondrocytes in the cartilage.

Figure 13C. Other cells.

Figure 13. Histogram showing aging changes of average labeling indices in respective cell types of the trachea of aging mice labeled with ^{3}H-thyidine. Mean ± Standard Deviation.

almost the same as the adult after 1 month (Figure 12F) to 2 to 6 months, and further to senescent stage over 12 months to 22 months. On electron microscopic radioautograms of the pulmonary tissues labeled with ^{3}H-thymidine, silver grains were observed over the nuclei of some pulmonary cells corresponding to the DNA synthesis in S-phase as observed by light microscopic radioautography (Sun et al. 1995a,b, 1997a). The DNA synthetic activity of respective pulmonary cells as expressed by labeling indices demonstrated increases from perinatal stage to developmental stage and decreased due to aging. We also studied inhalation of ^{3}H-thymidine in air by means of a nebulizer into the lungs of 1 week old young mice as experimental studies and observed by LM and EM RAG (Duan et al. 1994). After min. inhalation, the lung tissues were taken out and processed by either rapid freezing and freeze-substitution for dry-mounting radioautography or conventional chemical fixation for wet-mounting radioautography. By wet-mounting RAG silver grains were observed in the nuclei of a few alveolar type 2 cells and interstitial cells demonstrating DNA synthesis. By dry-mounting RAG, numerous silver grains were located diffusely over all the epithelial cells and interstitial cells demonstrating soluble compounds. The results showed that ^{3}H-thymidine inhaled into the lung distributed over all the pulmonary cells but only some of the alveolar type 2 cells and interstitial cells did synthesize DNA.

On the other hand, we studied the aging changes of DNA synthesis in the lungs of salamanders, *Hynobius nebolosus*, from larvae at 2 month after fertilization, juvenile at 1 month, adults at 10 and 12 months after metamorphosis, and finally to senescence at 5 years by LM RAG after ^{3}H-thymidine injections (Matsumura et at. 1994). The results showed that the labeling indices of in the ciliated cells and mucous cells in the suprerficial layer of young animals were higher than those of the basal cells and they decreased in adults, demonstrating aging changes in salamanders.

2.5.2.1. Mitochondrial DNA Synthesis of Aging Mouse Pulmonary Cells

On electron microscopic radioautograms of the pulmonary tissues labeled with ^{3}H-thymidine, silver grains were observed over not only the nuclei of some pulmonary cells corresponding to the DNA synthesis in S-phase as observed by LM radioautograpy (Sun et al. 1995a,b) but also over the mitochondria by EMRAG (Sun et al. 1995b). Some mitochondria in both S-phase cells and interphase cells which did not show any silver grains over their nuclei were labeled with silver grains showing intramitochondrial DNA synthesis. The intramitochondrial DNA synthesis was observed in all cell types, the type I epithelial cells, the type II epithelial cells (Figure 12E, 12F), the interstitial cell (Figure 12F) and the endothelial cell. Because enough numbers of electron photographs (more than 5) were not obtained from all the cell types in respective aging groups, only some cell types and some aging groups when enough numbers of electron photographs were available were used for quantitative analysis. The numbers of mitochondria per cell profile area, the numbers of labeled mitochondria per cell and the labeling indices of the type I epithelial cells in only a few aging groups were observed and counted. The labeling indices in respective aging stages were calculated from the number of labeled mitochondria and the number of total mitochondria per cellular profile area which were calculated, respectively. These results demonstrated that the labeling indices in these cell types decreased from prenatal stages at embryo day 16 to day 18 (20-25%), and further decreased to postnatal days up to senescent stages due to aging.

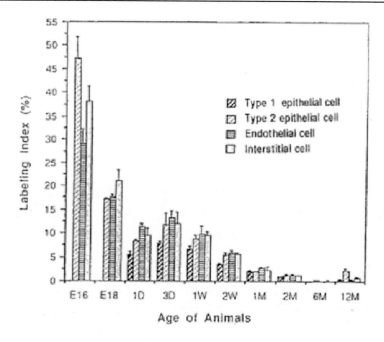

From Nagata, T. Radioautographology, General and Special.In, Progr.Histochem.Cytochem.Vol. 37, No. 2, p. 149, 2002, Urban & Fischer, Jena, Germany.

Figure 14. Histogram showing aging changes of average labeling indices in respective cell types of the tracheal epithelial cells of aging mice labeled with ^{3}H-thyidine. Mean ± Standard Deviation.

3.6. The DNA Synthesis inthe Urinary System

The urinary system consists of the kidney and the urinary tract. We studied only the macromolecular synthesis of the kidneys of several groups of aging miceby LM and EM RAG, while the localization of an anti-allergic agent was observed in the urinary bladders of adult rats (Nagata 2005).

3.6.1. The DNA Synthesis inthe Kidney

The kidneys of mammals microscopically consist of the nephrons, which can be divided into two portions, the renal corpuscles and the uriniferous tubules. The renal corpuscles are composed of the glomeruli which are covered with the Bowman's capsules. They are localized in the outer zone of the kidney, the renal cortex, while the uriniferous tubules are composed of two portions, the proximal portions and the distal portions which can further be divided into several portions which run from the outer zone of the kidney, the renal cortex, to the inner zone, the medulla. We studied the DNA synthesis by ^{3}H-thymidine radioautography in 3 groups of ddY mouse embryos from prenatal day 13 (Figure 15A), day 15 (Figure 15B) to day 19 in vitro, as well as perinatal mice from embryonic day 19 to postnatal day 1, 8, 30, 60 and 365 (1 year) in vivo (Hanai 1993, Hani et al. 1993, Hanai and Nagata 1994a,b). The labeling indices by LMRAG in glomeruli (28 to 32%) and uriniferous tubules (31 to 33%) in the superficial layer were higher than those of labeling indices (10 to 12%) and (8 to 16%) in the deeper layer from the late fetal to the suckling period, then decreased with aging from weaning to senescence (Figure 16). EMRAG revealed the same results (Hanai and Nagata 1994a,b,c). At the same time, immunocytochemical localization of PCNA/cyclin was carried out in the same animals in several aging groups as ^{3}H-thymidine RAG (Hanai 1993, Hanai et al. 1993). The results from the PCNA/cyclin positive indices in respective aging groups were almost the same as the labeling indices with ^{3}H-thymidine RAG. The incorporation of ^{3}H-thymidine was formerly observed by EMRAG in mitochondrial matrix of cultured kidney cells from chickens and mice in vitro demonstrating mitochondrial DNA synthesis (Nagata et al. 1967b).

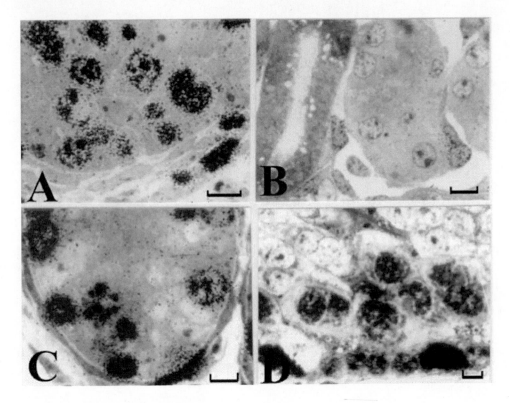

From Nagata, T.: Special Cytochemistry in Cell Biology, In, Internat. Rev. Cytol. Vol. 211, No. 1, p. 108, 2001, Academic Press, San Diego, USA, London, UK.

Figure 15A. LM RAG of the metanephros of a prenatal day 13.5 mouse embryo labeled with ^{3}H-thymidine, showing DNA synthesis. x1,200.

Figure 15B. LM RAG of the metanephros cortex of a prenatal day 15.5 mouse embryo labeled with ^{3}H-thymidine, showing DNA synthesis. x1,200.

Figure 15C. LM RAG of the testis of a postnatal day 7 male mouse labeled with ^{3}H-thymidine, showing DNA synthesis. x800.

Figure 15D. LM RAG of the testis of a postnatal year 1 male mouse labeled with ^{3}H-thymidine, showing DNA synthesis. x800.

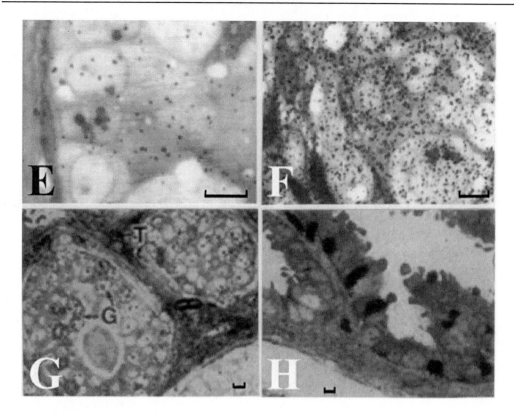

Figure 15E. LM RAG of the testis of a postnatal day 3 male mouse labeled with ³H-uridine, showing RNA synthesis. x1,500.

Figure 15F. LM RAG of the lung of a 1 month old mouse labeled with ³H-thymidine, showing DNA synthesis. x1,125.

Figure 15G. LM RAG of the ovary of a postnatal day 3 female mouse labeled with ³H-thymine, showing DNA synthesis in granulosa cells (G) and theca cells (T). x400.

Figure 15H. LM RAG of the oviduct of a postnatal day 30 female mouse labeled with ³H-thymine, showing DNA synthesis in epithelial cells. x400.

Figure 15. LM RAG of the uro-genital organs.

3.6.2. The DNA Synthesis inthe Urinary Tract

The urinary tract is composed of the ureter, the urinary bladder and the urethra. We studied the urinary bladder of adult rats by LMRAG after oral administration of ³H-tranilast, an anti-allergic agent produced by Kissei Pharmaceutical Co. (Momose et al. 1989, Nishigaki et al. 1987, 1990a,b). It was found that this agent specifically localized over the transitional epithelium and the endothelium of the veins in the mucosa of normal adult rats. However, any study on the DNA synthesis in the ureter, the urinary bladder and the urethra was not carried out.

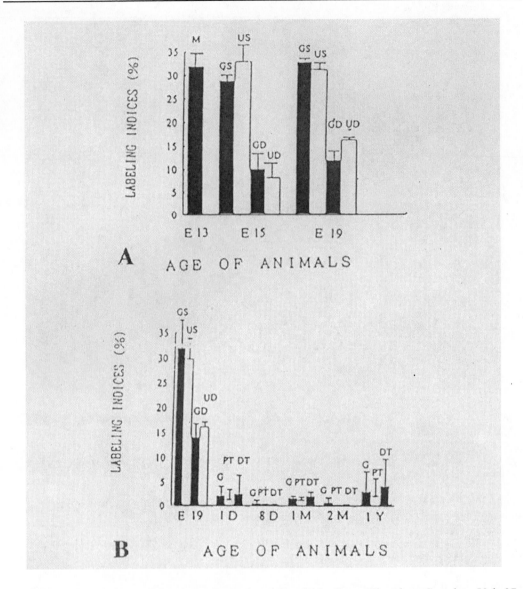

A   AGE OF ANIMALS

B   AGE OF ANIMALS

From Nagata, T. Radioautographology, General and Special.In, Progr.Histochem.Cytochem.Vol. 37, No. 2, p. 156, 2002, Urban & Fischer, Jena, Germany.

Figure 16A.Labeling indices of the glomeruli and the uriniferous tubules of of mouse embryo from prenatal day 13 and 19.

Figure 16B.Labeling indices of the glomeruli and the proximal and distal tubules of mouse embryo from prenatal day 19 to postnatal year 1.Abbreviations: GS=glomeruli of the superficial layer. US=uriniferous tubules of the superficial layer. GD=glomeruli of the deeper layer. UD=uriniferous tubules of the deeper layer. G=glomeruli. PT=proximal tubules. DT=distal tubules.

Figure 16. Histogram showing aging changes of average labeling indices in respective cell types of the kidneys of aging mice labeled with ³H-thyidine. Mean ± Standard Deviation.

3.7. The DNA Synthesis inthe Reproductive System

The reproductive system is divided into two parts, the male and female genital organs. We studied both the male and female genital organs of several groups of ddY aging mice by LM and EMRAG using macromolecular precursors.

3.7.1. The DNA Synthesis inthe Male Genital Organs

The male genital organ consists of the testis and its excretory ducts such as ductuli efferentes, ductus epididymidis, ductus deferens, ejaculatory ducts, auxiliary glands and penis. Among these organs, the testis was the main target of the scientific interests. Formerly, Clermont (1958, 1963) demonstrated using ^{3}H-thymidine radioautography that several stages of development of the spermatogonia were found at different levels in the germinal epithelium of mature man and rodents, with the most primitive germ cells found at the base and the more differentiated cells located at higher levels.

3.7.1.1. The DNA Synthesis inthe Testis

The structure of the testis of mammals is a compound tubular gland enclosed in tunica albuginea, a thick fibrous capsule. The parenchyma of the testis is composed of around 250 pyramidal compartments in men, named lobules. Each lobule is made of convoluted seminiferous tubules, consisting of many spermatogenic cells differentiatiang to sperms among the supporting cells of Sertoli in the seminiferous epithelium, surrounded by the interstitial cells of Leydig (Gao 1993, Gao et al. 1994,1995a,b).

We fifrst studied the macromolecular synthesis in the testis of aging male ddY mice at various ages. When testicular tissues were labeled with ^{3}H-thymidine and observed by LM and EM RAG, many spermatogonia and myoid cells as well as Leydig cells were labeled with ^{3}H-thymidine at various ages from embryonic day 19 to postnatal day 1, 3, 7 (Figure 15C), 14, month 1, 2, 6, 12 (Figure 15D) and 24 (2 years). Silver grains were localized over the nuclei and several mitochondria of the spermatogonia showing DNA synthesis. Among of the aging groups, we counted the numbers of mitochondria per cell profile area, the numbers of labeled mitochondria per cell of the spermatogonia from 4 aging groups, prenatal embryonic day 19, postnatal day 3, and adults at month 1 and 6, and the labeling indices were calculated. The results showed that the LI of the spermatogonia increased from embryonic day 19 (17%) to postnatal day 7 and month 1(30%), reaching the maximum, then decreased to month 6 (25%) to year 2.

At embryonic and neonatal stages, DNA synthesis of spermatocytes was weak and only a few labeled spermatogonia could be observed during the perinatal stages. The labeled spermatocytes were recognized at postnatal day 4 and 7 (Figure 15C) and the number of labeled spermatogonia and spermatocytes increased from 2 and 3 weeks, keeping high level from month 1 to year 1 and 2 until senescence (Figure 17A). However, the Sertoli's cells (Figure 17B) and myoid cells (Figure 17C) labeled with ^{3}H-thymidine were frequently observed at perinatal stages from embryo to postnatal day 7, while the labeling indices of both cells decreased from young adulthood (postnatal 2 weeks) to senescence (Gao 1993, Gao et al. 1994,1995a). The interstitial cells of Leydig in the testis surrounding the seminiferous tubules shall be described in the following section of the endocrine system in detail.

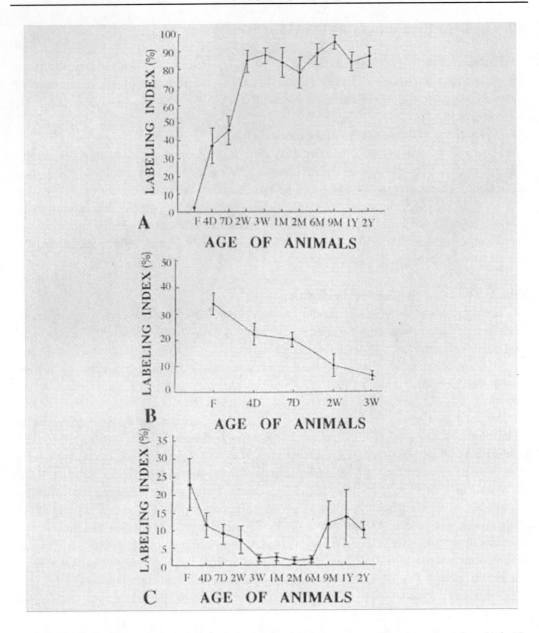

From Nagata, T. Radioautographology, General and Special.In, Progr.Histochem.Cytochem.Vol. 37, No. 2, p. 160, 2002, Urban & Fischer, Jena, Germany.

Figure 17A.The spermatogonia.

Figure 17B.The Sertoli cells.

Figure 17C. The myoid cells.

Figure 17. Transitional curves of the labeling indices of respective cell types in the testis of aging mice labeled with ^{3}H-thymidine, showing DNA synthesis. Mean ± Standard Deviation.

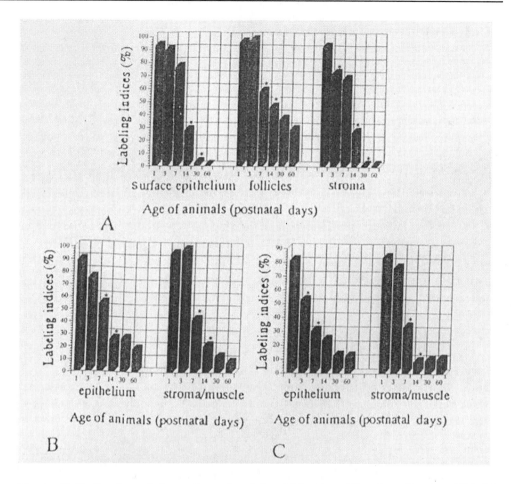

From Nagata, T. Radioautographology, General and Special.In, Progr.Histochem.Cytochem.Vol. 37, No. 2, p. 166, 2002, Urban & Fischer, Jena, Germany.

Figure 18A.The ovary.

Figure 18B.The uterus.

Figure 18C. The oviduct.

Figure 18. Histogram showing aging changes of average labeling indices in respective cell types of female genital organs of aging mice labeled with ^{3}H-thyidine. Mean ± Standard Deviation.

3.7.2. TheDNA Synthesis inthe Female Genital Organ

The female genital organ consists of the ovary, the oviduct, the uterus, the vagina and the external genitals. We studied the macromolecular synthesis in the ovary, oviduct and uterus of several litters of ddY mice in aging.

3.7.2.1. The DNA Synthesis inthe Ovary

The ovary consists of the germinal epithelium covering the surface and the stroma containing many developing ovarian follicles depending upon the age of animals.

The nucleic acids, DNA and RNA, syntheses in the developing virgin mice ovaries of 6 litters consisting of 36 female mice at various ages were studied by ^{3}H-thymidine and ^{3}H-uridine radioautography (Li 1994, Li and Nagata 1995, Li et al. 1992). The ^{3}H-thymidine incorporations were active in all surface epithelial cells, stromal and follicular cells of the ovary between postnatal days 1 to 7 and decreased from day 14 (Figure 15G) and maintained a lower level to day 60, while ^{3}H-uridine incorporations were active in all surface epithelial cells, stromal and follicular cells of the ovary between postnatal days 1 to 7 and maintained medium levels from day 14 on.

The labeling indices with ^{3}H-thymidine showing DNA synthetic activity were high in all the surface epithelial cells, follicular cells and stromal cells of mice at neonatal stage from postnatal day 1 to 7, but decreased from day 40 to day 60 at mature stage. The grain counts showing RNA synthetic activity were high at neonatal stage from day 1 to day 7, and maintained medium levels from day 14 to day 60 at mature stage.

3.7.2.2. The DNA Synthesis inthe Oviduct

The nucleic acids, DNA and RNA, syntheses in the oviducts of developing virgin mice at various ages were studied by ^{3}H-thymidine and ^{3}H-uridine radioautography (Li 1994, Li and Nagata 1995). The silver grains with ^{3}H-thymidine showing the DNA synthesis were observed over many nuclei in all surface epithelial cells, stromal and smooth muscle cells at neonatal stage between postnatal day 1 to 3 and decreased from day 7 to 30 (Figure 15H) and 60, while the silver grains showing the RNA synthesis with ^{3}H-uridine were observed over the nuclei and cytoplasm of all the epithelial and stromal cells from postnatal day 1 to day 60. The labeling indices with ^{3}H-thymidine were high at neonatal stage from postnatal day 1 to 3 but decreased from day 7 to day 60. The grain counts with ^{3}H-uridine were high at neonatal stage from postnatal day 1 to 3 and increased from day 7 to day 14 and decreased from day 30 to day 60. These results demonstrated an unparalleled alternation of DNA and RNA syntheses in the oviduct (Li and Nagata 1995).

3.7.2.3. The DNA Synthesis inthe Uterus

The silver grains with ^{3}H-thymidine showing DNA synthesis of the uterus was observed over some of the nuclei of all the cells in the epithelia, stroma and smooth muscles from postnatal day 1 to 60 (Li 1994, Li and Nagata 1995). The labeling indices with ^{3}H-thymidine were high (80-95%) at postnatal day 1 and decreased from day 3 to 60 (>10%). The silver grains showing RNA synthesis of the uterus were observed over all the nuclei and cytoplasm of all the cells in the uterine epithelia, stroma and smooth muscles from day 1 to 60. The number of silver grains in the uterine epithelium increased from postnatal day 1 to 7 and decreased from day 14 to 60, while they increased in the stroma from day 1 to 3 and decreased from day 7 to 60.

These results from the female genital organs showed that both DNA and RNA syntheses, as expressed by labeling indices and grain counting, were active in all kinds of cells, such as surface epithelial cells, stromal cells and follicular cells of the ovaries between postnatal days 1 to 7, then they decreased from day 14 to 60. However, the DNA synthesis in the epithelial cells and the stromal cells of both the uteri and the oviducts was active at postnatal day 1 and 3 and decreased from day 7 to 60. The RNA synthesis in the uteri and oviducts was active at postnatal day 1, increased from day 1 to day 14, and decreased from day 30 to 60. The unparalleled alteration of the DNA and RNA syntheses was shown between the ovary and the uterus or oviduct (Li and Nagata 1995).

We also studied PCNA/cyclin immunostaining in the ovary, oviduct and uterus (Li 1994). It was demonstrated that PCNA/cyclin positive cells were observed in the ovarian follicular epithelium, ovarian interstitial cells, tubal epithelial cells, tubal interstitial cells, uterine epithelial cells and uterine interstitial cells. The positive cells increased from postnatal 1 day to 3 and 7 days, then decreased from 14 days to senescence. These results accorded well with the results obtained from the ^{3}H-thymidine radioautography (Li 1994, Li and Nagata 1995). Moreover, the mucosubstance synthesis incorporating sulfuric acid was also carried out (Oliveira et al. 1991, Li et al. 1995).

3.7.2.4. The DNA Synthesis in Gametogenesis

The gametogenesis consists of both spermatogenesis in male germ cells and the oogenesis in female germ cells, leading to the implantation and further development of blastcysts. The macromolecular synthesis, DNA, RNA and protein synthesis, in both the testis and the ovary were already described in the sections of male and female reproductive systems (3.7.1. and 3.7.2.) previously.

3.7.2.5. The DNA Synthesis in Implantaion

In order to detect the changes of DNA, RNA and protein synthesis of the developing blastcysts in mouse endometrium during activation of the implantation, ovulations of female BALB/C strain mice were controlled by pregnant mare serum gonadotropin and human chorionic gonadotropin, then pregnant female mice were ovariectomized on the 4th day of pregnancy (Yamada 1993, Yamada and Nagata 1992a,b, 1993). The delay implantation state was maintained for 48 hrs and after 0 to 18 hrs of estrogen supply[3]H-thymidine was injected. The three regions of the endometrium, i. e. the interinplantation site, the antimesometrial and mesometrial sides of implantation site, were taken out and processed for LM and EM RAG. It was well known that the uterus of the rodent becomes receptive to blastcyst implantation only for a restricted period. This is called the implantation window which is intercalated between refractory states of the endometrium whose cycling is regulated by ovarian hormones (Yoshinaga 1988). We studied the changes of DNA synthesis by ^{3}H-thymidine (Yamada and Nagata 1992a,b) incorporations in the endometrial cells of pregnant-ovariectomized mice after time-lapse effect of nidatory estradiol. As the results, the endometrial cells showed topographical and chronological differences in the nucleic acid synthesis. The cells labeled with ^{3}H-thymidine increased after nidatory estradiol effects in the stromal cells around the blastocyst, but not in the epithelial cells. The results suggested that the presence of the blastocysts in the uterine lumen induced selective changes in the behavior of endometrial cells after nidatory estradiol effect showing the changes of DNAsynthesis.

As for a lower vertebrate, cell proliferation and migration of scleroblasts and their precursor cells during ethisterone-induced anal-fin process formation of the medaka, orizias latipes, was studied by LMRAG labeled with ^{3}H-thymidine(Uwa and Nagata 1976). The results showed that the labeling index in the posterior margin of the joint plate rapidly increased and the scleroblast population in the central portion increased simultaneously from the 3rd to 5th day of ethisterone treatment. These results indicated that the scleroblasts and their precursor cells migrated from the peripheral portion to the central portion along the proxi-distal axis of the joint plate.

3.8. The DNA Synthesis inthe Endocrine System

The endocrine system of mammals consists of the hypophysis, pineal body, thyroid, parathyroid, thymus, adrenal, pancreatic islet and genital glands. Among those organs, we studied macromolecular synthesis in the adrenal and steroid secreting cells of both sexes, the Leydig cells of the testis and the ovarian follicular cells in mice. On the other hand, incorporations of mercury chloride into human thyroid tissues as well as the intracellular localization of protein kinase C and keratin and vimentin were also studied (Nagata 2002).

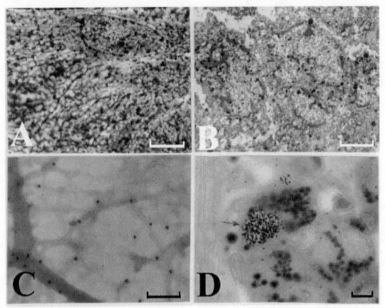

From Nagata, T. Special Cytochemistry in Cell Biology, In, Internat. Rev. Cytol. Vol. 211, No. 1, p. 115, 2001, Academic Press, San Diego, USA, London, UK.

Figure 19A. EM RAG of human thyroid cancer cells, labeled with ^{205}HgCl$_2$ in vitro, quick frozen, freeze-dried, embedded in Epoxy resin, dry-sectioned, and radioautographed by dry-mounting procedure for demonstrating soluble compounds, showing soluble ^{205}HgCl$_2$incorporations. x15,000.

Figure 19B. EM RAG of human thyroid cancer cells, labeled with ^{205}HgCl$_2$ in vitro, chemically fixed doubly in buffered glutaraldehyde and osmium tetroxide, dehydrated, embedded in Epoxy resin, wet-sectioned, and radioautographed by wet-mounting procedure for demonstrating insoluble compounds, showing insoluble ^{205}HgCl$_2$incorporations. x15,000.

Figure 19C. EM photo of a human thyroid cancer cell, fixed in paraformaldehyde and glutaraldehyde mixture, embedded in Lowicryl K4M, sectioned and immuno-stained with anti-keratin antibody by the protein-A gold technique, demonstrating keratin filaments.

Figure 19D. LM RAG of the zona glomelurosa of the adrenal cortex of a postnatal day 14 mouse labeled with ^{3}H-thymidine, showing DNA synthesis. x900.

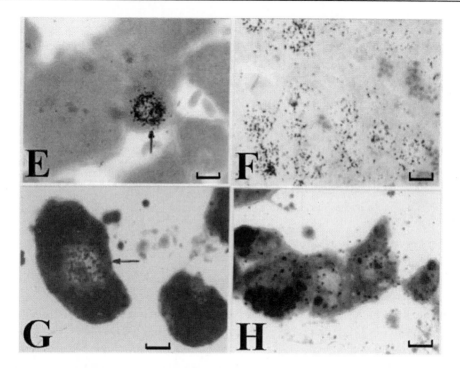

Figure 19E. LM RAG of the zona fasciculata of the adrenal cortex of a postnatal month 6 mouse labeled with ^{3}H-thymidine, showing DNA synthesis. x900.

Figure 19F. LM RAG of the zona gromeulosa of the adrenal cortex of a prenatal day 19 mouse labeled with ^{3}H-uridine, showing RNA synthesis. x1,000.

Figure 19G. LM RAG of the interstitial tissues of a postnatal month 12 male mouse labeled with ^{3}H-thymidine, showing DNA synthesis in Leydig cells. x1,000.

Figure 19H. LM RAG of the interstitial tissues of a postnatal day 3 male mouse labeled with ^{3}H-uridine, showing RNA synthesis in Leydig cells. x1,000.

Figure 19. LM and EM RAG of the endocrine organs.

3.8.1. The DNA Synthesis inthe Thyroid Gland

We first studied incorporations of mercury chloride into human thyroid tissues of both normal and cancer cells obtained from human patients (Nagata et al. 1977b, Nagata 1994a,b,c,d,e). The tissues were obtained surgically from human patients of both sexes in various ages suffering from the cancer of thyroids and the both normal and cancer cells were cut into small pieces (3x3 mm) aseptically which were incubated in a medium (Eagle's MEM) containing RI-mercury chloride (^{205}HgCl$_2$) and fixed either cryo-fixation at -196°C and freeze-dried or chemically fixed with buffered glutaraldehyde and osmium tetroxide. The former tissue blocks were processed for dry-mounting radioautography, while the latter were processed for conventional wet-mounting radioautography. The results revealed that the silver grains appeared much more in the cancer cells processed for freeze-fixation and dry-mounting radioatuography (Figure 19A) than the cancer cells processed for chemical fixation and wet-mounting radioatuography (Figure 19B), as well as much more in the cancer cells than the normal cells under the same conditions. On the other hand, PCNA/cycline and both keratin kinase C and vimentin were immunostained in connection to DNA synthetic activity. It was found that PCNA/cycline, keratin kinase C and vimentin antibodeis were localized around the filaments in the thyroid cancer cells (Figure 19C), demonstrating the relation between those antigens and DNA synthetic activity in cell cycle (Shimizu et al. 1993, Nagata 1994b,c, Gao et al. 1994).

3.8.2. The DNA Synthesis inthe Adrenal Gland

We studied the adrenal tissues of aging mice, both the adrenal cortex and the medulla, from embryo to postnatal 2 years. Some of the results were already published in several original articles (Ito 1996, Ito and Nagata 1996, Liang 1998, Liang et al. 1999, Nagata 1994, 1999c, 2000a,b, 2008a,b, 2009c,d,e,f,g,h,i,j, 2010a, Nagata et al. 2000b).

3.8.2.1. Structure of Aging Mouse Adreno-Cortical Cells

We studied the adrenal tissues of mice at various ages from embryo to postnatal 2 years (Ito 1996, Ito and Nagata 1996, Nagata 2008a,b, Nagata 2008a,b, 2009c,d,e,f,g,h,i,j). The adrenal tissues obtained from ddY strain mice at various ages from embryo day 19 to postnatal day 30 of both sexes, consisted of the adrenal cortex and the adrenal medulla. The former consisted of 3 layers, zona glomerulosa, zona fasciculata and zona reticularis, developing gradually with aging as observed by light microscopy (Figure 20). At embryonic day 19 and postnatal day 1, the 3 layers of the adreno-cortical cells, zona glomerulosa (Figure 19D), zona fasciculate (Figure 19E) and zona reticularis were composed mainly of polygonal cells, while the specific orientation of the 3 layers was not yet well established. At postnatal day 3, orientation of 3 layers, especially the zona glomerulosa became evident. At postnatal day 9 and 14, the specific structure of the 3 layers was completely formed and the arrangements of the cells in respective layers became typical especially at day 14 (Figure 19D) and month 1 (Figure 19E) to 24. Observing the ultrastructure of the adreno-cortical cells by electron microscopy, cell organelles including mitochondria were not so well developed at perinatal and early postnatal stages from embryonic day 19 to postnatal day 9. However, these cell organelles, mitochondria, endoplasmic reticulum, Golgi apparatus, appeared well developed similarly to the adult stages at postnatal day 14. The zona glomerulosa is the thinnest layer found at the outer zone, covered by the capsule, consisted of closely packed groups of columnar or pyramidal cells forming arcades of cell columns. The cells contained many spherical mitochondria and well developed smooth surfaced endoplasmic reticulum but a compact Golgi apparatus in day 14 animals. The zona fasciculate was the thickest layer, consisted of polygonal cells that were larger than the glomerulosa cells, arranged in long cords disposed radially to the medulla containing many lipid droplets. The mitochondria were less numerous and were more variable in size and shape than those of the glomerulosa cells, while the smooth surfaced endoplasmic reticulum were more developed and the Golgi apparatus was larger than the glomerulosa. In the zona reticularis, the parallel arrangement of cell cords were anastomosed showing networks continued to the medullar cells. The mitochondria were less numerous and were more variable in size and shape than those of the glomerulosa cells like the fasciculata cells, as well as the smooth surfaced endoplasmic reticulum were developed and the Golgi apparatus was large like the fasciculata cells. However, the structure of the adrenal cortex tissues showed changes due to development and aging at respective developmental stages.

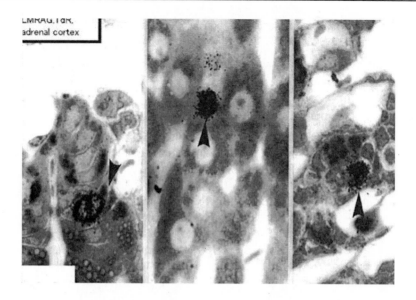

From Nagata, T. Annals Microsc.Vol. 10, p. 61, 2010, Microsc.Soc. Singapore.

Figure 20. LM RAG of a young mouse adrenal cortex, labeled with ^{3}H-thymine, showing DNA synthesis in 3 layers, zona glomerulosa (left), zona fasciculate (middle), zona reticularis (right).

3.8.2.2. The DNA Synthesis in the Adreno-Cortical Cells

Observing both LM and EM RAG labeled with ^{3}H-thymidine, demonstrating DNA synthesis, the silver grains were found over the nuclei of some adreno-cortical cells in S-phase of cell cycle mainly in perinatal stages at embryonic day 19, postnatal day 1 and day 3, while less at day 9 and day 14 to month 1-24 (Ito 1996, Ito and Nagata 1996, Nagata 2008a,b,c,d, 2009c,d,e). Those labeled cells were found in all the 3 layers (Figure 20), the zona glomerulosa (Figure 20 left), the zona fasciculata (Figure 20 middle) and the zona reticularis (Figure 20 right), at respective aging stages. In labeled adreno-cortical cells in the 3 layers the silver grains were mainly localized over the euchromatin of the nuclei and only a few or several silver grains were found over the mitochondria of these cells as observed by EM RAG (Figure 19D,E). To the contrary, most adreno-cortical cells were not labeled with any silver grains in their nuclei nor cytoplasm, showing no DNA synthesis after labeling with ^{3}H-thymidine. Among many unlabeled adreno-cortical cells, however, most cells in the 3 layers were observed to be labeled with several silver grains over their mitochondria due to the incorporations of ^{3}H-thymidine especially at the perinatal stages from embryonic day 19 to postnatal day 1, day 3, day 9 and 14 (Figure 19D). The ultrastructural localizations of silver grains over the mitochondria were mainly on the mitochondrial matrices and some over the cristae or membranes.

3.8.2.3. Number of Mitochondria of Aging Mouse Adreno-Corticl Cells

Preliminary quantitative analysis on the number of mitochondria in 10 adreno-cortical cells whose nuclei were labeled with silver grains and other 10 cells whose nuclei were not labeled in each aging group revealed that there was no significant difference between the number of mitochondria and the labeling indices (P<0.01). Thus, the number of mitochondria and the labeling indices were later calculated regardless whether their nuclei were labeled or not (Ito 1996, Ito and Nagata 1996, Nagata 2008a,b,c,d, 2009c,d,e). The results obtained from the number of mitochondria in adreno-cortical cells in the 3 layers of respective animals in 10 aging groups at perinatal stages, prenatal embryo day 19, postnatal day 1, 3, 9 and 14, showed an gradual increase from the prenatal day 19 (glomerulosa 12.5, fasciculata 14.9, reticularis 15/2/cell) to postnatal day 14 and month 1 and 2 (glomerulosa 62.2, fasciculata 64.0, reticularis 68.2/cell). The increase from embryo day 19 to postnatal month 2 was stochastically significant (P <0.01). Then, they did not change significanctly to month 12 and 24 (Figure 21A).

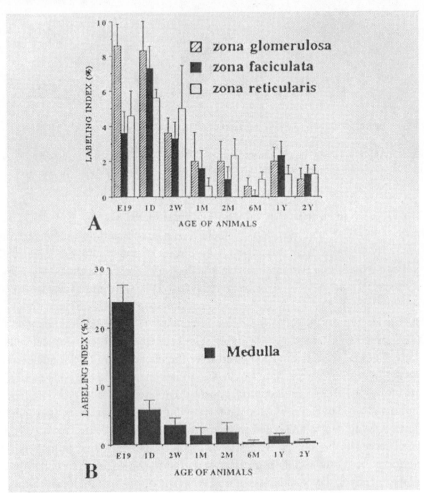

From Nagata, T. Radioautographology, General and Special.In, Progr.Histochem.Cytochem.Vol. 37, No. 2, p. 171, 2002, Urban & Fischer, Jena, Germany.

Figure 21A.The adrenal cortex.

Figure 21B.The adrenal medulla.

Figure 21. Histogram showing aging changes of average labeling indices in respective cell types of the adrenal glands of aging mice labeled with ^{3}H-thyidine showing DNA synthesis. Mean ± Standard Deviation.

3.8.2.4. Mitochondrial DNA Synthesis of Aging Mouse Adreno-Cortical Cells

The results of visual grain counts on the number of mitochondria labeled with silver grains obtained from 10 adreno-cortical cells in the 3 layers of each animal labeled with ^{3}H-thymidine demonstrating DNA synthesis in 10 aging groups at perinatal stages, prenatal embryo day 19, postnatal day 1, 3, 9 and 14, and month 1, 2, 6, 12 and 24 were reported previously (Ito 1996, Ito and Nagata 1996, Nagata 2008a,b,c,d, 2009c,d,e). The results demonstrated that the numbers of labeled mitochondria with ^{3}H-thymidine showing DNA synthesis gradually increased from prenatal embryo day 19 (glomerulosa 0.3, fasciculata 0.5, reticularis 0.4/cell) to postnatal day 14, month 1 and 2 (glomerulosa 5.3, fasciculata 5.0, reticularis 6.2/cell), reaching the maximum, then decreased to month 6, 12 and 24 (Figure 21A). The increase and decrease were stochastically significant (P <0.01).

3.8.2.5. The Labeling Index of DNA Synthesis in Mouse Adreno-Cortical Mitochondria

On the other hand, the labeling indices in respective aging stages were calculated from the number of labeled mitochondria, dividing by the number of total mitochondria per cell which were mentioned previously (Ito 1996, Ito and Nagata 1996, Nagata 2008a,b,c,d, 2009c,d,e,j, 2010e). The results showed that the labeling indices gradually increased from prenatal day 19 (glomerulosa 2.4, fasciculata 2.7, reticularis 2.6%) to postnatal day 14, month 1 and 2 (glomerulosa 8.5, fasciculata 7.8, reticularis 8.8%), reaching the maximum and decreased to month 6 (glomerulosa 4.1, fasciculata 4.2, reticularis 3.8%), 12 and 24 (Figure 21A). The increase and decrease were stochastically significant (P <0.01).

3.8.2.6. Structure of Aging Mouse Adreno-Medullary Cells

We studied the adrenal tissues of mice at various ages from embryo to postnatal 2 years. The adrenal tissues obtained from ddY strain mice at various ages from embryo day 19 to postnatal day 30 of both sexes, consisted of the adrenal cortex and the adrenal medulla (Ito 1996, Ito and Nagata 1996, Nagata 2008a,b,c,d, 2009c,d,e,g, 2010d,e,f,g). The former consited of 3 layers, zona glomerulosa, zona fasciculata and zona reticularis, developing gradually with aging as observed by light microscopy (Figure 19D, 19E), while the latter consisted of 2 cell types in one layer when observed by electron microscopy (Nagata 2009c,d,e, 2010d,e,f,g). At embryonic day 19 and postnatal day 1, the 3 layers of the adreno-cortical cells, zona glomerulosa, zona fasciculate and zona reticularis were composed mainly of polygonal cells, while the specific orientation of the 3 layers was not yet well established. However, the orientation of 3 layers became evident at day 3 and completely formed at day 14 (Figure 19D) and to month 1-24 (Figure 19E). On the other hand, the medulla consisted of only one layer containing 2 types of cells, adrenalin cell and noradrenalin cell. The former contains adrenalin granlules with low electron density, while the latter contains noradrenalin granules with high electron density.

3.8.2.7. The DNA Synthesis in the Adreno-Medullary Cells

The adrenal medulla is the deepest layer in the adrenal glands, surrounded by the 3 layers of the adrenal cortex as observed by light and electron microscopy (Ito 1996, Ito and Nagata 1996, Nagata 2008a,b,c,d, 2009c,d,e, 2010d,e,f,g), containing either adrenalin granules or noradrenalin granules. Quantitative analysis revealed that the numbers of mitochondria in both adrenalin and noradrenalin cells at various ages increased from fetal day 19 to postnatal month 1 due to aging of animals, respectively, but did not decrease to month 24, while the number of labeled mitochondria and the labeling indices of intramitochondrial DNA synthesis changed due to aging. When they were labeled with ^{3}H-thymidine silver grains appeared over some nuclei of both cell types at perinatal stages, but they appeared almost all the cell bodies containing mitochondria. Quantitative analysis revealed that the numbers of mitochondria in both adrenalin and noradrenalin cells at various ages increased from fetal day 19 to postnatal month 1 due to aging of animals, respectively, while the number of labeled mitochondria and the labeling indices of intramitochondrial DNA synthesis incorporating ^{3}H-thymidine increased from fetal day 19 to postnatal day 14 (2 weeks), reaching the maxima, and decreased to month 24. It was shown that the activity of intramitochnodrial DNA synthesis in the adrenal medullary cells in aging mice increased and decreased due to aging of animals (Figure 21B).

3.8.3. The DNA Synthesis in the Islet of Langerhans

When we studied macromolecular synthesis in the exocrine pancreatic cells of aging mice by LM and EMRAG we also studied the islet cells of Langerhans together with the exocrine cells, using RI labeled precursors such as ^{3}H-thymidine for DNA (Nagata and Usuda 1985, 1986, Nagata et al. 1986a,b), ^{3}H-uridine for RNA (Nagata and Usuda 1985, 1993b, Nagata et al. 1986a,b), ^{3}H-leucine for proteins (Nagata 2000, Nagata and Usuda 1993a, 1995), ^{3}H-glucosamine for glucides (Nagata et al. 1992), ^{3}H-glycerol for lipids (Nagata et al. 1988b, 1990). The results showed that the islets cells, A, B and C cells, incorporated those precursors to synthesize DNA, RNA, proteins and glucides. The labeling index of DNA synthesis and the densities of silver grains showing RNA, proteins and glucides syntheses were high at prenatal and earlier postnatal stages from day 1 to day 14, then decreased from 1 month to 1 years due to aging. However, the labeling indices by ^{3}H-thymidine and the grain counts by ^{3}H-uridine and ^{3}H-leucine in the endocrine cells were less than those in the exocrine cells at the same ages.

3.8.4. The DNA Synthesis in the Leydig Cells of the Testis

The cells of Leydig can be found in the interstitial tissues between the seminiferous tubules of the testis of mammals (Gao 1993, Gao et al. 1994, 1995a,b, Nagata et al. 2000b). They are identified as spherical, oval, or irregular in shape and their cytoplasms contain lipid droplets. We studied the macromolecular synthesis of the cells in the testis of several groups of litter ddY mice at various ages from fetal day 19 to postnatal aging stages up to 2 years senescence by LM and EMRAG using ^{3}H-thymidine, ^{3}H-uridine and ^{3}H-leucine incorporations.

The Leydig cells from embryonic stage to senescent stages were labeled with ^{3}H-thymidine as observed by LMRAG (Figure 19G). The changes of the numbers of labeled Leydig cells with the ^{3}H-thymidine incorporation into the nuclei showing the DNA synthesis were found in these cells at different aging stages. Only a few cells were labeled after ^{3}H-thymidine at embryonic day 19. At early postnatal stages, there was a slight increase of the number of labeled cells. The number of labeled cells from perinatal stage to postnatal 14 days and 1, 2, 6 months were similar to the values found at prenatal and early postnatal stages. The notable increases in the number of labeled cells of Leydig were found from 9 months to 2 years in senescence. The labeling indices with ^{3}H-thymidine in perinatal stages to postnatal 6 months were low (5-10%) but increased at 9 months and maintained high level (50-60%) to 2 years (Gao 1993, Gao et al. 1994, 1995a, Nagata et al. 2000b). The labeling indices at senescent stages still maintained a relatively high level and they were obviously higher than those of young animals. By electron microscopy, typical Leydig cells contained abundant cell organelles such as smooth surfaced endoplasmic reticulum, Golgi apparatus and mitochondria with tubular cristae. The silver grains were mainly localized over the euchromatin of labeled nucleus. Some of the grains were also localized over some of the mitochondria in both the nuclei labeled and unlabeled cells.

3.8.5. The DNA Synthesis inthe Ovarian Follicles

The ovarian follicles in the ovaries of mature mice are one of the steroid secreting organs in female animals. We studied the DNA and RNA synthesis of the follicular cells in the developing ddY mice ovaries in several aging groups at postnatal day 1, 3, 7, 14, 30 and 60 by LM and EMRAG using ^{3}H-thymidine and ^{3}H-uridine (Li 1994, Li and Nagata 1995). From the results it was shown that both DNA and RNA synthesis in the ovarian follicular cells were observed. Quantitative analysis, as expressed with labeling indices and grain counts, revealed that both increased significantly from postnatal day 1 to 3, then decreased from day 7 to 60 (Figure 18A). Comparing the results to other female genital cells, a paralleled alteration of both DNA and RNA synthesis was revealed between the ovarian follicular cells and other uterine or oviductal cells (Figure 18B,C). On the other hand, the glycoconjugate synthesis as shown by the uptake of ^{35}SO$_4$ in mouse ovary during the estrus cycle was also demonstrated (Li et al. 1992).

3.9. The DNA Synthesis inthe Nervous System

The nervous system consists of the central nervous system and the peripheral nervous system. The former is divided into the brains and the spinal cord, while the latter into the cerebrospinal system and the autonomous system. We studied macromolecular synthesis of the brains, the spinal cord in the cerebrospinal system and the autonomic peripheral nerves in the autonomous system by LM and EM RAG (Cui 1995, Cui et al. 1996, Izumiyama et al. 1987, Nagata 1965, 1967a, Nagata and Stegerda 1963, 1964, Nagata et al. 1999a).

3.9.1. The DNA Synthesis inthe Brains

The brains of mammals consist of the cerebrum, the cerebellum and the brain stem. We studied on DNA synthesis and protein synthesis in the cerebellum of aging mice as well as the glucose incorporation in the cerebrum of adult gerbils (Cui 1995, Cui et al. 1996, Izumiyama et al. 1987). The DNA synthesis was examined in the cerebella of 9 groups of aging ddY strain mice from fetal day 19, to postnatal day 1, 3, 8, 14 and month 1, 2, 6, 12, each consisting of 3 litter animals, using ^{3}H-thymidine, a DNA precursor, by LM and EMRAG (Cui 1995, Cui et al. 1996). The labeled nuclei, by the precursor, in both the neurons and glias, i.e., neuroblasts and glioblasts, were observed in the external granular layers of the cerebella of perinatal mice from embryonic day 19 (Figure 23A) to postnatal day 1, 3, 7 and day 14 by LMRAG and EMRAG. The labeled nuclei disappeared at postnatal 1 month. The peak of labeling index was at postnatal day 3 in both neuroblasts and glioblasts (Figure

24A,B). The glioblasts of the external granular layer migrated inward, some of them formed the Bergmann glia cells located between Purkinje cells. Labeled nuclei of neuroblasts and glioblasts in the internal granular layers were observed at perinatal stages. The maximum of the labeling index in the internal granular layer was at postnatal day 3, similarly to the external granular layer. The endothelial cells of the cerebellar vessels were progressively labeled from embryos to neonates, reaching the peak at 1 week after birth and decreasing thereafter.

3.9.2. The DNA Synthesis inthe Peripheral Nerves

We first studied the degeneration and regeneration of autonomous nerve cells in the plexuses of Auerbach and Meissner of the jejunums of 15 dogs which were operated upon to produce experimental ischemia of the jejunal loops by perfusing with Tyrode's solution via the mesenteric arteries for 1, 2, 3 and 4 hours (Nagata 1965, 1967, Nagata and Steggerda 1963, 1964). Tissue blocks were obtained from the deganglionated portions and the adjoining normal portions, which were fixed in Carnoy's fluid, embedded in paraffin, sectioned and stained with buffered thionine, methyl-green and pyronine and PAS. Some animals were injected with either ^{3}H-thymidine or ^{3}H-cytidine and the intestinal tissues obtained from ischemic portions and normal portions were processed for LMRAG. The results revealed that the ganglion cells in Auerbach's plexus showed various degenerative changes in accordance with the duration of ischemia. After 4 hours ischemia, most of the ganglion cells in Auerbach's plexus were completely destroyed. The degenerative changes in Auerbach's plexus after 4 hours ischemia were irreversible after 1 week recovery. The ganglion cells in the Meissner's plexus, on the other hand, were less sensitive to the ischemia. They recovered completely even after 4 hours ischemia. The PAS positive substances in degenerative ganglion cells in both plexuses decreased immediately after 4 hours ischemia. The DNA contents of ganglion cells in both Auerbach's and Meissner's plexus did not show any change before and after ischemia. The RNA contents decreased immediately after the ischemia (Nagata and Steggerda 1963, 1964). The number of binucleate cells in ganglion cells in both Auerbach's and Meissner's plexuses after 4 hours ischemia increased to 4.6% and 5.7% respectively. In contrast, in the non-ischemic normal control preparations, the binucleate cells occurred only 0.5% and 1.8% in Auerbach's and Meissner's plexus respectively. The high frequency of binucleate cells in the ganglion cells persisted for more than 100 days after the ischemia, indicating a possible regeneration of ganglion cells. The radioautographic study revealed that there was no evidence for DNA synthesis in both Auerbach's and Meissner's plexus from either ischemic or normal loops. The RNA synthesis was observed to be higher in gaglion cells in normal loop than ischemic loop and higher in Auerbach's than in Meissner's as expressed by grain counting. It was higher in binucleate ganglion cells than in mononucleate cells.

3.10. The DNA Synthesis in the Sensory System

The sensory system consists of five organs, i. e., the visual organ or the eye, stato-acoustic organ or the ear, gustatory organ or the tongue, olfactory organ or the nose, and the dermis or the skin. Among these sensory organs, we mainly studied the visual organ and the skin (Gunarso 1983a,b, Gunarso et al. 1996, 1997, Gao et al. 1992a,b, 1993, Kong 1993,

Kong and Nagata 1994, Kong et al. 1992a,b, Nagata 1998, 1999, 2000, Nagata et al. 1994, Toriyama 1995).

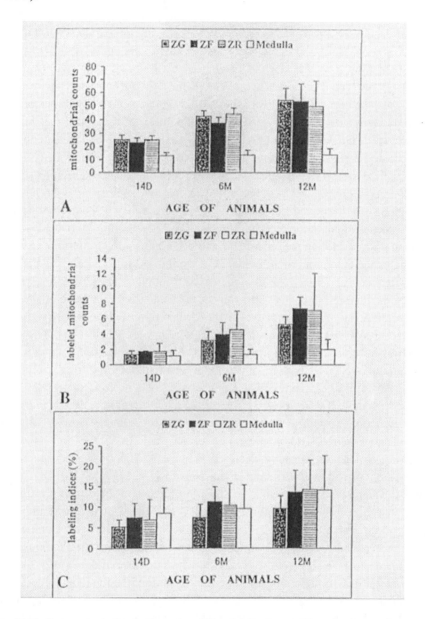

From Nagata, T. Radioautographology, General and Special. In, Progr. Histochem. Cytochem. Vol. 37, No. 2, p. 173, 2002, Urban & Fischer, Jena, Germany.

Figure 22A. The number of mitochondria per cell.

Figure 22B. The number of labeled mitochondria per cell.

Figure 22C. The mitochondrial labeling index.

Figure 22. Histogram showing aging changes of the respective cell types of the adrenal glands of aging mice labeled with ^{3}H-uridine showing RNA synthesis. Mean ± Standard Deviation.

3.10.1. The DNA Synthesis in the Eye

The visual organ consists of the eye and its accessory organs. The eye of mammals consists of the cornea, iris, cilliary body, lens, retina, choroid and sclera. We studied mainly the macromolecular synthesis in the retina of chickens and mice (Nagata 2000f). The nucleic acid syntheses, both DNA and RNA, were first studied in the ocular tissues of white Leghorn chick embryos from day 1 to day 14 incubation by LM and EMRAG (Gunarso 1984a,b, Gunarso et al. 1997, Gao et al. 1992a,b, 1993, Kong 1993, Kong and Nagata 1994, Kong et al. 1992a,b, Nagata et al. 1994). It was shown that the labeled cells with silver grains due to ^{3}H-thymidine were most frequently observed in the nuclei of the retinal cells in the posterior region of the day 2 chick embryo optic vesicle (Figure 23C) and the labeled cells moved from anterior to posterior regions. The number of labeled cells as expressed by labeling index (%), was more in the posterior regions than the anterior and the equatorial regions and more in the outer portions than in the inner portions at day 2, but the labeling index became more in the anterior regions than the equatorial and posterior regions at day 3, 4 and 7 and it became more in the inner portions than in the outer portions at day 7, decreasing from day 2 to 3, 4 and 7 in each regions (Figure 24). On the other hand, the silver grains due to ^{3}H-uridine were observed over the nuclei and cytoplasm of all retinal cells from day 2 to 7 (Figure 23D) and the number of silver grains incorporating ^{3}H-uridine increased from day 1 to day 7 and it was more in the anterior regions than in the posterior regions at the same stage (Gunarso et al. 1996). On the other hand, DNA and RNA syntheses in the ocular tissues of aging ddY mice were also studied (Gao et al. 1993, Kong and Nagata 1994, Kong et al. 1992a,b). The ocular tissues taken out from several groups of litter ddY mice at ages varying from fetal day 9, 12, 14, 16, 19 to postnatal day 1, 3, 7, 14 were labeled with ^{3}H-thymidine in vitro and radioautographed (Gao et al. 1992a,b, 1993, Kong 1993, Kong et al. 1992a,b, Toriyama 1995). Silver grains showing DNA synthesis were localized over the nuclei of retinal cells and pigment epithelial cells in the anterior, equatorial and posterior regions of perinatal animals (Figure 25). The labeling indices of the retina and pigment epithelium were higher in earlier stages than in later stages, during which they steadily declined (Figure 25AB). However, the retina and the pigment epithelium followed different courses in their changes of labeling indices during embryonic development. In the retina, the labeling indices in the vitreal portions were more than those in the scleral portions during the earlier stages. However, the indices of scleral portions were more than those in the vitreal portions in the later stages. Comparing the three regions of the retinae of mice, the anterior, equatorial and posterior regions, the labeling indices of the anterior region were generally higher than those of the equatorial and posterior regions (Figure 25A). In the pigment epithelium (Figure 25B), the labeling indices gradually increased in the anterior region, but decreased in the equatorial and the posterior regions through all developmental stages. These results suggest that the proliferation of both the retina and pigment epithelium in the central region occurred earlier than those of the peripheral regions (Nagata 1999a, Gao et al. 1992a,b, Kong 1993, Kong and Nagata 1994, Kong et al. 1992a, b). In the juvenile and adult stages, however, the labeled cells were localized at the middle of the bipolar-photoreceptor layer of the retina, where was supposed to be the undifferentiated zone.

In the corneas of aging mice, DNA synthesis was observed in all 3 layers, i. e., the epithelial, stromal and endothelial layers, at perinatal stages (Gao et al. 1993). The labeled cells with ^{3}H-thymidine were localized in the epithelial cells at prenatal day 19, postnatal day 1, 14 (Figure 23G) to 1 year, while the labeled cells in the stromal and endothelial layers were less. The labeling index of the corneal epithelial cells reached a peak at 1 month after birth and decreased to 1 year, while the indices of the stromal and endothelial cells were low and reached a peak at 3 days after birth and disappeared completely from postnatal 1 month to 1 year (Nagata 1999c).

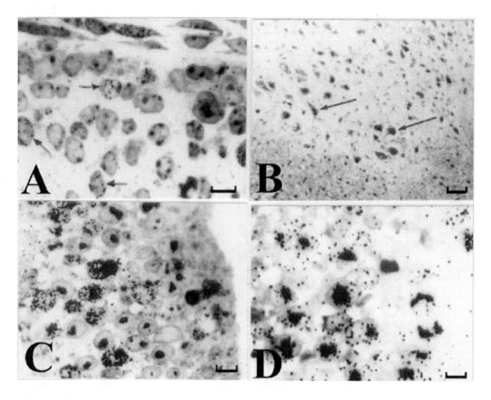

From Nagata, T. Special Cytochemistry in Cell Biology, In, Internat. Rev. Cytol. Vol. 211, No. 1, p. 122, 2001, Academic Press, San Diego, USA, London, UK.

Figure 23A. LM RAG of a prenatal day 19 mouse cerebellum labeled with ^{3}H-thymidine, showing DNA synthesis. x900.

Figure 23B. LM RAG of the spinal cord of a postnatal day 14 mouse immunostained with rabbit anti-TGF-β1 polyclonal IgG followed by ABC method, showing the ventral horn motoneurons are strongly positive. x70.

Figure 23C. LM RAG of the optic vesicle of a day 2 chick embryo labeled with ^{3}H-thymidine, showing DNA synthesis. x750.

Figure 23D. LM RAG of the optic vesicle of a day 2 chick embryo labeled with ^{3}H-uridine, showing RNA synthesis. x750.

Figure 23. (Continued)

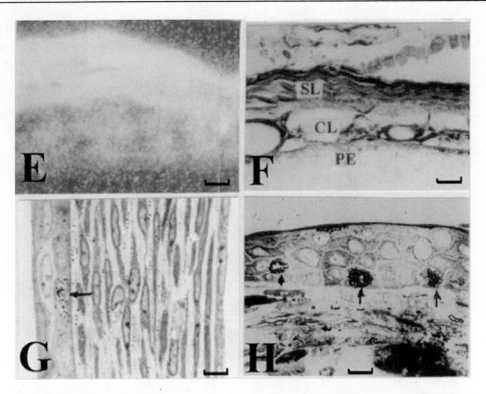

Figure 23E. Dark-field LM photo of the skleral layer (top), choroid layer (middle) and pigment epithelium (bottom) of an adult 1 month old mouse demonstrating intense silver grains by in situ hybridization for TGF-β1 mRNA. x450.

Figure 23F. Bright-field LM photo of the skleral layer (top), choroid layer (middle) and pigment epithelium (bottom) of an adult 1 month old mouse demonstrating intense silver grains by in situ hybridization for TGF-β1 mRNA. x450.

Figure 23G. LM RAG of the cornea of a postnatal day 14 mouse labeled with ^{3}H-thymine, showing DNA synthesis in the epithelial nucleus (arrow) as well as in the stroma. x900.

Figure 23H. LM RAG of the skin of the fore-limb of a salamander at 6weeks after hatching labeled with ^{3}H-thymidine, showing DNA synthesis. x900.

Figure 23. LM RAG of the neuro-sensory cells.

In the cilliary body, the labeled cells were located in the cilliary and pigment epithelial cells, stromal cells and smooth muscle cells from prenatal day 19 to postnatal 1 week, but no labeled cells were observed in any cell types from postnatal day 14 to 1 year (Nagata et al. 1994). The labeling indices of all the cell types in the cilliary body were at the maximum at prenatal day 19 and decreased gradually after birth reaching 0 at postnatal day 14. On the other hand, when the ocular tissues were labeled with ^{3}H-uridine, silver grains appeared over all cell types at all stages of development and aging (Toriyama 1995, Nagata 2000f). The grain counts in the retina (Figure 23A) and the pigment epithelium (Figure 23B) increased from prenatal day 9 to postnatal day 1 in the retinal cells, while they increased from prenatal day 12 to postnatal day 7 in the pigment epithelial cells (Nagata 1999a,b, Nagata et al. 1994).

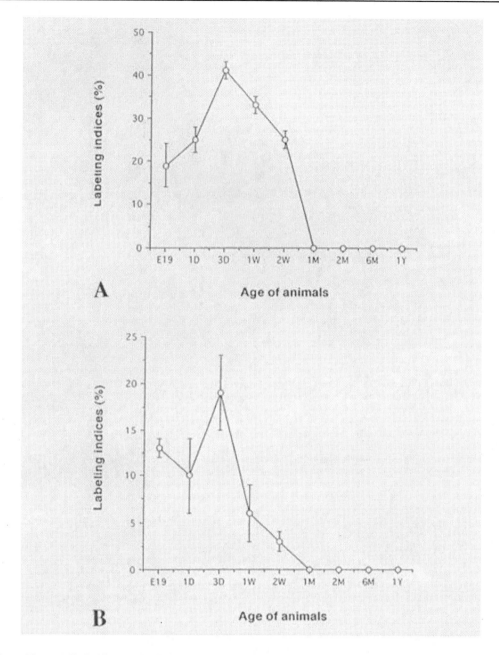

A   Age of animals

B   Age of animals

From Nagata, T. Radioautographology, General and Special.In, Progr.Histochem.Cytochem.Vol. 37, No. 2, p. 186, 2002, Urban & Fischer, Jena, Germany.

Figure 24A.The neuroblasts in the extragranular layer of the cerebella.

Figure 24B.The glioblasts in the extragranular layer of the cerebella.

Figure 24. Transitional curves of the labeling indices of respective cell types in the cerebella of aging mice labeled with ^{3}H-thymidine, showing DNA synthesis. Mean ± Standard Deviation.

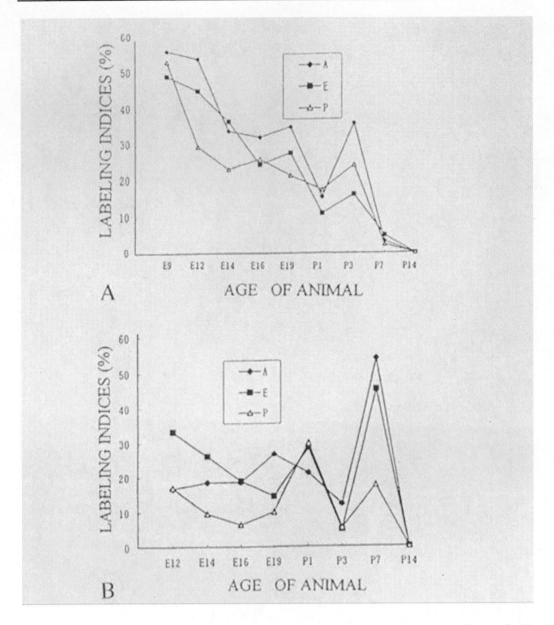

From Nagata, T. Radioautographology, General and Special.In, Progr.Histochem.Cytochem.Vol. 37, No. 2, p. 191, 2002, Urban & Fischer, Jena, Germany.

Figure 25A.The labeling index in the retina.

Figure 25B.The labeloing index in the pigment epithelium.

Figure 25. Transitional curves of the labeling indices in the three regions (A: anterior, E: eqauator, P: posterior) of the retina and the pigment epithelium aging mice labeled with[3]H-thymidine, showing DNA synthesis.Mean ± Standard Deviation.

3.10.2. The Skin

The skin which covers the surface of the animal body can be divided into 3 layers, the epidermis, the dermis and the hypodermis. We studied only the epidermal cells of young salamanders after hatching by radioautography (Nagata 1998c). The fore-limbs and hind-limbs of salamanders were composed of skeletons consisting of bones and cartilages which were covered with skeletal muscles, connective tissues and epidermis consisting of stratified squamous epithelial cells in the outermost layer. We observed both the cartilage cells in the bone and the epithelial cells in the epidermis to compare the two cell populations. The skin of a salamander consisted of epidermis and dermis or corium which was lined with connective tissue layers designated as the subcutaneous layer. The former consisted of stratified squamous epithelium, while the latter consisted of dense connective tissues. The epithelial cells in the juvenile animals at 4 weeks after hatching were cuboidal in shape and not keratinized. Radioautograms labeled with ^{3}H-thymidine at this stage showed that many cells were labeled demonstrating DNA synthesis at both the superficial and deeper layers (Figure 23H), resulting very high labeling index. At 6 weeks after hatching, the superficial cells changed their shape from cuboidal to flattened squamous, while the deeper and basal cells remained cuboidal. The numbers of labeled cells were almost the same as the previous stage at 4 weeks, but they were localized at the basal layer. The shape of epithelial cells in juvenile animals at 8, 9, 10, and 11 weeks differentiated gradually forming the superficial corneum layer which appeared keratinized and the deeper basal layer. Radioautograms at these stages showed that the labeled cells remarkably reduced as compared with that of 4 and 6 weeks. In the adult salamanders at 8 months up to 12 months, the dermal and epidermal cells showed complete mature structure and examination of radioautograms revealed that the labeled cells were localized at only the basal cell layer and their number reached very low but at constant level. No difference was found on the morphology and labeling between the fore-limbs and hind-limbs at any stages. Comparing the labeling indices of both epidermal cells and the cartilage cells in the limbs, the labeling index of the epidermal cells was higher than the cartilage cells. The index of the dermal cells in the hind-limbs was at its maximum about 25% at 4 weeks, and fell down markedly with time from 6 weeks to 9 weeks. The labeling index of epidermal cells of the hind-limbs, on the other hand, had its maximum about 23% at 6 weeks, increasing from 20% at 4 weeks, and decreasing to about 18% at 8 weeks, then fell progressively with time, dropped to 5% at 9 weeks. The labeling indices of the epidermal cells of both fore-limbs and hind-limbs were almost the same from 9 weeks to 12 months, keeping low constant level about 4-5%, but never reaching 0. These results indicated that the cutaneous cell belonged to the renewing cell population(Nagata 1998c).

3.11. The DNA Synthesis in the Tumor Cells

We carried out several experiments dealing with the nucleic acid synthesis in some malignant tumor cells by means of LM and EM RAG. The DNA and RNA syntheses in nuclei and mitochondria of cultured HeLa cells, an established cell line obtained from the carcinoma of the human uterus, or IgG myeloma cells from a human patient, labeled with either ^{3}H-thymidine or ^{3}H-uridine were demonstrated (Nagata 1972b,c,d, Fujii et al. 1980). The incorporations of these precursors increased or decreased depending upon the aging of isolated cells in vitro.

4. The RNA Synthesis

The RNA (ribonucleic acid) contained in cells can bedemonstrated either by morphological histochemical techniques staining tissue sections such as methyl green-pyronin staining or by biochemical techniques homogenizing tissues and cells. To the contrary, the synthetic RNA or newly synthesized RNA but not all the RNA in the cells can be detected as macromolecular synthesis together with other macromolecules such as DNA or proteins in various organs of experimental animals by either morphological or biochemical procedures employing RI-labeled precursors. We have studied the sites of macromolecular synthesis in almost all the organs of mice during their aging from prenatal to postnatal development to senescence by means of microscopic radioautography, one of the morphological methods (Nagata 1992, 1994b,c, 1996a,b,c, 1997a). The results obtained from RNA synthesis should be here described according to the order of organ systems in anatomy or histology. In contrast to the results obtained from DNA synthesis of almost all the organs, we have studied only several parts of the organ systems. The skeletal system, the muscular system or the circulatory system were not so much studied.

4.1. The RNA Synthesis in the Digestive System

We have mainly studied the digestive system, but not all the digestive organs yet concerning to the RNA synthesis. Study on RNA synthesis was carried out on the small intestines and the liver and the pancreas.

4.1.1. The RNA Synthesis in the Intestines

We studied the RNA synthesis of the small intestines of mice after feeding or refeeding under the restricted conditions (Nagata 1966). Five groups of ddY mice, each consisting of 5 individuals, total 25, were injected with ^{3}H-uridine, an RNA precursor, and sacrificed at different time intervals after feeding. The animals of the first group were injected with ^{3}H-uridine at 9 a.m. and fed at 10 a.m. for 30 min. and sacrificed at 11 a.m. 1 hour after the feeding and 2 hours after the injection, the 2nd group was sacrificed at 1 p.m. 3 hours after feeding and 4 hours after the injection, the 3rd group at 5 p.m., 7 and 8 hours later, the 4th group at 9 a.m. on the next day 23 and 24 hours later, and finally the 5th group at 1. p.m. on the next day 3 hours after refeeding and 28 hours after the injection. Then, the jejunums obtained from each animal were prepared for isolated cell radioautograms according to Nagata et al. (1961). The results demonstrated that the grain counts in mononucleate villus cells reached the maximum (20-30 grains per cell) 4 hours after injection and decreased (10-20/cell) after 28 hours, while the counts in mononucleate villus cells only increased gradually from 4 hours (10/cell) to 28 hours (20/cell). In contrast to this, the grain counts of binucleate cells which appeared in villus cells increased parallely to the mononucleate villus cells (10-20/cell). It was concluded that the RNA synthesis in the jejunal epithelial cells was high in the following order: mononucleate crypt cells, binucleate cells and mononucleate villus cells. These results revealed that the feeding or refeeding affected the RNA synthesis of the intestinal epithelial cells (Nagata 1966).

We also studied intracellular localization of mRNA in adult rat hepatocytes localizing over the peroxisomes by means of in situ hybridization technique (Usuda and Nagata 1992, 1995, Usuda et al. 1992). However, its relationship to the aging of animals was not yet studied.

4.1.2. The RNA Synthesis in the Liver

The RNA synthesis in the liver was studied by ^{3}H-uridine RAG. When the RI-labeled precursor ^{3}H-uridine was administered to experimental animals, or cultured cells were incubated in a medium containing ^{3}H-uridine in vitro and LMRAG was prepared, silver grains first appeared over the chromatin of the nucleus and nucleolus of all the cells within several minutes, then silver grains spread over the cytoplasm within 30 minutes showing messenger RNA and ribosomal RNA (Nagata 1966, Nagata and Nawa 1966a,b, Nagata et al.1969). We studied quantitative changes of RNA synthesis in the livers of adult mice before and after feeding by incorporations of ^{3}H-uridine. Five groups of ddY mice, each consisting of 5 individuals, total 25, were injected with ^{3}H-uridine and sacrificed at different time intervals. The animals of the first group were injected with ^{3}H-uridine at 9 a.m. and fed at 10 a.m. for 30 min. and sacrificed at 11 a.m. 1 hour after the feeding and 2 hours after the injection, the 2nd group was sacrificed at 1 p.m. 3 hours after feeding and 4 hours after the injection, the 3rd group at 5 p.m., 7 and 8 hours later, the 4th group at 9 a.m. on the next day 23 and 24 hours later, and finally the 5th group at 1. p.m. on the next day 3 hours after refeeding and 28 hours after the injection. Then, the livers were taken out from each animal, prepared for isolated cell radioautograms according to Nagata et al. (1961). The results demonstrated that the grain counts in both mononucleate and binucleate hepatocytes before feeding (15-25 grains per cell) increased 4 hours after feeding (30-40 grains per cell), reached the maximum in 24 hours (40-50 grains per cell), then decreased on the next day (30-40 grains per cell). It was concluded that the RNA synthesis in the binucleate hepatocytes was a little higher than the mononucleate hepatocytes at the same stages and both increased and decreased after feeding. These results revealed that the feeding or refeeding affected the RNA synthesis of the livers (Nagata 1966).

Then, we studied aging changes of ^{3}H-uridine incorporation in the livers and pancreases of aging mice at various ages from prenatal embryos to postnatal aged mice by LM and EMRAG (Ma and Nagata 1990b, Nagata 1999c). When aged mice were injected with ^{3}H-uridine, LM and EMRAG showed that silver grains were localized over the nucleoli, nuclear chromatin (both euchromatin and heterochromatin), mitochondria and rough surfaced endoplasmic reticulum of hepatocytes (Figure 8C) and other types of cells such as sinusoidal endothelial cells, Kupffer's cells, Ito's fat-storing cells (Figure 8D), ductal epithelial cells, fibroblasts and haematopoietic cells in the livers at various ages. By quantitative analysis, the total number of silver grains in nucleus, nucleolus and cytoplasm of each hepatocyte increased gradually from fetal day 19 to postnatal days, reached the maximum at postnatal day 14 (30%), then decreased to 24 months (5%). The number of silver grains in nucleolus, when classified into two compartments, grains over granular components and those over fibrillar components both increased paralelly after birth, reached the maxima on day 14 (granular 6-7, fibrillar 1-2/per cell), then decreased to 24 months with aging. However, when the ratio (%) of silver grains over euchromatin, heterochromatin of the nuclei and granular components and fibrillar components of the nucleoli are calculated, the ratio remained constant at each aging point.

4.4.5.2. Mitochondrial RNA synthesis in hepatocytes

The intramitochondrial RNA synthesis was first found in the cultured HeLa cells and the cultured liver cells in vitro using EM RAG (Nagata 1972c, d). Then, it was also found in any other cells in either in vitro or vivo (Nagata et al. 1977c, Nagata 2002). Observing light microscopic radioautograms labeled with ^{3}H-uridine, the silver grains were found over both the karyoplasm and cytoplasm of almost all the cells not only at the perinatal stages from embryo day 19 to postnatal day 1, 3, 9, 14, but also at the adult and senescent stages from postnatal month 1 to 2, 6, 12 and 24 (Nagata 2007a,c,d,e.f. Nagata and Ma 2005b). By electron microscopic observation, silver grains were detected in most mononucleate hepatocytes in respective aging groups localizing not only over euchromatin and nucleoli in the nuclei but also over many cell organelles such as endoplasmic reticulum, ribosomes, and mitochondria as well as cytoplasmic matrices from perinatal stage at embryonic day 19, postnatal day 1, 3, 9, 14, to adult and senescent stages at postnatal month 1, 2, 12 and 24. The silver grains were also observed in binucleate hepatocytes at postnatal day 1, 3, 9, 14, month 1, 2, 6, 12 and 24. The localizations of silver grains over the mitochondria were mainly on the mitochondrial matrices but a few over the mitochondrial membranes and cristae when observed by high power magnification.

As the results, it was found that almost all the hepatocytes were labeled with silver grains showing RNA synthesis in their nuclei and mitochondria. Preliminary quantitative analysis on the number of mitochondria in 10 mononucleate hepatocytes whose nuclei were intensely labeled with many silver grains (more than 10 per nucleus) and other 10 mononucleate hepatocytes whose nuclei were not so intensely labeled (number of silver grains less than 9) in each aging group revealed that there was no significant difference between the number of mitochondria, number of labeled mitochondria and the labeling indices in both types of hepatocytes ($P<0.01$). Thus, the number of mitochondria and the labeling indices were calculated in 10 hepatocytes selected at random in each animal in respective aging stages regardless whether their nuclei were very intensely labeled or not. The results obtained from the number of mitochondria in mononucleate hepatocytes per cellular profile area showed an increase from the prenatal day (mean ± standard deviation 26.2± /cell) to postnatal day 1 to day 14 (38.4-51.7/cell), then to postnatal month 1-2 (53.7-89.2/cell), reaching the maximum, then decreased to year 1-2 (83.7-80.4/cell) and the increase was stochastically significant ($P<0.01$). The results of visual grain counts on the number of mitochondria labeled with silver grains obtained from 10 mononucleate hepatocytes of each animal labeled with ^{3}H-uridine demonstrating RNA synthesis in 10 aging groups at perinatal stages, prenatal embryo day 19, postnatal day 1, 3, 9 and 14, month 1, 6 and year 1 and 2, were counted. The labeling indices in respective aging stages were calculated from the number of labeled mitochondria and the number of total mitochondria per cellular profile area, respectively. The results showed that the numbers of labeled mitochondria with ^{3}H-uridine showing RNA synthesis increased from prenatal embryo day 19 (3.3/cell) to postnatal month 1 (9.2/cell), reaching the maximum, and then decreased to month 6 (3.5/cell) and again increased to year 1 (4.0/cell) and year 2 (4.3/cell), while the labeling indices increased from prenatal day 19 (12.4%) to postnatal month 1 (16.7%), reaching the maximum, then decreased to year 1 (4.8%) and year 2 (5.3%). Stochastical analysis revealed that the increases and decreases of the number of labeled mitochondria from the perinatal stage to the adult and senescent stage were significant in contrast that the increases and decreases of the labeling indices were not significant ($P<0.01$). As for the binucleate hepatocytes, on the other hand, because the appearances of binucleate hepatocytes were not so many in the embryonic stage, only several binucleate cells (5-8 at least) at respective stages when enough numbers of binucleate cells available from postnatal day 1 to year 2 were analyzed. The results of visual counts on the number of mitochondria labeled with silver grains obtained from several (5 to 8) binucleate hepatocytes labeled with ^{3}H-uridine demonstrating RNA synthesis in 8 aging groups at perinatal stages, postnatal day 1, 9, 14, and month 1, 2, 6, and year 1 and 2, were counted and the labeling indices in respective aging stages were calculated from the number of labeled mitochondria and the number of total mitochondria per cellular profile area calculated, respectively. The results showed that the number of labeled mitochondria increased from postnatal day 1 (2.3/cell) to day 9 (5.2/cell) and remained almost constant around 4-5, but the labeling indices increased from postnatal day 1 (2.1%) to postnatal day 9 (13.6%), remained almost constant around 13% (12.5-13.6%) from postnatal day 9 to month 1, then decreased to month 2 (6.1%) to month 6 (3.9%), and slightly increased to year 1 (6.3%) and 2 (5.3%). The increases and decreases of the number of labeled mitochondria and the labeling indices in binucleate hepatocytes were stochastically not significant (P<0.01).

4.3. The RNA Synthesis in the Pancreas

On the other hand, LM and EMRAG of pancreas of mouse injected with ^{3}H-uridine demonstrated its incorporation into exocrine and then in endocrine cells, and more in pancreatic acinar cells (Figures. 10C,D) than in ductal or centro-acinar cells (Nagata and Usuda 1986a,b, Nagata et al. 1986). Among the acinar cells, the number of silver grains increased after birth to day 14 and then decreased with aging. Quantification of silver grains in the nucleoli, chromatin, and cell body were carried out by X-ray microanalysis (Nagata 1991, 1993, 2004, Nagata and Usuda 1985), which verified the results obtained by visual grain counting. In EMRAG obtained from the pancreas of fetal day 19 embryos, newborn day 1 and newborn day 14 mice labeled with ^{3}H-uridine, demonstrating RNA synthesis, the number of silver grains in the nucleoli, nuclear chromatin and cytoplasm increased (Nagata 1985, 1991, 1993a,b, 2004, Nagata and Usuda 1985, 1986). In order to quantify the silver contents of grains observed over the nucleoli, nuclei and cytoplasm, X-ray spectra were recorded by energy dispersive X-ray microanalysis (JEM-4000EX TN5400), demonstrating Ag-Kα peaks at higher energies. Thus, P/B ratios expressing relative silver contents were determined and compared between the two age groups. The results obtained by X-ray microanalysis in different cell compartments at postnatal day 1 and day 14 and he results obtained by visual grain counting in different cell compartments in day 1 day and day 14 animals were compared. The number of silver grains was calculated to express the counts per unit area to be compared with the XMA counts. These two results, the silver content analyzed by X-ray microanalysis and the results obtaining from visual grain counting were in good accordance with each other.

4.4. The RNA Synthesis in the Respiratory System

We studied the lungs of aging mice among the respiratory organs after administration of ^{3}H-uridine at various ages from prenatal embryonic day 16 to postnatal senescent month 22 as observed by LM and EM RAG (Figures. 12G,H).

4.4.1. RNA Synthesis of Aging Mouse Pulmonary Cells

When the lung tissues of mice were labeled with ^{3}H-uridine, RNA synthesis was observed in all cells of the lungs at various ages by RAG (Sun 1995, Sun et al. 1997b, Nagata 2002). Observing the light microscopic radioautograms labeled with ^{3}H-uridine, the silver grains were found over both the karyoplasm and cytoplasm of almost all the cells not only at the perinatal stages from embryo day 16, 18, to postnatal day 1, 3, 7, 14, but also at the adult and senescent stages from postnatal month 1 to 2, 6, 12 and 22. The number of silver grains,by electron micrography, changed with aging. The grain counts in type I epithelial cells increased from the 1st day after birth and reached a peak at 1 week and decreased gradually to month 22, while the counts in type II epithelial cells, interstitial cells and endothelial cells increased from embryo day 16 and reached peaks at 1 week after birth, then decreased to senescence.

4.4.2. Mitochondrial RNA Synthesis of Aging Mouse Pulmonary Cells

By electron microscopic radioautography, silver grains were observed in most pulmonary cells in respective aging groups localizing not only over euchromatin and nucleoli in the nuclei but also over many cell organelles such as endoplasmic reticulum, ribosomes, and mitochondria as well as cytoplasmic matrices from perinatal stage at embryonic day 16, 18, to postnatal day 1, 3, 7, 14, to adult and senescent stages at postnatal month 1, 2, 6, 12 and 22 (Sun 1995, Sun et al. 1997b, Nagata 2002). The localizations of silver grains over the mitochondria were mainly on the mitochondrial matrices but a few over the mitochondrial membranes and cristae when observed by high power magnification. However, quantitative analyses on the number of mitochondria, the number of labeled mitochondria and the labeling index were not performed because enough number of EMRAG was not obtained.

4.5. The RNA Synthesis in the Urinary system

We studied the RNA synthesis of the kidneys of aging mice among the urinary organs after the administration of ^{3}H-uridine at various ages.

4.5.1. The RNA Synthesis in the Kidney

The RNA synthesis by incorporation of ^{3}H-uridine into the kidneys of aging mice was studied by LM and EM RAG (Hanai and Nagata 1994a,b, Nagata 2002). When the kidneys of several groups of aging mice from embryo to postnatal 1 year were radioautographed with ^{3}H-uridine either in vitro (embryonic day 15, 19 and postnatal day 1) and in vivo (embryonic day 19, postnatal day 1, 8, month 1, 2, 12), RNA synthesis was observed in all the cells of the kidney at various ages. The numbers of silver grains demonstrating the incorporation of ^{3}H-uridine in glomeruli (34.6 per cell) and uriniferous tubules (56.4 per cell) were higher in the superficial layer than those (15.6 and 18.6 per cell) in the deeper layer at embryonic day 15 and decreased gradually with aging. These results demonstrated the aging changes of RNA synthesis in the kidney.

5. THE RNA SYNTHESIS IN THE REPRODUCTIVE SYSTEM

We studied the RNA synthesis of both the male and female reproductive organs of aging mice after the administration of ^{3}H-uridine at various ages.

5.1. The RNA Synthesis in the Testis

We studied the RNA syntheses in aging mouse testis by LM and EM RAG, demonstrating the incorporations of ^{3}H-uridine into various cells of the seminiferous tubules (Gao 1993, Gao et al. 1994, Nagata 2002). The RNA synthesis of various cells in the seminiferous tubules was studied using ^{3}H-uridine, while the protein synthesis was by ^{3}H-leucine incorporations. Silver grains due to ^{3}H-uridine demonstrating RNA synthesis were observed over the nuclei and cytoplasm of all spermatogonia, spermatocytes, Sertoli's cells, myoid cells of immature mice at perinatal stages (Figure 15E) at day 1 and 3, as well as in mature and senescent mice from month 1, 6 to year 1 and 2. The synthetic activities of spermatogonia, Sertoli's cells and myoid cells as shown by grain counting with ^{3}H-uridine, as expressed by grain counting, were low (2-8 grain counts per 10 μm^2 at the embryonic and neonatal stages but increased at adult stages and maintained high levels (10-20 grain counts per 10 μm^2) until senescence. These results showed that DNA synthesis in myoid cells and Sertoli's cells increase at the perinatal stages and decrease from postnatal 2 weeks as described previously (Figure 15), while the RNA synthesis in spermatogonia increase from postnatal 2 weeks together with DNA (Figure 15E) and protein syntheses (Figure 15F) to senescence.

5.2. The RNA Synthesis in the Implantation

In order to detect the changes of RNA synthesis of the developing blastcysts in mouse endometrium during activation of the implantation, ovulations of female BALB/C strain mice were controlled by pregnant mare serum gonadotropin and human chorionic gonadotropin, then pregnant female mice were ovariectomized on the 4th day of pregnancy (Yamada 1993, Yamada and Nagata 1992a,b, 1993). The delay implantation state was maintained for 48 hrs and after 0 to 18 hrs of estrogen supply, ^{3}H-uridine was injected. The three regions of the endometrium, i. e. the interinplantation site, the antimesometrial and mesometrial sides of implantation site, were taken out and processed for LM and EMRAG. It was well known that the uterus of the rodent becomes receptive to blastcyst implantation only for a restricted period. This is called the implantation window which is intercalated between refractory states of the endometrium whose cycling is regulated by ovarian hormones (Yoshinaga 1988). We studied the changes of RNA synthesis by[3]H-uridine (Yamada and Nagata 1993) incorporations in the endometrial cells of pregnant-ovariectomized mice after time-lapse effect of nidatory estradiol. As the results, the endometrial cells showed topographical and chronological differences in the nucleic acid synthesis. The cells labeled with ^{3}H-uridine demonstrating RNA synthesis were observed in both epithelial cells and stromal cells. Quantitative analysis revealed that the labeling indices with ^{3}H-thymidine in the stromal cells at antimesometrial and mesometrial side of implantation sites and interimplantation sites increased from 0 hr to 3 (10% in antimesometrial site), 6, 12 and 18 hr, reaching a peak (60% in antimesometrial site) at 18 hrs after estrogen induction, while the number of silver grains as expressed by grain counting per μm^2 in both the stromal and epithelial cells on the antimesometrial side due to ^{3}H-uridine increased from 0 hr to 3 and 6 hr, reaching a peak (4 grain count per μm^2 in luminal epithelial cells) at 6 hr and decreased from 12 to 18 hr (2 grain count per μm^2 in luminal epithelial cells). These results suggested that the presence of the blastocysts in the uterine lumen induced selective changes in the behavior of endometrial cells after nidatory estradiol effect showing the changes of DNA and RNA synthesis. The time coincident peak of RNA synthesis detected in the endometrial cells at the anti-mesometrial side of the implantation site, probably reflected the activation moment of the implantation window.

6. THE RNA SYNTHESIS IN THE ENDOCRINE SYSTEM

We studied the RNA synthesis of the adrenal glands and the cells of Leydig in the testis of aging mice among the endocrine organs after the administration of ^{3}H-uridine at various ages.

6.1. The RNA Synthesis in the Adrenal Glands

The RNA synthesis in the adrenal glands was studied in both the adrenal cortex and the adrenal medulla after administration of ^{3}H-uridine in aging mice at various ages from perinatal stages to senescence at year 2.

6.1.1. The RNA Synthesis in Aging Mouse Adreno-Cortical Cells

Observing both LM and EM RAG labeled with ^{3}H-uridine, demonstrating RNA synthesis, the silver grains were found over the nuclei and cytoplasm of almost all the adreno-cortical cells from perinatal stages to postnatal month 1-24 (Liang 1998, Liang et al. 1999, Nagata et al. 2000b, Nagata 2010a). Those labeled cells were found in all the 3 layers, the zona glomerulosa (Figure 19F), the zona fasciculata and the zona reticularis, at respective aging stages. In labeled adreno-cortical cells in the 3 layers the silver grains were mainly localized over the euchromatin of the nuclei and several silver grains were found over the endoplasmic reticulum, ribosomes and mitochondria of these cells. The ultrastructural localizations of silver grains over the mitochondria were mainly on the mitochondrial matrices and some over the cristae or membranes.

6.1.2. Mitochondrial RNA Synthesis of Aging Mouse Adreno-Cortical Cells

The results of visual grain counts on the number of mitochondria labeled with silver grains obtained from 10 adreno-cortical cells in the 3 layers of each animal labeled with ^{3}H-uridine demonstrating RNA synthesis in 10 aging groups at perinatal stages, prenatal embryo day 19, postnatal day 1, 3, 9 and 14, and month 1, 2, 6, 12 and 24 were reported previously (Liang 1998, Liang et al. 1999, Nagata et al. 2000b, Nagata 2010a). The results demonstrated that the numbers of labeled mitochondria with ^{3}H-uridine showing RNA synthesis gradually increased from prenatal embryo day 19 (glomerulosa 0.3, fasciculata 0.5, reticularis 0.4/cell) to postnatal day 14, month 1 and 2 (glomerulosa 5.3, fasciculata 5.0, reticularis 6.2/cell), reaching the maximum, then decreased to month 6, 12 and 24 (Figure 22A). The increase and decrease were stochastically significant (P <0.01).

6.1.3. The Labeling Index of RNA Synthesis in Mouse Adreno-Cortical Mitochondria

On the other hand, the labeling indices in respective aging stages were calculated from the number of labeled mitochondria, dividing by the number of total mitochondria per cell which were mentioned previously (Liang 1998, Liang et al. 1999, Nagata et al. 2000b, Nagata 2010a). The results showed that the labeling indices gradually increased from prenatal day 19 (glomerulosa 10.4, fasciculata 12.1, reticularis 13.1%) to postnatal day 1 (glomerulosa 12.6, fasciculata 11.4, reticularis 11.1%), 3, 9 (glomerulosa 16.6, fasciculata 18.0, reticularis 18.0%), reaching the maximum and decreased to day 14, month 1 (glomerulosa 11.4, fasciculata 11.0, reticularis 10.7%) 2 (glomerulosa 8.5, fasciculata 7.8, reticularis 8.8%), month 6 (glomerulosa 4.1, fasciculata 4.2, reticularis 3.8%), 12 and 24 (Figure 21A). The increase and decrease were stochastically significant (P <0.01).

6.1.4. RNA Synthesis in Aging Mouse Adreno-Medullary Cells

The adrenal medulla consists of 2 cell types, the adrenalin cells and noradrenalin cells. When they were labeled with ^{3}H-uridine, an RNA precursor, silver grains appeared over almost all the cells, both nuclei and cytoplasm containing mitochondria (Ligng et al. 1999, Nagata et al. 2000b, 2010b). Quantitative analysis revealed that the numbers of mitochondria in both adrenalin and noradrenalin cells at various ages increased from fetal day 19 to postnatal month 1 due to aging of animals, respectively, but did not decrease to month 24, while the number of labeled mitochondria and the labeling indices of intramitochondrial RNA synthesis incorporating ^{3}H-uridine increased from fetal day 19 to postnatal month 1, reaching the maxima, but din not decrease to month 24. It was shown that the activity of intramitochnodrial RNA synthesis in the adrenal medullary cells in aging mice increased but did not decrease due to aging of animals in contrast to DNA synthesis (Nagata 2010b).

6.2. The Leydig Cells of the Testis

The cells of Leydig can be found in the interstitial tissues between the seminiferous tubules of the testis of mammals (Gao 1993, Gao et al. 1994, 1995a, Nagata et al. 2000b). They are identified as spherical, oval, or irregular in shape and their cytoplasms contain lipid droplets. We studied the macromolecular synthesis of the cells in the testis of several groups of litter ddY mice at various ages from fetal day 19 to postnatal aging stages up to 2 years senescence by LM and EMRAG using ^{3}H-thymidine, ^{3}H-uridine and ^{3}H-leucine incorporations.

The incorporation of ^{3}H-uridine into RNA was observed in almost all the Leydig cells in the interstitial tissues of the testis from embryonic day 19 to 2 years after birth. A few silver grains over the nuclei and cytoplasm of the Leydig cells labeled with ^{3}H-uridine were observed at embryonic day 19. The silver grains over those cells slightly decreased at postnatal day 1, 3 (Figure 19H), 7 and 14. The number of the silver grains over the nuclei increased from postnatal 1 months onwards. The average number of silver grains over the cytoplasm increased gradually and reached the maximum at 12 months after birth. At each stage, the activity of RNA synthesis was specifically localized over the euchromatin in the nucleus and nucleolus as observed by EMRAG. From adult to senescent stages, the activity of RNA synthesis maintained a high level in their nuclei as compared to the cytoplasm. In the cytoplasm of Leydig cells in respective aging groups some of the mitochondria and endoplasmic reticulum were also labeled with silver grains. It is noteworthy that the average grain counts increased prominently in the senescent aging groups at 1 and 2 years after birth.

6.3. The RNA Synthesis in the Nervous System

We studied only messenger RNA in the spinal cords of aging mice from perinatal to postnatal adult stages by means of in situ hybridizaion.

6.3.1. The RNA Synthesis in the Spinal Cord

The localization of TGF-β1 mRNA in the segments of the spinal cords of mice was investigated by means of in situ hybridization techniques together with immunohistochemical staining (Nagata et al. 1999). The tissues of lower cervical segments of the spinal cords of BALB/c mice, from embryonic day 12, 14, 16, 19 and postnatal day 1, 3, 7, 14, 21, 28, 42 and 70, were used. For in situ hybridization, ^{35}S-labeled oligonucleotide probes for TGF-β1 were used to detect their messenger RNA. Cryosections were incubated under silicon cover slides with 100 μl of pre-incubation solution plus final concentration of 2.4×10^6 cpm/ml probes and 100 mM DTT for 16 hours. After washing with SSC and DTT, the slides were dried and processed for radioautography by dipping in Konica NR-M2 emulsion, which were exposed and developed. The results showed that TGF-β1 mRNA was detectable in the meninges surrounding the spinal cord, but scarcely detected in spinal cord parenchyma (Figure 23B). The localization of TGF-β1 mRNA in the spinal cord suggested that TGF- β1 acted through paracrine mechanism in the morphogenesis of the spinal cord in mice. The localization of TGF-β1 and its mRNA in the segments of the spinal cords of mice was also investigated with immunocytochemical techniques (Nagata et al. 1999). The tissues of lower cervical segments of the spinal cords of BALB/c mice, from embryonic day 12, 14, 16, 19 and postnatal day 1, 3, 7, 14, 21, 28, 42 and 70, the same as in situ hybridization were used. For immunocytochemistry, transverse cryosections of the spinal cords were cut and stained with rabbit anti-TGF-β1 polyclonal antibody followed with ABC method. The results showed that positive immunoreactivities arose in the ventral horn motoneurons from the embryonic stage to postnatal neonates (Figure 23B) up to the adults. The extracellular matrix of the white matter, however, showed positive immunocytochemical staining from postnatal day 14, and thereafter, and the immunoreactivity remained with aging. The whole white matter showed only background level of staining before postnatal day 14. The results indicated that TGF-β1 regulates motoneuron growth and differentiation as well as they were probably correlated with formation, differentiation and regeneration of myelin of nerve tracts. The immunostaining with βFGF antibody presented the same basal pattern as shown in TGF-β1 immunocytochemistry (Nagata and Kong 1988). The positive immunoreactivities were detected in ganglion cell layer, inner and outer plexiform layers, retinal pigment epithelial layer, choroidal and scleral layers. Since TGF-β1 mRNA was detectable in the meninges surrounding the spinal cord by in situ hybridization but scarcely detected in spinal cord parenchyma, the disparate localization of TGF-β1 polypeptide and TGF-β1 mRNA in the spinal cord suggest that TGF-β1 acts through paracrine mechanism in the morphogenesis of the spinal cord in mice. The negative control abolished virtually all reactivity when using the normal rabbit serum instead of primary antibody or using avidin-biotin-peroxidase complex solution only.

6.4. The RNA Synthesis in the Sensory System

We studied only the RNA synthesis in the chicken and mouse eyes among of the sensory organs.

6.4.1. The RNA Synthesis in the Eye

The RNA synthesis in the chicken eyes was studied with the ocular tissues of chicken embryos in incubation (Figure 26D). Silver grains due to the incorporations of ^{3}H-uridine were observed over all the nuclei, cell organelles, cytoplasm of all the cells in the optic cups in development showing the RNA synthesis (Gunarso 1984a,b, Gunarso et al. 1969). Grain counting revealed that the counts gradually increased from day 2 to 7 and the numbers of silver grains were the most in the nuclei, while the numbers between the 3 portions of the optic cups, the anterior, equator and the posterior portions decreased from the anterior to the posterior at the same developmental stages (Gunarso 1984a).

On the other hand, the ocular tissues of aging mice were also labeled with ^{3}H-uridine. The silver grains demonstrating RNA synthesis appeared over all the cell types at all the stages of development and aging. The grain counts in the retina and the pigment epithelium increased from prenatal day 9 to postnatal day 1 in the retinal cells, while they increased from prenatal day 12 to postnatal day 7 in the pigment epithelial cells (Kong et al. 1992b).

On the other hand, the distribution and localization of TGF-β1 and βFGF and their mRNA in the ocular tissues of aging mice were also studied (Nagata and Kong 1998). The posterior segment of BALB/c mouse eyes from embryonic day 14, 16, 19 and postnatal 1, 3, 5, 7, 14, 28, 42 and 70 were used. For in situ hybridization, ^{35}S-labeled oligonucleotide probes for TGF-β1 and βFGF were used to detect their mRNA. Cryo-sections were picked up on glass slides which were processed for in situ hybridization and for radioautography. As the results, silver grains mainly located in the scleral layers and some in the choroidal and pigment epithelial layers, but only background level of grains were found in the whole retina.

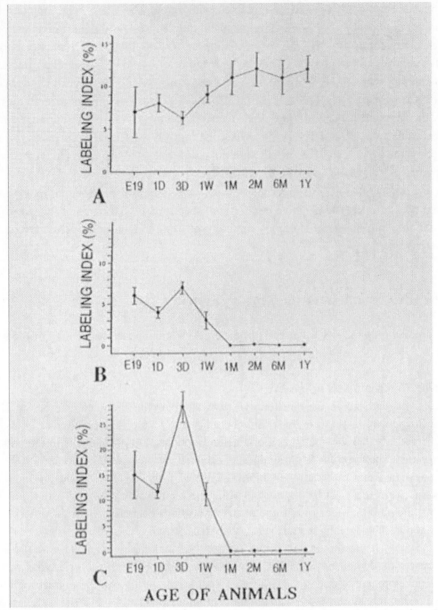

From Nagata, T. Radioautographology, General and Special.In, Progr.Histochem.Cytochem.Vol. 37, No. 2, p. 192, 2002, Urban & Fischer, Jena, Germany.

Figure 26A.The labeling index in the epithelium.

Figure 26B.The labeling index in the stroma.

Figure 26C. The labeling index in the endothelium.

Figure 26. Transitional curves of the labeling indices in the three layers of the central area of the cornea of aging mice labeled with[3]H-thymidine, showing DNA synthesis.Mean ± Standard Deviation.

In the radioautograms from embryonic day 14 to adult mice at week 10 (day 70), the significant distribution of silver grains representing TGF-β1 mRNA was not detected in the whole retina. However, the significant silver grains were detected in scleral and chorioidal layers and mesenchymal cells at embryonic day 14, then the number of grains increased in these layers particularly in sclera from prenatal to postnatal neonate until adult (Figure 22E). These results suggest that mRNA for TGF-β1 and βFGF were synthesized in scleral, choroidal and pigment epithelial layers, but their proteins were transferred to the target cells of the retina and elsewhere. Furthermore, it is suggested that TGF-β1 and βFGF may play important roles on retinal differentiation, development and aging, particularly during the late embryonic and newborn stages (Nagata and Kong 1998).

These results showed that RNA synthetic activities in the ocular cells changed due to the aging of individual animals.

7. THE PROTEIN SYNTHESIS

The proteins found in animal cells are composed of various amino-acids which initially form low molecular polypeptides and finally macromolecular compounds designated as proteins.They are chemically classified into two, simple proteins and conjugated proteins. Therefore, the proteins can be demonstrated by showing specific reactions to respective amino-acids composing any proteins. Thus, the proteins contained in cells can bedemonstrated either by morphological histochemical techniques staining tissue sections such as Millon reaction or tetrazonium reaction or otherwise by biochemical techniques homogenizing tissues and cells. To the contrary, the newly synthesized proteins but not all the proteinsin the cells can be detected as macromolecular synthesis together with other macromolecules such as DNA or RNA in various organs of experimental animals by either morphological or biochemical procedures employing RI-labeled precursors. We have studied the sites of macromolecular synthesis in almost all the organs of mice during their aging from prenatal to postnatal development to senescence by means of microscopic radioautography (Nagata 1992, 1994b,c, 1996a,b, 1997a, 2002). The results obtained from protein synthesis should be described according to the order of organ systems in anatomy or histology. In contrast to the results obtained from DNA synthesis of almost all the organs, we have studied only several parts of the organ systems.

7.1.The Protein Synthesis in the Muscular System

We studied only ^{3}H-taurin incorporation in the skeletal muscles of both normal and dystrophy adult mice incubated in Eagle's medium containing ^{3}H-taurin in vitro at varying time intervals from 1 min to 5, 10, 30 and 60 min (1 hour). The silver grains were observed over the skeletal muscle cells as well as over the smooth muscle cells and the endothelial cells in the arteries, both the nuclei and cytoplasm, by LM and EM RAG, showing taurin incorporation into the proteins. However, the aging changes were not studied.

7.2. The Protein Synthesis in the Circulatory System

Among the circulatory organs, we first studied the protein synthesis in the spleens of aging mice at various ages. Several groups of litter mates, each 3, from fetal day 19 to postnatal day 1, 14, and month 6 to 12 (year 1) were administered with ^{3}H-leucine and sacrificed, the spleens were taken out and processed for LM and EM RAG (Nagata and Olea 1999). The results demonstrated that the sites of incorporations were hematopoietic cells, i.e., lymphoblasts, myeloblasts, erythroblasts and littoral cells in the splenic tissues at every aging stage. In most labeled cells silver grains were observed over the nuclei, nucleoli, endoplasmic reticulum, ribosomes, Golgi apparatus and mitochondria. Quantitative analysis revealed that grain counts in respective cells were higher in young animals than adult aged animals. The grain counts and the labeling index increased from prenatal to postnatal day 14, reaching the maximum, then decreased to month 12. These results showed the increase and decrease due to aging of animals.

7.3. The Protein Synthesis in the Digestive System

We studied the protein synthesis of the stomach and the intestines in the digestive tracts as well as the liver and the pancreas among the digestive glands of rats and mice.

7.3.1. Protein Synthesis in the Stomach

Weobserved the secretion process in G-cells by EMRAG using ^{3}H-amino acid (Sato 1978, Sato et al. 1977, Komiyama et al. 1978). When the stomach tissues were taken out from the adult Wistar rats at postnatal month 1 and were labeled with either ^{3}H-glutamic acid or ^{3}H-glycine in vitro at varying time intervals, silver grains in the EM radioautograms appeared first over the Golgi zones, then migrated to secretory granules and were stored in the cytoplasm, suggesting the secretory kinetics. We also studied the mechanism of serum albumin passing through the gastric epithelial cells into the gastric cells by EMRAG (Sato et al. 1977). When adult Wistar rat stomach tissues were labeled with ^{132}I-albumin in vitro at varying time intervals, silver grains in the radioautograms appeared over rough endoplasmic reticulum within 3 min, then moved to the Golgi apparatus in 10 min, and on to secretory granules and into the lumen in 30 min, suggesting the pathway of serum albumin absorption from the blood vessels through the gastric mucous epithelial cells into the gastric lumen (Komiyama et al. 1978). These results demonstrated that the stomach cells of adult rats synthesized proteins and secreted. However, aging changes of these protein synthesis between the young and senescent animal were not yet completed.

7.3.2. Protein Synthesis in the Intestines

We first studied the incorporations of ^{3}H-leucine and ^{3}H-tryptophane in mouse small intestines in connection to the binuclearity before and after feeding (Nagata 1967b). The results showed that the incorporations of both amino acids were greater in binucleate intestinal epithelial columnar cells than mononucleate villus and crypt cells at both before and after feeding. However, the aging changes of these incorporations were not studied.

7.3.3. The Protein Synthesis in the Liver

As for the protein synthesis in the liver, we first studied the incorporations of ^{3}H-leucine and ^{3}H-tryptophane in mouse hepatocytes in connection to the binuclearity before and after feeding (Nagata 1967b, Nagata et al. 1967a, Ma et al. 1991). The results showed that the incorporations of both amino acids were greater in binucleate hepatocytes than mononucleate. When ^{3}H-leucine was injected into several groups of mice at various ages and the liver tissues were processed for LM and EMRAG, silver grains were observed over all cell types of the liver, i.e., hepatocytes (Figure 8E), sinusoidal endothelial cells (Figure 8F), ductal epithelial cells, Kupffer's cells, Ito's fat storing cells, fibroblasts and haematopoietic cells. In hepatocytes, number of silver grains in cytoplasm and karyoplasm increased from perinatal animals to postnatal 1 month young adult animals and decreased with aging to senescence at 24 months. Number of silver grains observed over respective cell organelles, the Golgi apparatus, mitochondria and endoplasmic reticulum, changed with aging, reaching the maxima at 1 month but the ratio remained constant at each point. When ^{3}H-proline was injected into mice at various ages from prenatal embryos to postnatal senescence and quantitative changes of collagen and protein synthesis in the livers were studied by electron microscopic radioautography (Ma and Nagata 2000, Nagata 2006a). The silver grains due to ^{3}H-proline showing collagen synthesis were localized over the nuclei, cytoplasmic matrix, endoplasmic reticulum, the Golgi apparatus, mitochondria and peroxisomes of almost all the cells such as hepatocytes (Figure 8F), sinusoidal endothelial cells, Kupffer's cells (Figure 8G), Ito's fat-storing cells, ductal epithelial cells, fibroblasts and haematopoietic cells at various ages. The number of silver grains in the cell bodies and nuclei, cytoplasmic matrix, endoplasmic reticulum, mitochondria, the Golgi apparatus and peroxisomes of hepatocytes gradually increased from embryo, reaching the maxima at postnatal month 1 and 6, and decreased with aging until 24 months. The grain counts of the cell bodies reached the maximum at month 6 and the nuclei at month 2, while that of endoplasmic reticulum at month 6 and mitochondria at month 1. The number of silver grains localized over the extracellular collagen fibrils and matrices was not so many in respective aging groups and did not show any remarkable changes with aging. From the results, it was concluded that ^{3}H-proline was incorporated not only into collagen but also into the structural proteins of hepatocytes and increased and decreased due to aging under normal aging conditions.

7.3.3.1. Mitochondrial protein synthesis in hepatocytes

When the aging mice at various agesfrom embryo to senescence were injected with ^{3}H-leucine, it was found that almost all the hepatocytes, from embryonic day 19, postnatal day 1, 3, 9, 14, to adult and senescent stages at postnatal month 1, 2, 12 and 24, incorporated silver grains (Figure 8E). The silver grains were also observed in binucleate hepatocytes at postnatal day 1, 3, 9, 14, month 1, 2, 6, 12 and 24 (Nagata 2007a,b,c,d, 2006a,b, 2007b,c,e). The localizations of silver grains observed over the mitochondria were mainly on the mitochondrial matrices but a few over their nuclei, cytoplasmic matrix, endoplasmic reticulum, ribosomes, Golgi apparatus and mitochondria (Nagata 2006a,b, 2007b,c,e). In the mitochondria the silver grains were localized over the mitochondrial membrans and cristae when observed by high power magnification. Preliminary quantitative analysis on the number of mitochondria in 20 mononucleate hepatocytes whose nuclei were intensely labeled with many silver grains (more than 10 per nucleus) and other 20 mononucleate hepatocytes whose nuclei were not so intensely labeled (number of silver grains less than 9) in each aging group revealed that there was no significant difference between the number of mitochondria, number of labeled mitochondria and the labeling indices in both types of hepatocytes (P<0.01).

On the other hand, the numbers of mitochondria, the numbers of labeled mitochondria and the labeling indices were calculated in 10 binucleate hepatocytes selected at random in each animal in respective aging stages, regardless whether their nuclei were very intensely labeled or not, except the prenatal stage at embryonic day 19, because no binucleate cell was found at this stage, resulted in no significant difference between them. Thus, the numbers of mitochondria, the numbers of labeled mitochondria and the labeling indices were calculated in 20 hepatocytes selected at random in each animal in respective aging stages regardless whether their nuclei were very intensely labeled or not. The results obtained from the total numbers of mitochondria in mononucleate hepatocytes showed an increase from the prenatal day (34.5/cell) to postnatal days 1 (44.6/cell), 3 (45.8/cell), 9 (43.6/cell), 14 (48.5/cell), to postnatal months 1 (51.5/cell), 2 (52.3/cell), reaching the maximum at month 6 (60.7/cell), then decreased to years 1 (54.2/cell) and 2 (51.2/cell). The increase and decrease were stochastically significant (P<0.01). The results obtained from visual counting on the numbers of mitochondria labeled with silver grains from 20 mononucleate hepatocytes of each animal labeled with ^{3}H-leucine demonstrating protein synthesis in 10 aging groups at perinatal stages, prenatal embryo day 19, postnatal day 1, 3, 9 and 14, month 1, 2, 6 and year 1 and 2, were counted. The labeling indices in respective aging stages were calculated from the numbers of labeled mitochondria and the numbers of total mitochondria per cell. The results showed that the numbers of labeled mitochondria with ^{3}H-leucine showing protein synthesis increased from prenatal embryo day 19 (8.3/cell) to postnatal days 1 (9.6/cell), 3 (8.1/cell), 9 (8.9/cell), 14 (9.5/cell), and month 1 (11.2/cell), reaching the maximum, and then decreased to months 2 (9.1/cell), 6 (8.8/cell) to years 1 (6.7/cell) and 2 (2.2/cell), while the labeling indices increased from prenatal day 19 (20.1%) to postnatal days 1 (21.2%), 3 (21.6%), 9 (22.2%), 14 (23.1%), reaching the maximum, then decreased to month 1 (21.7%), 2 (17.4%), 6 (14.6%), and years 1 (12.4%) and 2 (4.4%). Stochastical analysis revealed that the increases and decreases of the numbers of labeled mitochondria as well as the labeling indices from the perinatal stage to the adult and senescent stages were significant (P<0.01).

The results obtained from the numbers of mitochondria in binucleate hepatocytes showed an increase from the postnatal days 1 (66.2/cell), to 3 (66.4/cell), 14 (81.8/cell), to postnatal months 1 (89.9/cell), 2 (95.1/cell), and 6 (102.1), reaching the maximum at month 12 (128.0/cell), then decreased to years 2 (93.9/cell). The increase and decrease were stochastically significant (P<0.01). The results obtained from visual counting on the numbers of mitochondria labeled with silver grains from 10 binucleate hepatocytes of each animal labeled with ^{3}H-leucine demonstrating protein synthesis in 10 aging groups at postnatal day 1, 3, and 14, month 1, 6 and year 1 and 2, were counted. The labeling indices in respective aging stages were calculated from the numbers of labeled mitochondria and the numbers of total mitochondria per cell which showed that the numbers of labeled mitochondria with ^{3}H-leucine showing protein synthesis increased from postnatal day 1 (7.3/cell) to day 3 (6.8/cell), 14 (10.2/cell), and month 1 (15.0/cell), 2 (15.9/cell), reaching the maximum at month 6 (19.6/cell), then decreased to year 1 (8.3/cell) and 2 (5.1/cell), while the labeling indices increased from postnatal day 1 (11.8%) to 3 (10.2%), 14 (12.5%), month 1 (18.3%) and 2 (18.7%), reaching the maximum at month 6 (19.2%), then decreased to year 1 (6.4%) and 2 (5.5%). Stochastical analysis revealed that the increases and decreases of the numbers of labeled mitochondria as well as the labeling indices from the newborn stage to the adult and senescent stages were significant (P<0.01).

7.3.4. The Protein Synthesis in the Pancreas

As for the protein synthesis in the pancreas, ^{3}H-leucine incorporation into endoplasmic reticulum, Golgi apparatus and to secretory granules of pancreatic acinar cells was first demonstrated by Jamieson and Palade (1967). We first studied ^{3}H-glycine incorporation into these cell organelles of mouse pancreatic acinar cells in connection with soluble compounds by EMRAG (Nagata 2000c, 2007a). It was demonstrated that soluble ^{3}H-glycine distributed not only in these cell organelles but also in the karyoplasm and cytoplasm diffusely. Then, the quantitative aspects of protein synthesis with regards the aging from fetal day 19, to postnatal day 1, 3, 7, 14 and 1, 2, 5 and 12 months were alsoclarified (Nagata and Usuda 1993a, Nagata 2000c). The results showed an increase of silver grain counts labeled with ^{3}H-leucine after birth, reaching a peak from postnatal 2 weeks to 1 month (Figure 10E), and decreasing from 2 months to 1 year (Figure 10F).

On the other hand, we also studied ^{3}H-leucine incorporations into the pancreatic acinar cells of both normal adult rats and experimentally pancreatitis induced rats with either ethionine or alcohol (Yoshizawa et al. 1974, 1977). The results showed that the incorporations as indicated silver grain counts in the pancreatitis rats were less than normal control rats. However, its relation to the aging was not yet studied.

7.4. The Protein Synthesis in the Respiratory System

We studied the protein synthesis only in the lungs of aging mice at various ages among the respiratory organs including the respiratory tract.

7.4.1. Protein Synthesis of Aging Mouse Pulmonary Cells

When the lung tissues of mice at various ages were labeled with ^{3}H-leucine, protein synthesis was observed in all types of cells, type I and type II epithelial cells, interstitial cells and endothelial cells, of the lung at various ages from embryo to senescence (Sun et al. 1997b). Observing the light microscopic radioautograms, the silver grains were detected over both the karyoplasm and cytoplasm of almost all the cells not only at the perinatal stages from embryo day 16, 18, to postnatal day 1 to 14, but also at the adult and senescent stages from postnatal month 1 to 2, 6, 12 and 22. The number of silver grains,by light microscopic radioautography, changed with aging. The grain counts in type I epithelial cells were the highest on postnatal day 1, decreased on day 3 and increased again at 1 week, then decreased gradually to month 22, while the counts in type II epithelial cells, interstitial cells and endothelial cells reached the highest levels on fetal day 16, declined progressively with aging from fetal day 18 to postnatal day 3, then increased again from postnatal to day 7, reached peaks at 1 week (day 7) after birth, then decreased to senescence. However, grain counting on cell organelles by electron microscopy was not performed because enough numbers of EM RAG were not obtained at that time.

7.5. The Protein Synthesis in the Reproductive System

We studied the protein synthesis of the reproductive system in both the male and female reproductive organs.

7.5.1. The Protein Synthesis in the Testis

We studied the protein syntheses in aging mouse testis by LM and EM RAG, demonstrating the incorporations of ^{3}H-leucine into various cells of the seminiferous tubules (Gao 1993, Gao et al. 1994, Nagata 2002). The protein synthesis of various cells in the seminiferous tubules was first studied after administration of ^{3}H-leucine into aging male mice at various ages from perinatal to senescence at postnatal 2 years. Silver grains due to ^{3}H-leucine incorporation demonstrating protein synthesis were observed over the nuclei and cytoplasm of all the cells, spermatogonia, spermatocytes, Sertoli's cells, myoid cells of all male mice at respective stages from perinatal to senescence. The synthetic activities of spermatogonia, Sertoli's cells and myoid cells as shown by the number of silver grainsdue to ^{3}H-leucine, as expressed by grain counting, were low at the embryonic and neonatal stages but increased at adult stages and maintained high levels until senescence. These results showed that DNA synthesis in myoid cells and Sertoli's cells that increased at the perinatal stages and decreased from postnatal 2 weeks (Figure 17), while the DNA synthesis in spermatogonia increase from postnatal 2 weeks together with RNA (Figure 17E) and protein syntheses (Figure 17F) to senescence.

7.5.2. The Protein Synthesis in the Implantation

In order to detect the changes of DNA, RNA and protein synthesis of the developing blastocysts in female mouse endometrium during activation of the implantation, ovulations of female BALB/C strain mice were controlled by pregnant mare serum gonadotropin and human chorionic gonadotropin, then pregnant female mice were ovariectomized on the 4th day of pregnancy (Yamada 1993, Yamada and Nagata 1992a,b, 1993). The delay implantation state was maintained for 48 hrs and after 0 to 18 hrs of estrogen supply and ^{3}H-leucine was injected. The three regions of the endometrium, i. e. the interinplantation site, the antimesometrial and mesometrial sides of implantation site, were taken out and processed for LM and EM RAG. It was well known that the uterus of the rodent becomes receptive to blastocyst implantation only for a restricted period. This is called the implantation window which is intercalated between refractory states of the endometrium whose cycling is regulated by ovarian hormones (Yoshinaga 1988). We studied the changes of protein synthesis by ^{3}H-leucine (Yamada 1993, Yamada and Nagata 1992a) incorporations in the endometrial cells of pregnant-ovariectomized mice after time-lapse effect of nidatory estradiol. As the results, the endometrial cells showed topographical and chronological differences in the nucleic acid and protein synthesis. The cells labeled with ^{3}H-leucine were observed in both epithelial cells and stromal cells. Quantitative analysis revealed that the number of silver grains as expressed by grain counting per $\mu\mu^2$ in both the stromal and epithelial cells on the antimesometrial side with ^{3}H-leucine increased from 0 hr to 3 and 6 hr, reaching the peak at 6 hr and decreased from 12 to 18 hr. These results suggested that the presence of the blastocysts in the uterine lumen induced selective changes in the behavior of endometrial cells after nidatory estradiol effect showing the changes of DNA, RNA and protein synthesis. The time coincident peak of

RNA and protein synthesis detected in the endometrial cells at the anti-mesometrial side of the implantation site, probably reflected the activation moment of the implantation window. The protein synthesis in the decidual cells of pregnant mice uteri was compared to the endometrial cells of virgin mice uteri using ^{3}H-proline and ^{3}H-tryptophane incorporations. The results demonstrated that silver grains were localized over the endoplasmic reticulum and the Golgi apparatus of fibroblasts and accumulated over collagen fibrils in the extracellular matrix suggesting that the decidual cells produced collagen in the matrix. On the other hand, collagen synthesis in the mouse decidual cells was studied by LM and EMRAG using ^{3}H-proline (Oliveira et al. 1991, 1995). Silver grains were localized over the endoplasmic reticulum and Golgi apparatus of fibroblasts and accumulated over collagen fibrils in the extracellular matrix. The results suggested that the decidual cells produced collagen into the matrix. The quantitative analysis showed that both incorporations in the decidual cells and the matrix increased in the pregnant mice than the endometrial cells in virgin mice.

7.6. The Protein Synthesis in the Endocrine System

Among the endocrine organs, we studied only the cells of Leydig in the testis of male mice.

7.6.1. The Protein Synthesis in the Leydig Cells of the Testis

We studied the macromolecular synthesis of the cells in the testis of several groups of litter ddY mice at various ages from fetal day 19 to postnatal aging stages up to 2 years senescence by LM and EMRAG using ^{3}H-thymidine, ^{3}H-uridine and ^{3}H-leucine incorporations (Gao 1993, Gao et al. 1994, 1995a, Nagata 2000b). The incorporation of ^{3}H-leucine into proteins was observed in almost all the Leydig cells in the interstitial tissues of the testis. The silver grains were located over the nuclei and cytoplasm of respective Leydig cells. The aging change of protein synthesis of Leydig cells among different aging groups was also found (Nagata 2001c, 2002). At embryonic day 19, the silver grains of Leydig cells labeled with ^{3}H-leucine was observed in both nucleus and cytoplasm and there was no obvious difference between the number of silver grains on the cytoplasm and the nucleus. The number of silver grains decreased at postnatal day 1 and then increased at day 3 and 7. However, the number of silver grains on both the nucleus and cytoplasm decreased from 1 month to 3 months and increased again from 6 months onwards maintaining a high level from adult to senescent stages. Some of the silver grains were also localized over some of the mitochondria in respective aging groups as observed by EMRAG. These results indicate that the DNA, RNA and protein syntheses in Leydig cells are maintained at rather high level even at senescent stages at postnatal 1 and 2 years when the animals survived for longer lives.

7.7. The Protein Synthesis in the Nervous System

We studied the protein synthesis in the nervous system of aging mice in the cerebella and the spinal cords of mice at various ages from prenatal to postnatal 2 years.

7.7.1. The Protein Synthesis in the Cerebellum of Aging Mouse

When 10 groups of aging ddY mice from fetal day 19, to postnatal day 1, 3, 7, 14 and month 1, 2, 6, 12and 24, each consisting of 3 litter mates, using ^{3}H-leucine, protein syntheses of both neuroblasts and glioblasts were observed by LM and EM RAG in the extragranular layers of perinatal animals (Cui 1995, Cui et al. 1996, 2000, Nagata et al. 2001). The silver grains due to ^{3}H-leucine demonstrating protein synthesis were localized over the nuclei and cytoplasm of neuroblasts and glioblasts of embryos at fetal day 19 and the number of silver grains increased after birth from postnatal day 1, 3 to day 7 and onward. On day 3, some Purkinje cells were recognized incorporating silver grains. The number of silver grains in these cells increased from neonatal stages to mature adult stage at postnatal day 14 and 30, then decreased from month 1, 2, 6, 12 to 24. The increase and decrease of the silver grains were due to the aging changes of protein synthesis in the cerebella due to development and senescence of individual animals.

7.8. The Protein Synthesis in the Sensory System

Among the several sensory organs, we studied only the ocular tissues of aging mice at various ages from perinatal to senescent stages.

7.8.1. The Protein Synthesis of the Eye

The protein synthesis in the ocular tissues of aging mouse were studied in all the 3 layers of the eye, the tunica fibrosa, the tunica vasculosa and the tunica intima, or the cornea, cilliary bodies and the retina of the aging mouse at various stages after the administration of several precursors (Toriyama 1995, Nagata 1997c, 1999b,c,d, 2000e,f, 2001c, Nagata and Kong 1998, Cui et al. 2000).

The protein synthesis of the retina in aging mouse as revealed by ^{3}H-leucine incorporation demonstrated that number of silver grains in bipolar cells and photoreceptor cells was the most intense at embryonic stage and early postnatal days. The peak was 1 day after birth and decreased from 14 days to 1 year after birth. (Toriyama 1995). The protein synthesis of the cornea as revealed by ^{3}H-leucine incorporation (Nagata 1997c, 1999d, 2000f, 2001c, Nagata and Kong 1998, Cui et al. 2000) and the glycoprotein synthesis demonstrated by ^{3}H-glucosamine (Nagata et al. 1995) were also studied in several groups of aging ddY mice. Silver grains of both ^{3}H-leucine and ^{3}H-glucosamine incorporations were located in the epithelial cells, the stromal fibroblasts and the endothelial cells from prenatal day 19 to postnatal 6 months. No silver grains were observed in the lamina limitans anterior (Bowman's membrane) and the lamina limitans posterior (Descemet's membrane). The grain densities by ^{3}H-leucine incorporation in 3 layers, i. e., epithelial, stromal and endothelial layers, increased from embryonic stage to postnatal day 3 and 7, then decreased to 2 weeks and 1 year. The grain densities due to the glycoprotein synthesis with[3]H-glucosamine were more observed in the endothelial cells of prenatal day 19 animals, but more in the epithelial cells of postnatal day 1, 3 and 7 animals. From the results, it was shown that the glycoprotein synthetic activity in respective cell types in the cornea of mouse changed with aging of the animals.

The collagen synthesis in the ocular tissues was also demonstrated by the incorporation of ^{3}H-proline in 4 groups of mice at various ages, from prenatal day 20, postnatal day 3, 7 and

30. The results showed that the sites of ^{3}H-proline incorporation were located in the stromal fibroblasts in both cornea and the trabecular meshworks in the iridocorneal angle in prenatal and postnatal newborn mice. No silver grains were observed in the epithelial and endothelial cells. On EMRAG, silver grains were localized over the endoplasmic reticulum and Golgi apparatus of fibroblasts and over intercellular matrices consisting of collagen fibrils. From the quantitative analysis, the grain densities were more observed in the fibroblasts in postnatal day 7 animals than younger animals at fetal day 20 and postnatal day 3, 7 and 30. In the same aging groups, the grain densities were more in the cornea than the iridocorneal angle. It was concluded that the collagen synthetic activity was localized in the fibroblasts in the cornea and the trabecular meshworks in the iridocorneal angle and the activity changed with aging, reaching the maximum at postnatal day 7.

On the other hand, the distributions of some of the ophthalmological drugs used for the treatment of human glaucoma patients were examined in the ocular tissues by LM and EM RAG (Nagata 2000f). However, its relationship to the aging was not studied.

8. THE GLUCIDE SYNTHESIS

The glucides found in animal cells and tissues are composed of various low molecular sugars such as glucose or fructose called monosaccharides whichform compounds of polysaccharides or complex mucopolysaccharides connecting to sulfated compounds. The former are called simple polysaccharides, while the latter mucopolysubstances. Thus, the glucides are chemically classified into 3 groups, monosaccharides such as glucose or fructose, disaccharides such as sucrose and polysaccharides such as mucosubstances. However, in most animal cells polysaccharides are much more found than monosaccharides or disaccharides. The polysaccharides can be classified into 2, i.e. simple polyscaccharides and mucosubstances. Anyway, they are composed of various low molecular sugars that can be demonstrated by either histochemical reactions or biochemical techniques. To the contrary, the newly synthesized glucides but not all the glucides in the cells and tissues can be detected as macromolecular synthesis together with other macromolecules such as DNA, RNA or proteins in various organs of experimental animals by either morphological or biochemical procedures employing RI-labeled precursors. We have studied the sites of macromolecular synthesis in almost all the organs of mice during their aging from prenatal to postnatal development to senescence by means of microscopic radioautography (Nagata 1992, 1994a,b,c, 1996a, 1997a, 2002). The results obtained from glucides synthesis are described according to the order of organ systems in anatomy or histology. In contrast to the results obtained from DNA synthesis of almost all the organs, we have studied only several parts of the organ systems. The skeletal system, the muscular system and the circulatory system were not yet studied.

8.1. The Glucide Synthesis in the Digestive System

Among several digestive organs in the digestive tracts and the digestive glands, we studied the glucide synthesis of the stomach and the intestines in the digestive canals as well as the salivary gland, the liver and the pancreas in the digestive glands of aging mice.

8.1.1. The Glucide Synthesis in the Oral Cavity

In the oral cavity, we studied the incorporations of ^{3}H-glucosamine in the submandibular glands of 10 groups of litter mice at various ages. The animals from embryonic day 19, postnatal day 1, 3, 7, 14, and 1, 3, 6 months to 1 and 2 years were sacrificed after administration of ^{3}H-glucosamine and the submandibular glands were processed for LM and EMRAG (Watanabe et al. 1997, Nagata 2002). The results showed that the silver grains appeared over the endoplasmic reticulum, Golgi apparatus and the secretory granules of the acinar cells, demonstrating the glycoprotein synthesis in these cells. Grain counting revealed that the counts increased from the fetal stage at embryonic day 19 to postnatal day 1 to 3, 7, 14, reaching the peak at day 14, then decreased to month 1, 3, 6, to year 1 and 2, showing the aging changes, inverse proportion to DNA synthesis of these cells.

The sulfate uptake and accumulation in sulfomucin in several digestive organs of mice were also studied by light microscopic radioautography (Nagata and Kawahara 1999, Nagata et al. 1999b). Two litters of normal ddY mice 30 days after birth, each consisting of 3 animals, were studied. One litter of animals was sacrificed at 30 min after the intraperitoneal injections with phosphate buffered $Na_2{}^{35}SO_4$, and the other litter animals were sacrificed at 12 hr after the injections. Then the submandibular glands and the sublingual glands were taken out, fixed, embedded in epoxy resin, sectioned, radioautographed and analyzed by light microscopy. As the results, many silver grains were observed on serous cells of the salivary glands at 30 min and 12 hr after the injections (10-20/cell). The numbers of silver grains at 30 min were less than those at 12 hr. From the results, it was concluded that glycoprotein synthesis was demonstrated in both the submandibular and sublingual glands by radiosulfate incorporation. In the salivary glands the silver grains were more observed in serous cells than mucous cells at 30 min, while in mucous cells more at 12 hr than 30 min after the injection. These results show the time difference of glycoprotein synthesis in the two salivary glands, showing inverse proportion to DNA synthesis of these cells (Watanabe et al. 1997, Nagata 2002).

8.1.2. The Glucide Synthesis of the Stomach

When incorporation of radiosulfate into sulfated complex carbohydrate in rat stomach was studied by labeling with $^{35}SO_4$ in vivo, silver grains appeared over the glandular cells of the pyloric gland but not those of the fundic gland, demonstrating the mucous synthesis in the former glands (Nagata et al. 1988, Nagata and Kawahara 1999). The radiosulfate uptake and accumulation in the stomach of mouse were also studied by light microscopic radioautography (Nagata et al. 1999b). Two litters of normal ddY mice 30 days after birth, each consisting of 3 animals, were studied. One litter animals were sacrificed at 30 min after the intraperitoneal injections with phosphate buffered $Na_2{}^{35}SO_4$, and the other litter animals were sacrificed 12 hr after the injections. Then the antrum and the fundus tissues of the stomachs were taken out. The tissues were fixed, dehydrated, embedded in epoxy resin, sectioned, radioautographed and analyzed. As the results, many silver grains were observed on the mucosa and submucosa of the stomach at 30 min after the injection. Then at 12 hr after the injection silver grains were observed on some of the fundic glands. The numbers of silver grains observed in the stomach especially over the pyloric glands at 30 min (a few per cell) were less than those (several per cell) at 12 hr. The results showed the time difference of glycoprotein synthesis in the stomach, showing inverse proportion to DNA synthesis (Nagata and Kawahara 1999, Nagata 2002).

8.1.3. The Glucide Synthesis in the Intestines

We also studied the aging changes of glucide synthesis by ^{3}H-glucosamine uptake in the small intestines of mouse (Morita 1993), and found that the silver grains in the ileum columnar epithelial cells were mainly localized over the brush borders and the Golgi regions in these cells (Figure 5F). The grain counting revealed that the numbers of silver grains over the brush borders and cytoplasm of the columnar epithelial cells increased in the villi (10-15/cell) than in the crypts (1-2/cell) from 6 months up to 2 years due to aging. The grain counting in other cell types also revealed that the number of silver grains in goblet cells, basal granulate cells, Paneth cells increased by aging, but did not in the undifferentiated cells.

The glycoprotein synthesis in goblet cells as well as in absorptive epithelial cells was also studied using ^{35}SO$_4$ incorporation in the duodenums, the jejunums and the colons of adult mice at varying time intervals at 30, 60, and 180 min after the administration (Nagata et al. 1988a, Nagata and Kawahara 1999, Nagata et al. 1999b). Silver grains were localized over the columnar absorptive cells and the goblet cells, especially over the Golgi regions and mucous granules of the goblet cells. By EMRAG the intracellular localization of silver grains in goblet cells was clearly shown in the Golgi apparatus. The results from grain counting revealed that the average grain counts were different in the upper and deeper regions of the crypts in the 4 portions and it was shown that silver grains over goblet cells in the lower region of the crypt transferred rapidly from 30 min to 180 min, while they transferred slowly in goblet cells in the upper region of the colonic crypt, leading to the conclusion that the rates of transport and secretion of mucous products of the goblet cells at these two levels in the crypts were different. By EMRAG silver grains first appeared over the Golgi zone at 30 min. and then moved to the secretory granules at 60 and 180 min. The incorporation of Na$_2$^{35}SO$_4$ into sulfated complex carbohydrate was investigated in the mouse small and large intestines by LM and EMRAGas well as in the submandibular glands and the stomachs. Quantitative differences have been observed in the relative uptake of radiosulfate in the various labeled cells of each organ. Incorporation by the colon in goblet cells exceeded that elsewhere in the deep goblet cells of the colonic crypts migration of label progressed during the time tested from the supranuclear Golgi region to the deep position of the goblet and then extended throughout the mucosubstance in the goblet in the superficial goblet cells of the colon. The radioautographic and cytochemical staining differences between secretory cells in the deeper region compared with the upper region of the colonic crypts are considered to reflect differences in the rate of transport of secretory products in the theca and the rate of secretion at the low levels in the crypt (Figures. 5GH). These results showed the time differences of glycoprotein synthesis in respective organs. The sulfate uptake and accumulation in several mouse digestive organs were also studied by LM RAG. Two litters of normal ddY mice 30 days after birth, each consisting of 3 animals, were studied. One litter of animals was sacrificed 30 min after the intraperitoneal injections with phosphate buffered Na$_2$^{35}SO$_4$, and the other litter animals were sacrificed 12 hr after the injections. Then several digestive organs, the parotid gland, the submandibular gland, the sublingual gland, antrum and fundus of the stomach, the duodenum, the jejunum, the ileum, the caecum, the ascending colon and the descending colon were taken out and radioautographed. As the results, many silver grains were observed on villous cells and crypt cells of the small intestines and whole mucosa of the large intestines at 30 min after the injection. Then at 12 hr after the injection silver grains were observed on mucigen granules of goblet cells in the small intestines and the large intestines. The numbers of silver grains observed in respective organs at 30 min were less than those at 12 hr. From the results, it was concluded that the time difference of the glycoprotein synthesis was demonstrated in several digestive organs by radiosulfate incorporation, in reverse proportion to DNA synthesis. The total S contents in colonic goblet cells in upper and deeper regions of colonic crypts in aging mice were also analyzed by X-ray microanalysis (Nagata et al. 2000b, Nagata 2004). The results accorded well with the results from RAG (Nagata 2002) showing increase and decrease of mucosubstances in these cells due to development and aging to senescence.

8.1.4. The Glucide Synthesis in the Liver

We first studied ^{3}H-glucose incorporation into glycogen in the livers of adult mice, in connection to soluble compounds (Nagata and Murata 1977, Nagata et al. 1977a,d). Soluble ^{3}H-glucose, which was demonstrated by cryo-fixation (at -196°C) in combination with dry-mounting radioautography, was localized over the nuclei, nucleoli, all the cell organelles and cytoplasmic ground substance of all the hepatocytes diffusely. On the other hand, by conventional chemical fixation and wet-mounting radioautography, silver grains were localized only over glycogen granules, endoplasmic reticulum and Golgi apparatus showing glycogen synthesis. However, the relationship of glycogen synthesis to aging has not yet been fully clarified.

8.1.5. The Glucide Synthesis in the Pancreas

Concerning the glucide synthesis of the pancreas, we first studied the incorporation of ^{3}H-glucose into the pancreatic acinar cells of mouse in connection with soluble compounds by EMRAG (Nagata et al. 1977a). It was demonstrated that soluble ^{3}H-glucose distributed not only in such cell organelles as endoplasmic reticulum, Golgi apparatus, mitochondria but also in the karyoplasm and cytoplasm diffusely. Then, the incorporation of ^{3}H-glucosamine into the pancreases of aging mice at various ages was studied by LM and EMRAG (Nagata et al. 1992). When perinatal baby mice received ^{3}H-glucosamine injections and the pancreatic tissues were radioautographed, silver grains were observed over exocrine and endocrine pancreatic cells. However, the number of silver grains was not so many (Figure 10G). When juvenile mice at the age of 14 days after birth were examined, many silver grains appeared over the exocrine pancreatic acinar cells (Figure 10H). Less silver grains were observed over endocrine pancreatic cells and ductal epithelial cells. The grains in the exocrine pancreatic acinar cells were localized over the nucleus, endoplasmic reticulum, Golgi apparatus and secretory granules, demonstrating glycoprotein synthesis. Adult mice at the ages of postnatal 1 month, 6 month or senile mice at the ages of 12 months or 24 months showed very few silver grains on radioautograms. Thus, the glucide synthesis in the pancreatic acinar cells of mice revealed quantitative changes, increase and decrease of ^{3}H-glucosamine incorporation with aging (Nagata 1994a,b,c,d,e, Nagata et al. 1992).

8.2. The Glucide Synthesis in the Respiratory System

The incorporation of $^{35}SO_4$ in the trachea of aging mice was studied among the repiratory organs (Nagata 2000d). As the results, silver grains indicating the incorporations of radiosulfate were found over the cartilage matrices and the cartilage capsules in the hyaline cartilages of the tracheae of fetal (Figure 12C) and postnatal newborn mice (Figure 12D). The grain density as analyzed by grain densitometry was the maximum at the fetal day 19(1200 grains per unit area). The grain density then decreased from the fetal day 19 to the postnatal day 1, 3, 7 (600/area), 14 (200/area) and reached 0 on day 30, and no silver grain was found in the animals aged from 1 to 12 months. The silver grains in the perinatal animals aged at postnatal day 1 and 3, disappeared from the internal layer to the external layer of the cartilage and from the interterritorial matrix to the territorial matrix and the cartilage capsule. In the juvenile animals aged at postnatal day 9 and 14, intense incorporations were observed disseminatedly over several groups of cartilage capsules in the external layer. The results indicated that the glycoproteins constituting the cartilage matrix were synthesized from prenatal to postnatal day 30. To the contrary, no incorporation of silver grains was observed in the aging animals from postnatal 1 to 12 months by both LM and EMRAG. These results demonstrated the aging changes of glycoprotein synthesis in the cartilage matrix of mice at various ages during development and aging.

8.3. The Glucide Synthesis in the Urinary System

The incorporations of ^3H-glucosamine in the kidneys of aging micewere studies by LMRAG (Joukura 1996, Joukura and Nagata 1995) and EMRAG (Joukura et al. 1996). Silver grains were observed over all the cell type nephrons at embryonic day 19, i.e., glomerular epithelial cells, endothelial cells, mesangial cells, Bowman's capsular cells and tubule cells. In newborn and suckling stages, from postnatal day 1, 3, 8 to 14, both the renal corpuscles and urinary tubules were well differentiated and the number of silver grains increased (Figures. 36CDEFGH in Nagata 2002). The results from grain counting revealed that the numbers of silver grains in both the renal corpuscles and the uriniferous tubules were less in the embryonic stage, but increased postnatally and reached peaks at day 1 and 3, then decreased to senescence at 1 year. These results showed that glucidesynthesis in the kidney cells also changed with aging of animals.

8.4. The Glucide Synthesis in the Reproductive System

Among the reproductive organs, only the mucosubstance synthesis with radiosulfate, $^{35}SO_4$, was studied in the ovaries of mice during the estrus cycle. Litter mate groups of female ddY mice, aged 8-10 weeks, were divided into 4 groups, diestrus, proestrus, estrus and metestrus according to the vaginal smears. The ovaries were taken out, labeled with $^{35}SO_4$ in vitro and radioautographed. In all the animals, silver grains were localized over the granulosa and theca cells. Almost all compartments of the ovaries were labeled. The grain counts per cell changed according to cell cycle. From the results, it was concluded that all the cells of the ovary incorporated mucosubstances throughout the estrus cycle (Li et al. 1992).

8.5. The Glucide Synthesis in the Nervous System

The incorporation of ^{3}H-deoxyglucose was studied in the adult gerbil brains among the nervous system of experimental animals (Izumiyama et al. 1987). The changes of soluble deoxyglucose uptake in the hippocampus were studied after ^{3}H-deoxyglucose injections by means of cryo-fixation, freeze-substitution and dry-mounting radioautography to demonstrate soluble compounds under normal and post-ischemic conditions. The results demonstrated that the neurons in the hippocampus subjected to ischemia revealed higher uptake of soluble glucose than normal control. The concentration of soluble ^{3}H-deoxyglucose was higher than the chemically fixed and wet-mounted radioautograms that demonstrated only insoluble compounds. However, the relation of glycogen synthesis to aging has not yet been fully clarified.

8.6. The Glucide Synthesis in the Sensory System

We studied the aging changes of glucide synthesis by ^{3}H-glucosamine uptake in the ocular tissues of aging mice.

8.6.1. The Glucide Synthesis in the Eye

The glycoprotein synthesis of the cornea in aging mouse as revealed by ^{3}H-glucosamine incorporation was studied in several groups of aging mice at various ages from prenatal stages to senescence (Nagata et al. 1995). Silver grains were located in the epithelial cells, the stromal fibroblasts and the endothelial cells from prenatal day 19 to postnatal 6 months. No silver grains were observed in the lamina limitans anterior (Bowman's membrane) and the lamina limitans posterior (Descemet's membrane). On the other hand, the grain densities by ^{3}H-leucine incorporation in 3 layers, i.e., epithelial, stromal and endothelial layers, increased from embryonic stage to postnatal day 3 and 7, then decreased to week 2 and year 1. The grain densities due to the glycoprotein synthesis with ^{3}H-glucosamine were more observed in the endothelial cells of prenatal day 19 animals, but more in the epithelial cells of postnatal day 1, 3 and 7 animals. From the results, it was shown that the glycoprotein synthetic activity in respective cell types in the cornea of mouse changed with aging of the animals.

9. THE LIPIDS SYNTHESIS

The lipids found in animal cells are chemically composed of various low molecular fatty acids. They are esters of high fatty acids and glycerol that can biochemically be classified into simple lipids and compound lipids such as phospholipids, glycolipids or proteolipids. The simple lipids are composed of only fatty acids and glycerol, while the latter composed of lipids and other components such as phosphates, glucides or proteins. In order to demonstrate intracellular localization of total lipids, we can employ either histochemical reactions or biochemical techniques. To the contrary, the newly synthesized lipids but not all of the lipids in the cells can be detected as macromolecular synthesis similarly to the other macromolecules such as DNA, RNA, proteins or glucides in various organs of experimental animals by either morphological or biochemical procedures employing RI-labeled precursors. We have studied the sites of macromolecular synthesis in almost all the organs of mice during their aging from prenatal to postnatal development to senescence by means of microscopic radioautography (Nagata 1992, 1994a,b,c,d,e, 1996a, 1997a, 2002). However, we have not studied the lipids synthesis so much as compared to other compounds. We have studied only a·few organs of the digestive system.

9.1. The Lipids Synthesis in the Digestive System

We studied only the livers and the pancreases of aging mice at various ages but only demonstrating soluble compounds by means of cryo-fixation and dry-mounting radioautography.

9.1.1. The Lipids Synthesis in the Liver

We observed lipid synthesis in the liver using ^{3}H-glycerol in connection to soluble compounds (Nagata 1994a,d, Nagata and Murata 1977a, Nagata et al. 1977a,d). When adult mice were injected with ^{3}H-glycerol and the livers were taken out, cryo-fixed in liquid nitrogen at -196°C, then freeze-substituted, embedded in epoxy resin, dry-sectioned, and prepared for dry-mounting radioautography, many silver grains appeared over the nuclei and cytoplasm of hepatocytes diffusely. However, when the same liver tissues were fixed chemically in buffered glutaraldehyde and osmium tetroxide at 4°C, dehydrated, embedded, wet-sectioned and radioautographed by conventional wet-mounting procedures, very few silver grains were observed only over the endoplasmic reticulum and the lipid droplets, which demonstrated insoluble macromolecular lipid synthesis accumulating into the lipid droplets. However, the aging change of the lipid synthesis in the liver has not yet been fully clarified.

9.1.2. The Lipids Synthesis in the Pancreas

In order to demonstrate the lipids synthesis in the pancreas, several litters of ddY mice aged fetal day 19, postnatal day 1, 3, 7, 14, and 1, 2, 6 up to 12 months, were injected with ^{3}H-glycerol and the pancreas tissues were prepared for LM and EMRAG. The silver grains were observed in both exocrine and endocrine cells of respective ages (Nagata 1995, Nagata et al. 1988, 1990). In perinatal animals from fetal day 19 to postnatal 1, 3, and 7 days, cell organelles were not well developed in exocrine and endocrine cells and number of silver grains was very few. In 14 day old juvenile animals, cell organelles such as endoplasmic reticulum, Golgi apparatus, mitochondria and secretory granules were well developed and many silver grains were observed over these organelles and nuclei in both exocrine and endocrine cells. The number of silver grains was more in exocrine cells than endocrine cells. In 1, 2, 6 month old adult animals, number of silver grains remained constant. In 12 month old senescent animals, silver grains were fewer than younger animals. It was demonstrated that the number of silver grains expressed the quantity of lipids synthesis, which increased from perinatal atages to adult and senescent stages and finally decreased to senescence.

10. THE INTRACELLULAR LOCALIZATION OF THE OTHER SUBSTANCES

The other substances than macromolecules that can also be demonstrated by radioautography are target tracers not the precursors for the macromolecular synthesis. They are hormones such as ^{3}H-methyl prednisolone (Nagata et al. 1978b), neurotransmitters and inhibitors such as ^{14}C-bupranolol, a beta-blocking agent (Tsukahara et al. 1980) or ^{3}H-befunolol (Nagata and Yamabayashi 1983, Yamabayashi et al. 1981), vitamins, drugs such as synthetic anti-allergic agent ^{3}H-tranilast (Nagata et al. 1986b, Nishigaki et al. 1987, 1990a.b, Momose et al. 1989), hypolipidemic agent bezafibrate (Momose et al. 1993a,b, 1995), calmodulin antagonist (Ohno et al. 1982, 1983) or anti-hypertensive agent[3]H-benidipine hydrochloride (Suzuki et al. 1994), toxins, inorganic substances such as mercury (Nagata et al. 1977b) and others such as laser beam irradiation (Nagata 1984). The details are referred to the previous publication on the radioatuographology (Nagata 2002). However, their relationships to the cell aging were not studied.

CONCLUSION

From the results obtained, it was concluded that almost all the cells in various organs of all the organ systems of experimental animals at various ages from prenatal to postnatal development and senescence during the aging of cells and individual animals demonstrated to incorporate various macromolecular precursors such as ^{3}H-thymidine, ^{3}H-uridine, ^{3}H-leucine, ^{3}H-glucose or glucosamine, ^{3}H-glycerol and others localizing in the nuclei, cytoplasmic cell organelles showing silver grains due to DNA, RNA, proteins, glucides, lipids and others those which the cells synthesized during the cell aging. Quantitative analysis carried out on the numbers of silver grains in respective cell organelles demonstrated quantitative changes, increases and decreases, of these macromolecular synthesis in connection to cell aging of respective organs. In general, DNA synthesis with ^{3}H-thymidine incorporations in most organs showed maxima at perinatal stages and gradually decreased due to aging. To the contrary, the other synthesis such as RNA, proteins, glucides and lipids increased due to aging and did not remarkably decrease until senescence. Anyway, these results indicated that macromolecular synthetic activities of respective compounds in various cells were affected from the aging of the individual animals.

Thus, the results obtained from the various cells of various organs should form a part of special radioautographology that I had formerly proposed (Nagata 1999e, 2002), i.e., application of radioautography to the aging of cells, as well as a part of special cytochemistry (Nagata 2001), as was formerly reviewed. We expect that such special radioautographology and special cytochemistry should be further developed in all the organs in the future.

ACKNOWLEDGMENTS

This study was supported in part by Grant-in-Aids for Scientific Research from the Ministry of Education, Science and Culture of Japan (No. 02454564) while the author worked at Shinshu University School of Medicine as well as Grants for Promotion of Characteristic Research and Education from the Japan Foundation for Promotion of Private Schools (No. 1997, 1998 1999, 2000) while the author worked at Nagano Women's Jr. College. The author is also grateful to Grant-in-Aids for Scientific Research from the Japan Society for Promotion of Sciences (No. 18924034, 19924204, 20929003) while the author has been working at Shinshu Institute of Alternative Medicine and Welfare since 2005 up to the present time. The author thanks Dr. Kiyokazu Kametani, Technical Official, Department of Instrumental Analysis, Research Center for Human and Environmental Sciences, Shinshu University, for his technical assistance in electron microscopy during the course of this study.

REFERENCES

Chen, S., Gao, F., Kotani, A., Nagata, T.: Age-related changes of male mouse submandibular gland: A morphometric and radioautographic study. *Cell.Mol. Biol.* 41, 117-124, 1995.

Clermont Y.: The contractime elements in the limiting membrane of the seminiferous tubules of rats. *Exp. Cell Res.* 15, 438-342, 1958.

Clermont, Y.: Renewal of spermatogonia in man. *Amer. J. Anat.* 112, 35-51, 1963.

Cui, H.: Light microscopic radioautographic study on DNA synthesis of nerve cells in the cerebella of aging mice. *Cell.Mol. Biol.* 41, 1139-1154, 1995.

Cui, H., Gao, F., Nagata, T.: Light microscopic radioautographic study on protein synthesis in perinatal mice corneas. *Acta Histochem.Cytochem.* 33, 31-37, 2000.

Duan, H., Gao, F., Li, S., Hayashi, K., Nagata, T.: Aging changes and fine structure and DNA synthesis of esophageal epithelium in neonatal, adult and old mice. *J. Clin. Electron Microsc.* 25, 452-453, 1992.

Duan, H., Gao, F., Li, S., Nagata, T.: Postnatal development of esophageal epithelium in mouse: a light and electron microscopic radioautographic study. *Cell.Mol. Biol.* 39, 309-316, 1993.

Duan, H., Gao, F., Oguchi, K., Nagata, T.: Light and electron microscopic radioautographic study on the incorporation of ^{3}H-thymidine into the lung by means of a new nebulizer. Drug Res. 44, 880-883, 1994.

Feulgen, R., Rossenbeck, H.: Mikroskopische-chemischer Nachweis einer Nukeinsaeure von Thymus der Thymonukeinsaeure Z. Physik. Chem. 135: 203-248, 1924.

Fujii, Y., Ohno, S., Yamabayashi, S., Usuda, N., Saito, H., Furuta, S., Nagata, T.: Electron microscopic radioautography of DNA synthesis in primary cultured cells from an IgG myeloma patient. -*J. Clin. Electr.Microsc.* 13, 582-583, 1980.

Gao, F.: Study on the macromolecular synthesis in aging mouse seminiferous tubules by light and electron microscopic radioautography. *Cell.Mol. Biol.* 39, 659-672, 1993.

Gao, F., Toriyama, K., Nagata, T.: Light microscopic radioautographic study on the DNA synthesis of prenatal and postnatal aging mouse retina after labeled thymidine injection. *Cell.Mol. Biol.* 38, 661-668, 1992a.

Gao, F., Li, S., Duan, H., Ma, H., Nagata, T.: Electron microscopic radioautography on the DNA synthesis of prenatal and postnatal mice retina after labeled thymidine injection. *J. Clin. Electron Microsc.* 25, 721-722, 1992b.

Gao, F., Toriyama, K., Ma, H., Nagata, T.: Light microscopic radioautographic study on DNA synthesis in aging mice corneas. *Cell.Mol. Biol.* 39, 435-441, 1993.

Gao, F., Ma, H., Sun, L., Jin, C., Nagata, T.: Electron microscopic radioautographic study on the nucleic acids and protein synthesis in the aging mouse testis. *Med. Electron Microsc.*27, 360-362, 1994.

Gao, F., Chen, S., Sun, L., Kang, W., Wang, Z., Nagata, T.: Radioautographic study of the macromolecular synthesis of Leydig cells in aging mice testis. *Cell.Mol. Biol.* 41, 145-150, 1995a.

Gao, F., Jin, C., Ma, H., Sun, L., Nagata, T.: Ultrastructural and radioautographic studies on DNA synthesis in Leydig cells of aging mouse testis. *Cell.Mol. Biol.* 41, 151-160, 1995b.

Gunarso, W.: Radioautographic studies on the nucleic acid synthesis of the retina of chick embryo. I. Light microscopic radioautography. *Shinshu Med. J.* 32, 231-240, 1984a.

Gunarso, W.: Radioautographic studies on the nucleic acid synthesis of the retina of chick embryo. II. Electron microscopic radioautography. *Shinshu Med. J.* 32, 241-248, 1984b.

Gunarso, W., Gao, F., Cui, H., Ma, H., Nagata, T.: A light and electron microscopic radioautographic study on RNA synthesis in the retina of chick embryo. *Acta Histochem.*98, 309-32, 1996.

Gunarso, W., Gao, F., Nagata, T.: Development and DNA synthesis in the retina of chick embryo observed by light and electron microscopic radioautography. *Cell.Mol. Biol.* 43, 189-201, 1997.

Hanai, T.: Light microscopic radioautographic study of DNA synthesis in the kidneys of aging mice. *Cell.Mol. Biol.* 39, 81-91, 1993.

Hanai, T., Nagata, T.: Electron microscopic radioautographic study on DNA and RNA synthesis in perinatal mouse kidney. In, Radioautography in Medicine, Nagata, T., Ed., pp. 127-131, Shinshu University Press, Matsumoto, 1994a.

Hanai, T., Nagata, T.: Study on the nucleic acid synthesis in the aging mouse kidney by light and electron microscopic radioautography. In, Radioautography in Medicine, Nagata, T..Ed., pp. 209-214, Shinshu University Press, Matsumoto, 1994b.

Hanai, T., Nagata, T.: Electron microscopic study on nucleic acid synthesis in perinatal mouse kidney tissue. *Med. Electron Microsc.*27, 355-357, 1994c.

Hanai, T., Usuda, N., Morita, T., Shimizu, T., Nagata, T.: Proliferative activity in the kidneys of aging mice evaluated by PCNA/cyclin immunohistochemistry. *Cell.Mol. Biol.* 39, 181-191, 1993.

Hayashi, K., Gao, F., Nagata, T.: Radioautographic study on ^{3}H-thymidine incorporation at different stages of muscle development in aging mice. *Cell.Mol. Biol.* 39, 553-560, 1993.

Ito, M.: The radioautographic studies on aging change of DNA synthesis and the ultrastructural development of mouse adrenal gland. *Cell.Mol. Biol.* 42, 279-292, 1996.

Ito, M., Nagata, T.: Electron microscopic radioautographic studies on DNA synthesis and ultrastructure of aging mouse adrenal gland. *Med. Electron Microsc.*29, 145-152, 1996.

Izumiyama, K., Kogure, K., Kataoka, S., Nagata, T.: Quantitative analysis of glucose after transient ischemia in the gerbil hippocampus by light and electron microscope radioautography. *Brain Res.* 416, 175-179, 1987.

Jamieson, J. D., Palade, G. E.: Intracellular transport of secretory proteins in the pancreatic exocrine cells. *J. Cell Biol.* 34, 577-615, 1967.

Jin, C.: Study on DNA synthesis of aging mouse colon by light and electron microscopic radioautography. *Cell.Mol. Biol.* 42, 255-268, 1996.

Jin, C., Nagata, T.: Light microscopic radioautographic study on DNA synthesis in cecal epithelial cells of aging mice. *J. Histochem. Cytochem.*43, 1223-1228, 1995a.

Jin, C., Nagata, T.: Electron microscopic radioautographic study on DNA synthesis in cecal epithelial cells of aging mice. *Med. Electron Microsc.*28, 71-75, 1995b.

Johkura, K.: The aging changes of glycoconjugate synthesis in mouse kidney studied by ^{3}H-glucosamine radioautography. *Acta Histochem.Cytochem.*29, 57-63, 1996.

Johkura, K., Nagata, T.: Aging changes of ^{3}H-glucosamine incorporation into mouse kidney observed by radioautography. *Acta Histochem.Cytochem.*28, 494-494, 1995.

Johkura, K., Usuda, N., Nagata, T.: Quantitative study on the aging change of glycoconjugates synthesis in aging mouse kidney. Proc. Xth Internat. Cong. Histochem. Cytochem.,*Acta Histochem. Cytochem.*29, Suppl. 507-508, 1996.

Kobayashi, K., Nagata, T.: Light microscopic radioautographic studies on DNA, RNA and protein syntheses in human synovial membranes of rheumatoid arthritis patients. *J. Histochem. Cytochem.*42, 982-982, 1994.

Komiyama, K., Iida, F., Furihara, R., Murata, F., Nagata, T.: Electron microscopic radioautographic study on 125I-albumin in rat gastric mucosal epithelia. *J. Clin. Electron Microsc.*11, 428-429, 1978.

Kong, Y.: Electron microscopic radioautographic study on DNA synthesis in perinatal mouse retina. *Cell.Mol. Biol.* 39, 55-64, 1993.

Kong, Y., Nagata, T.: Electron microscopic radioautographic study on nucleic acid synthesis of perinatal mouse retina. *Med. Electron Microsc.*27, 366-368, 1994.

Kong, Y., Usuda, N., Nagata, T.: Radioautographic study on DNA synthesis of the retina and retinal pigment epithelium of developing mouse embryo. *Cell.Mol. Biol.* 38, 263-272, 1992a.

Kong, Y., Usuda, N., Morita, T., Hanai, T., Nagata, T.: Study on RNA synthesis in the retina and retinal pigment epithelium of mice by light microscopic radioautography. *Cell.Mol. Biol.* 38, 669-678, 1992b.

Leblond, C. P.: Localization of newly administered iodine in the thyroid gland as indicated by radioiodine. *J. Anat.* 77, 149-152, 1943.

Leblond, C. P.: The life history of cells in renewing systems. *Am. J. Anat.* 160, 113-158, 1981.

Leblond, C. P., Messier, B.: Renewal of chief cells and goblet cells in the small intestine as shown by radioautography after injection of thymidine-3H into mice. *Anat. Rec.* 132: 247-259. 1958.

Li, S.: Relationship between cellular DNA synthesis, PCNA expression and sex steroid hormone receptor status in the developing mouse ovary, uterus and oviduct. *Histochemistry*, 102, 405-413, 1994.

Li, S., Nagata, T.: Nucleic acid synthesis in the developing mouse ovary, uterus and oviduct studied by light and electron microscopic radioautography. *Cell.Mol. Biol.* 41, 185-195, 1995.

Li, S., Gao, F., Duan, H., Nagata, T.: Radioautographic study on the uptake of [35]SO$_4$ in mouse ovary during the estrus cycle. *J. Clin. Electron Microsc.*25, 709-710, 1992.

Liang, Y.: Light microscopic radioautographic study on RNA synthesis in the adrenal glands of aging mice. *Acta Histochem.Cytochem.*31, 203-210, 1998.

Liang, Y., Ito, M., Nagata, T.: Light and electron microscopic radioautographic studies on RNA synthesis in aging mouse adrenal gland. Acta Anat. Nippon.74, 291-300, 1999.

Ma, H.: Light microscopic radioautographic study on DNA synthesis of the livers in aging mice. *Acta Anat. Nippon*.63, 137-147, 1988.

Ma, H., Nagata, T.: Electron microscopic radioautographic study on DNA synthesis of the livers in aging mice. *J. Clin. Electron Microsc*.21, 335-343, 1988a.

Ma, H., Nagata, T.: Studies on DNA synthesis of aging mice by means of electron microscopic radioautography. *J. Clin. Electron Microsc*.21, 715-716, 1988b.

Ma, H., Nagata, T.: Electron microscopic radioautographic studies on DNA synthesis in the hepatocytes of aging mice as observed by image analysis. *Cell.Mol. Biol*. 36, 73-84, 1990a.

Ma, H., Nagata, T.: Study on RNA synthesis in the livers of aging mice by means of electron microscopic radioautography. *Cell.Mol. Biol*. 36, 589-600, 1990b.

Ma, H., Nagata, T.: Collagen and protein synthesis in the livers of aging mice as studied by electron microsopic radioautography. *Ann. Microsc*. 1, 13-22, 2000.

Ma, H., Gao, F., Olea, M. T., Nagata, T.: Protein synthesis in the livers of aging mice studied by electron microscopic radioautography. *Cell.Mol. Biol*. 37, 607-615, 1991.

Matsumura, H., Kobayashi, Y., Kobayashi, K., Nagata, T.: Light microscopic radioautographic study of DNA synthesis in the lung of aging salamander, Hynobius nebulosus. *J. Histochem. Cytochem*.42, 1004-1004, 1994.

Momose, Y., Nagata, T.: Radioautographic study on the intracellular localization of a hypolipidemic agent, bezafibrate, a peroxisome proliferator, in cultured rat hepatocytes. *Cell.Mol. Biol*. 39, 773-781, 1993a.

Momose, Y., Naito, J., Nagata, T.: Radioautographic study on the localization of an anti-allergic agent, tranilast, in the rat liver. *Cell.Mol. Biol*. 35, 347-355, 1989.

Momose, Y., Shibata, N., Kiyosawa, I., Naito, J., Watanabe, T., Horie, S., Yamada, J., Suga, T., Nagata, T.: Morphometric evaluation of species differences in the effects of bezafibrate, a hypolipidemic agent, on hepatic peroxisomes and mitochondria. *J. Toxicol. Pathol*.6, 33-45, 1993b.

Momose, Y., Naito, J., Suzawa, H., Kanzawa, M., Nagata, T.: Radioautographic study on the intracellular localization of bezafibrate in cultured rat hepatoctyes. *Acta Histochem.Cytochem*.28, 61-66, 1995.

Morita, T.: Radioautographic study on the aging change of ^{3}H-glucosamine uptake in mouse ileum. *Cell.Mol. Biol*. 39, 875-884, 1993.

Morita, T., Usuda, N. Hanai, T., Nagata, T.: Changes of colon epithelium proliferation due to individual aging with PCNA/cyclin immunostaining comparing with ^{3}H-thymidine radioautography. *Histochemistry*, 101, 13-20, 1994.

Murata, F., Momose,Y. , Yoshida, K., Nagata, T.: Incorporation of ^{3}H-thymidine into the nucleus of mast cells in adult rat peritoneum. *Shinshu Med. J*. 25, 72-77, 1977a.

Murata, F., Momose, Y., Yoshida, K., Ohno, S., Nagata, T.: Nucleic acid and mucosubstance metabolism of mastocytoma cells by means of electron microscopic radioautography. *Acta Pharmacol.Toxicol*.41, 58-59, 1977b.

Murata, F., Yoshida, K., Ohno, S., Nagata, T.: Ultrastructural and electron microscopic radioautographic studies on the mastocytoma cells and mast cells. *J. Clin. Electron Microsc*.11, 561-562, 1978.

Murata, F., Yoshida, K., Ohno, S., Nagata, T.: Mucosubstances of rabbit granulocytes studied by means of electron microscopic radioautography and X-ray microanalysis. *Histochemistry*, 61, 139-150, 1979.

Nagata, T.: On the relationship between cell division and cytochrome oxidase in the Yoshida sarcoma cells. *Shinshu Med. J.* 5: 383-386, 1956.

Nagata, T.: Studeis on the amitosis in the Yoshida sarcoma cells. I. Observation on the smear preparation under normal conditions. *Med. J. Shinshu Univ.* 2: 187-198, 1957a.

Nagata, T.: Studeis on the amitosis in the Yoshida sarcoma cells. II. Phase-contrast microscopic observations under normal conditions. *Med. J. Shinshu Univ.* 2: 199-207, 1957b.

Nagata, T.: Cell divisions in the liver of the fetal and newborn dogs. *Med. J. Shinshu Univ.* 4: 65-73, 1959.

Nagata, T.: A radioautographic study of the DNA synthesis in rat liver, with special reference to binucleate cells. *Med. J. Shinshu Univ.* 7, 17-25, 1962.

Nagata, T.: A quantitative study on the ganglion cells in the small intestine of the dog. *Med. J. Shinshu Univ.* 10, 1-11, 1965.

Nagata, T.: A radioautographic study on the RNA synthesis in the hepatic and the intestinal epithelial cells of mice after feeding with special reference to binuclearity. *Med. J. Shinshu Univ.* 11, 49-61, 1966.

Nagata, T.: On the increase of binucleate cells in the ganglion cells of dog small intestine due to experimental ischemia. *Med. J. Shinshu Univ.* 12, 93-113, 1967a.

Nagata, T.: A radioautographic study on the protein synthesis in the hepatic and the intestinal epithelial cells of mice, with special reference to binucleate cells. *Med. J. Shinshu Univ.* 12, 247-257, 1967b.

Nagata, T.: Chapter 3. Application of microspectrophotometry to various substances.In , Introduction to Microspectrophotometry. S. Isaka, T. Nagata, N. Inui, Eds., pp. 49-155, Olympus Co., Tokyo, 1972a.

Nagata, T.: Electron microscopic dry-mounting autoradiography. *Proc. 4th Internat. Cong. Histochem. Cytochem.Kyoto*, pp. 43-44, 1972b.

Nagata, T.: Electron microscopic radioautography of intramitochondrial RNA synthesis of HeLa cells in culture. *Histochemie*, 32, 163-170, 1972c.

Nagata, T.: Quantitative electron microscope radioautography of intramitochondrial nucleic acid synthesis. *Acta Histochem.Cytochem.*5, 201-203, 1972d.

Nagata, T.: Electron microscopic observation of target cells previously observed by phase-contrast microscopy: Electron microscopic radioautography of laser beam irradiated cultured cells. *J. Clin. Electron Microsc.*17, 589-590, 1984.

Nagata, T.: Principles and techniques of radioautography. In, Histo- and Cyto-chemistry 1985, Japan Society of Histochemistry and Cytochemistry, Ed., pp. 207-226, Gakusai Kikaku Co., Tokyo, 1985.

Nagata, T.:.Electron microscopic radioautography and analytical electron microscopy.*J. Clin. Electron Microsc.*24, 441-442, 1991.

Nagata, T.: Radiolabeling of soluble and insoluble compounds as demonstrated by light and electron microscopy. Recent Advances in Cellular and Molecular Biology, Wegmann, R. J., Wegmann, M. A., Eds. Peters Press, Leuven, Vol. 6, pp. 9-21, 1992.

Nagata, T.: Quantitative analysis of histochemical reactions: Image analysis of light and electron microscopic radioautograms. *Acta Histochem.Cytochem.*26, 281-291, 1993a.

Nagata, T. Quantitative light and electron microscopic radioautographic studies on macromolecular synthesis in several organs of prenatal and postnatal aging mice. *Chinese J. Histochem.Cytochem.* 2: 106-108, 1993b.

Nagata, T.: Electron microscopic radioautography with cryo-fixation and dry-mounting procedure. *Acta Histochem.Cytochem.* 27: 471-489, 1994a.

Nagata, T.: Application of electron microscopic radioautography to clinical electron microscopy. *Med. Electron Microsc.*27; 191-212, 1994b.

Nagata, T.: Radioautography in Medicine. Shinshu University Press, 268pp, Matsumoto, 1994c.

Nagata, T.: Radioautography, general and special. In, Histo- and Cyto-chemistry 1994, Japan Society of Histochemistry and Cytochemistry, ed, pp. 219-231, Gakusai Kikaku Co., Tokyo, 1994d.

Nagata, T., Application of electron microscopic radioautography to clinical electron microscopy.*Med. Electron Microsc.*27, 191-212, 1994e.

Nagata, T.: Light and electron microscopic radioautographic study on macromolecular synthesis in digestive organs of aging mice. *Cell.Mol. Biol.* 41, 21-38, 1995a.

Nagata, T.: Histochemistry of the organs: Application of histochemistry to anatomy. *Acta Anat. Nippon.*70, 448-471, 1995b.

Nagata, T.: Three-dimensional observation of whole mount cultured cells stained with histochemical reactions by ultrahigh voltage electron microscopy. *Cell.Mol. Biol.* 41, 783-792, 1995c.

Nagata, T.: Morphometry in anatomy: image analysis on fine structure and histochemical reactions with special reference to radioautography. *Ital. J. Anat.* 100 (Suppl. 1), 591-605, 1995d.

Nagata, T.: Technique and application of electron microscopic radioautography. *J. Electron Microsc.* 45, 258-274, 1996a.

Nagata, T.: Techniques of light and electron microscopic radioautography. In, Histochemistry and Cytochemistry 1996.Proc. Xth Internat.Congr.Histochem. Cytochem. *Acta Histochem.Cytochem.*29 (Suppl.), 343-344, 1996b.

Nagata, T.: Remarks: Radioautographology, general and special. *Cell.Mol. Biol.* 42 (Suppl.), 11-12, 1996c.

Nagata, T.: On the terminology of radioautography vs. autoradiography. *J. Histochem. Cytochem.*44, 1209-1209, 1996d.

Nagata, T.: Techniques and applications of microscopic radioautography. *Histol. Histopathol.*12, 1091-1124, 1997a.

Nagata T.: Three-dimensional observation on whole mount cultured cells and thick sections stained with histochemical reactions by high voltage electron microscopy. In, Recent Advances in Microscopy of Cells, Tissues and Organs, P. Motta, Ed., pp. 37-44, Antonio Delfino Editore, Roma, 1997b.

Nagata, T.: Radioautographic study on collagen synthesis in the ocular tissues. *J. Kaken Eye Res.* 15, 1-9, 1997c.

Nagata, T.: Techniques of radioautography for medical and biological research. *Braz. J. Biol. Med. Res.* 31, 185-195, 1998a.

Nagata, T.: Radioautographology, the advocacy of a new concept. *Braz. J. Biol. Med. Res.* 31, 201-241, 1998b.

Nagata, T.: Radioautographic studies on DNA synthesis of the bone and skin of aging salamander. *Bull. Nagano Women's Jr. College*, 6, 1-14, 1998c.

Nagata, T.: 3D observation of cell organelles by high voltage electron microscopy. Microscopy and Analysis, Asia Pacific Edition, 9, 29-32, 1999a.

Nagata, T.: Application of histochemistry to anatomy: Histochemistry of the organs, a novel concept. Proc. XV Congress of the International Federation of Associations of Anatomists, Ital. *J. Anat. Embryol.*104 (Suppl. 1), 486-486, 1999b.

Nagata, T.: Aging changes of macromolecular synthesis in various organ systems as observed by microscopic radioautography after incorporation of radiolabeled precursors. *Methods Find. Exp. Clin. Pharmacol.*21, 683-706, 1999c.

Nagata, T.: Radioautographic study on protein synthesis in mouse cornea. *J. Kaken Eye Res.* 8, 8-14, 1999d.

Nagata, T.: Radioautographology, general and special: a novel concept. *Ital. J. Anat. Embryol.* 104 (Suppl. 1), 487-487, 1999e.

Nagata, T.: Three-dimensional observations on thick biological specimens by high voltage electron microscopy. *Image Analysis Stereolog.*19, 51-56, 2000a.

Nagata, T.: Biological microanalysis of radiolabeled and unlabeled compounds by radioautography and X-ray microanalysis. *Scanning Microscopy International*, 14, on line, 2000b.

Nagata, T.: Electron microscopic radioautographic study on protein synthesis in pancreatic cells of perinatal and aging mice. *Bull. Nagano Women's Jr. College*, 8: 1-22, 2000c.

Nagata, T.: Light microscopic radioautographic study on radiosulfate incorporation into the tracheal cartilage in aging mice. *Acta Histochem.Cytochem.*32, 377-383, 2000d.

Nagata, T.: Introductory remarks: Special radioautographology. Cell.Mol. Biol. 46 (Congress Suppl.), 161-161, 2000e.

Nagata, T.: Special radioautographology: the eye. J. Kaken Eye Res. 18, 1-13, 2000f.

Nagata, T.: Three-dimensional high voltage electron microscopy of thick biological specimens. *Micron.*32, 387-404, 2001a.

Nagata, T.: Three-dimensional and four-dimensional observation of histochemical and cytochemical specimens by high voltage electron microscopy. *Acta Histochem.Cytochem.*34, 153-169, 2001b.

Nagata, T. : Special cytochemistry in cell biology. In, Internat. Rev. Cytol. Jeon, K.W., ed., Vol. 211, Chapter 2, pp. 33-154, Academic Press, New York, 2001c.

Nagata, T. : Radioautographology General and Special, In, Prog. Histochem. Cytochem., Graumann, W., Ed., Urban & Fischer, Jena, Vol. 37 No. 2, pp. 57-226, 2002.

Nagata T.: Light and electron microscopic study on macromolecular synthesis in amitotic hepatocyte mitochondria of aging mice. *Cell.Mol. Biol.* 49, 591-611, 2003.

Nagata, T.: X-ray microanalysis of biological specimens by high voltage electron microscopy. In, Prog.Histochem. Cytochem., Graumann, W., Ed., Urban & Fischer Verlag, Jena, Vol. 39, No. 4, pp. 185-320, 2004.

Nagata T.: Aging changes of macromolecular synthesis in the uro-genital organs as revealed by electron microscopic radioautography. *Ann. Rev. Biomed.Sci.* 6, 13-78, 2005.

Nagata T.: Electron microscopic radioautographic study on protein synthesis in hepatocyte mitochondria of developing mice. *Ann. Microsc.* 6, 43-54, 2006a.

Nagata T.: Electron microscopic radioautographic study on nucleic acids synthesis in hepatocyte mitochondria of developing mice. *Sci. World J.* 6: 1583-1598, 2006b.

Nagata T.: Macromolecular synthesis in hepatocyte mitochondria of aging mice as revealed by electron microscopic radioautography. I: Nucleic acid synthesis. In, Modern Research and Educational Topics in Microscopy. Mendez-Vilas, A. and Diaz, J. Eds., Formatex Micrscopy Series No. 3, Vol. 1, pp. 245-258, Formatex, Badajoz, Spain, 2007a.

Nagata T.: Macromolecular synthesis in hepatocyte mitochondria of aging mice as revealed by electron microscopic radioautography. II: Protein synthesis. In, Modern Research and Educational Topics in Microscopy. Mendez-Vilas, A. and Diaz, J. eds., Formatex Micrscopy Series No. 3, Vol. 1, pp. 259-271, Formatex, Badajoz, Spain, 2007b.

Nagata, T.: Electron microscopic radioautographic study on macromolecular synthesis in hepatocyte mitochondria of aging mouse. *J. Cell Tissue Res.* 7, 1019-1029, 2007c.

Nagata, T.; Electron microscopic radioautographic study on nucleic acids synthesis in hepatocyte mitochondria of developing mice. *Trends Cell Molec.Biol.*, 2, 19-33, 2007d.

Nagata, T.; Aging changes of macromolecular synthesis in the mitochondria of mouse hepatocytes as revealed by microscopic radioautography.*Ann. Rev. Biomed.Sci.* 9, 30-36, 2007e.

Nagata, T.: Radioautographology, Bull. Shinshu Institute Alternat. Med., 2, 3-32, 2007f.

Nagata, T.: Electron microscopic radioautographic study on mitochondrial DNA synthesis in adrenal cortical cells of developing mice. *J. Cell.Tis. Res.* 8, 1303-12, 2008a.

Nagata T.: Electron microscopic radioautographic study on mitochondrial DNA synthesis in adrenal cortical cells of developing and aging mice. *The Sci. World J.* 8, 683-97. 2008b.

Nagata, T.: Sexual difference between the macromolecular synthesis of hepatocyte mitochondria in male and female mice in aging as revealed by electron microscopic radioautography. Chapter 22. In, Women and Aging: New Research, H. T. Bennninghouse, A. D. Rosset, eds. Nova Biomed. Books, New York, pp. 461-487, 2009a

Nagata, T.: Protein synthesis in hepatocytes of mice as revealed by electron microscopic radioautography. In, Protein Biosynthesis.Esterhouse, T. E. and Petrinos, L. B., eds., Nova Biomed.Books, New York, pp. 133-161, 2009b.

Nagata, T.: Electron microscopic radioatuographic studies on macromolecular synthesis in mitochondria of various cells. 18EMSM Conference Proc. 9th Asia-Pacific Microscopy Conference (APMC9), Kuala Lumpur, Malaysia, pp. 48-50, 2009c.

Nagata, T.: Recent studies on macromolecular synthesis labeled with ^{3}H-thymidine in various organs as revealed by electron microscopic radioautography. *Current Radiopharmaceutics*, 2, 118-128, 2009d.

Nagata, T.: Electron microscopic radioautographic study on mitochondrial DNA synthesis in adrenal medullary cells of developing and aging mice. *J. Cell Tissue Res.* 9, 1793-1802, 2009e.

Nagata, T.: Applications of high voltage electron microscopy to thick biological specimens. *Ann. Microsc.* 9, 4-40, 2009f.

Nagata, T.: Electron microscopic radioautographic study on DNA synthesis of mitochondria in adrenal medullary cells of aging mice. *Open Anat. J.* 1, 14-24, 2009g.

Nagata, T.: Electron microscopic radioautographic studies on macromolecular synthesis in mitochondria of animal cells in aging. *Ann. Rev. Biomed.Sci.* 11, 1-17, 2009h.

Nagata, T.: Electron microscopic radioautographic studies on macromoleclular synthesis in mitochondria of some organs in aging animals. *Bull. Shinshu Inst. Alternat. Med. Welfare*, 4, 15-38, 2009i.

Nagata, T.: Electron microscopic radioautographic study on mitochondrial DNA synthesis in adreno-cortical cells of aging ddY mice. *Bull. Shinshu Inst. Alternat. Med. Welfare*, 4, 51-66, 2009j.

Nagata T.: Electron microscopic radioautographic study on mitochondrial RNA synthesis in adrenocortical cells of aging mice. *Open Anat J.* 2, 91-97, 2010a.

Nagata T. Electron microscopic radioautographic study on mitochondrial RNA synthesis in adrenal medullary cells of aging and senescent mice.*J Cell Tissue Res.* 10, 2213-2222, 2010b.

Nagata, T.: Macromolecular synthesis in the livers of aging mice as revealed by electron microscopic radioautography. In, Prog.Histochem. Cytochem., Sasse, D., ed., Elsevier, Amsterdam, Boston, London, New York, Oxford, Paris, Philadelphia, San Diego, St. Louis, Vol. 45, No. 1, pp. 1-80, 2010c.

Nagata, T.: Electron microscopic radioautographic study on protein synthesis of mitochondria in adrenal medullary cells of aging mice. *Bulletin Shinshu Inst Alternat Med Welfare*, 5: in press, 2010d.

Nagata, T.: Electron microscopic radioautographic study on mitochondrial RNA synthesis in adrenal cortical and medullary cells of aging mice. *J. Biomed. Sci. Enginer.* 1, in press, 2010e.

Nagata, T.: Electron microscopic radioautographic study on protein synthesis of mitochondria in adrenal cortical cells of aging mice. *Bulletin Shinshu Inst Alternat Med Welfare*, 5: in press, 2010f.

Nagata T.: Electron microscopic radioautographic study on mitochondrial DNA, RNA and protein synthesis in adrenal cells of aging mice. Formatex Micrscopy Series No. 3, Vol. 3, Formatex, Badajoz, Spain, in press, 2010g.

Nagata, T.: Electron microscopic radioautographic studies on macromolecular synthesis in mitochondria of animal cells in aging. *Ann. Rev. Biomed.Sci.* 12, 1-29, 2010h.

Nagata, T., Cui, H., Gao, F.: Radioautographic study on glycoprotein synthesis in the ocular tissues. *J. Kaken Eye Res.* 13, 11-18, 1995.

Nagata, T., Cui, H., Kong, Y.: The localization of TGF-□1 and its mRNA in the spinal cords of prenatal and postnatal aging mice demonstrated with immunohistochemical and in situ hybridization techniques. *Bull. Nagano Women's Jr. College*, 7, 75-88, 1999a.

Nagata, T., Cui, H., Liang, Y.: Light microscopic radioautographic study on the protein synthesis in the cerebellum of aging mouse. *Bull. Nagano Women's Jr. College*, 9, 41-60 2001.

Nagata, T., Fujii, Y., Usuda, N.: Demonstration of extranuclear nucleic acid synthesis in mammalian cells under experimental conditions by electron microscopic radioautography.*Proc. 10th Internat.Congr.Electr.Microsc.*2, 305-306, 1982b.

Nagata, T., Hirano, I., Shibata, O., Nagata, T.: A radioautographic study on the DNA synthesis in the hepatic and the pancreatic acinar cells of mice during the postnatal growth, with special reference to binuclearity. Med. J. Shinshu Univ. 11, 35-42, 1966.

Nagata, T., Ito, M., Chen, S.: Aging changes of DNA synthesis in the submandibular glands of mice as observed by.light and electron microscopic radioautography. *Ann. Microsc.* 1, 4-12, 2000a.

Nagata, T. Ito, M., Liang, Y.: Study of the effects of aging on macromolecular synthesis in mouse steroid secreting cells using microscopic radioautography. *Methods Find. Exp. Clin. Pharmacol.*22, 5-18, 2000b.

Nagata, T., Iwadare, I., Murata, F.: Electron microscopic radioautography of nucleic acid synthesis in cultured cells treated with several carcinogens. *Acta Pharmacol.Toxicol.*41, 64-65, 1977c.

Nagata, T., Kawahara, I.: Radioautographic study of the synthesis of sulfomucin in digestive organs of mice. *J. Trace Microprobe Analysis*, 17, 339-355, 1999.

Nagata, T., Kawahara, I., Usuda, N., Maruyama, M., Ma, H.: Radioautographic studies on the glycoconjugate synthesis in the gastrointestinal mucosa of the mouse. In, Glycoconjugate in Medicine, Ohyama, M., Muramatsu, T., Eds, pp. 251-256, Professional Postgrad. Service, Tokyo, 1988a.

Nagata, T., Kong, Y.: Distribution and localization of TGF□1 and □FGF, and their mRNAs in aging mice. *Bull. Nagano Women's Junior College*, 6, 87-105, 1998.

Nagata, T., Ma, H., Electron microscopic radioautographic study on mitochondrial DNA synthesis in hepatocytes of aging mouse.*Ann. Microsc.* 5, 4-1, 2005a.

Nagata, T., Ma, H., Electron microscopic radioautographic study on RNA synthesis in hepatocyte mitochondria of aging mouse.*Microsc.Res. Tech.* 67, 55-64, 2005b.

Nagata, T., Momoze, S.: Aging changes of the amitotic and binucleate cells in dog livers. *Acta Anat. Nipponica*, 34, 187-190, 1959.

Nagata, T., Morita, T., I. Kawahara, I.: Radioautographic studies on radiosulfate incorporation in the digestive organs of mice. *Histol.Histopathol.* 14, 1-8, 1999b.

Nagata, T., Murata, F.: Electron microscopic dry-mounting radioautography for diffusible compounds by means of ultracryotomy. *Histochemistry*, 54, 75-82, 1977.

Nagata, T., Murata, F., Yoshida, K., Ohno, S., Iwadare, N.: Whole mount radioautography of cultured cells as observed by high voltage electron microscopy. *Proc. Fifth Internat.Conf. High Voltage Electron Microsc.*347-350, 1977d.

Nagata, T., Nawa, T.: A modification of dry-mounting technique for radioautography of water-soluble compounds. *Histochemie*, 7, 370-371, 1966a.

Nagata, T., Nawa, T.: A radioautographic study on the nucleic acids synthesis of binucleate cells in cultivated fibroblasts of chick embryos. *Med. J. Shinshu Univ.* 11, 1-5, 1966b.

Nagata, T., Nawa, T., Yokota, S.: A new technique for electron microscopic dry-mounting radioautography of soluble compounds. *Histochemie*, 18, 241-249, 1969.

Nagata, T., Nishigaki, T., Momose, Y.: Localization of anti-allergic agent in rat mast cells demonstrated by light and electron microscopic radioautography. *Acta Histochem.Cytochem.*19, 669-683, 1986b.

Nagata, T., Ohno, S., Kawahara, I., Yamabayashi, S., Fujii, Y., Murata, F.: Light and electron microscopic radioautography of nucleic acid synthesis in mitochondria and peroxisomes of rat hepatic cells during and after DEHP administration. *Acta Histochem.Cytochem.*16, 610-611, 1979.

Nagata, T., Ohno, S., Murata, F.: Electron microscopic dry-mounting radioautography for soluble compounds. -*Acta Phamacol. Toxicol.*41, 62-63, 1977a.

Nagata, T., Ohno, S., Yoshida, K., Murata, F.: Nucleic acid synthesis in proliferating peroxisomes of rat liver as revealed by electron microscopical radioautography. *Histochem. J.* 14, 197-204, 1982a.

Nagata, T., Olea, M. T.: Electron microscopic radioautographic study on the protein synthesis in aging mouse spleen. Bull. Nagano Women's Jr. College 7, 1-9, 1999.

Nagata, T., Shibata, O., Nawa, T.: Simplified methods for mass production of radioautograms. -*Acta Anat. Nippon.*42, 162-166, 1967a.

Nagata, T., Shibata, O., Nawa, T.: Incorporation of tritiated thymidine into mitochondrial DNA of the liver and kidney cells of chickens and mice in tissue culture. *Histochemie*, 10, 305-308, 1967b.

Nagata, T., Shimamura, K., Onozawa, M., Kondo, T., Ohkubo, K., Momoze, S.: Relationship of binuclearity to cell function in some organs. I. Frequencies of binucleate cells in some organs of toads in summer and winter. *Med. J. Shinshu Univ.* 5, 147-152, 1960a.

Nagata, T., Shimamura, K., Kondo, T., Onozawa, M., Momoze, S., Okubo, M.: Relationship of binuclearity to cell function in some organs. II. Variation of frequencies of binucleate cells in some organs of dogs owing to aging. *Med. J. Shinshu Univ.* 5, 153-158, 1960b.

Nagata, T., Steggerda, F. R.: Histological study on the deganglionated small intestine of the dog. *Physiologist.*6, 242-242, 1963.

Nagata, T., Steggerda, F. R.: Observations on the increase of binucleate cells in the ganglion cells of the dog's intestine due to experimental ischemia. *Anat. Rec.* 148, 315-315, 1964.

Nagata, T., Toriyama, K., Kong, Y., Jin, C., Gao, F.: Radioautographic study on DNA synthesis in the ciliary bodies of aging mice. *J. Kaken Eye Res.*12, 1-11, 1994.

Nagata, T., Usuda, N.: Image processing of electron microscopic radioautograms in clinical electron microscopy. *J. Clin. Electron.Microsc.*18, 451-452, 1985.

Nagata, T., Usuda, N.: Studies on the nucleic acid synthesis in pancreatic acinar cells of aging mice by means of electron microscopic radioautography. *J. Clin. Electron Microsc.*19, 486-487, 1986.

Nagata, T., Usuda, N.: Electron microscopic radioautography of protein synthesis in pancreatic acinar cells of aging mice. *Acta Histochem.Cytochem.*26, 481-481, 1993a.

Nagata, T., Usuda, N.: In situ hybridization by electron microscopy using radioactive probes. *J. Histochem. Cytochem.*41, 1119-1119, 1993b.

Nagata, T., Usuda, N., Ma, H.: Electron microscopic radioautography of nucleic acid synthesis in pancreatic acinar cells of prenatal and postnatal aging mice. *Proc. XIth Intern. Cong. Electr. Microsc.*3, 2281-2282, 1984.

Nagata, T., Usuda, N., Ma, H., Electron microscopic radioautography of nucleic acid synthesis in pancreatic acinar cells of prenatal and postnatal aging mice.*Proc. XIth Intern. Cong. Electr. Microsc.*3, 2281-2282, 1986.

Nagata, T., Usuda, N., Ma, H.: Electron microscopic radioautography of lipid synthesis in pancreatic cells of aging mice. *J. Clin. Electr.Microsc.*23, 841-842, 1990.

Nagata, T., Usuda, N., Maruyama, M., Ma, H.: Electron microscopic radioautographic study on lipid synthesis in perinatal mouse pancreas. *J. Clin. Electr.Microsc.*21, 756-757, 1988b.

Nagata, T., Usuda, N., Suzawa, H., Kanzawa, M.: Incorporation of ^{3}H-glucosamine into the pancreatic cells of aging mice as demonstrated by electron microscopic radioautography. *J. Clin. Electron Microsc.*25, 646-647, 1992.

Nagata, T., Yamabayashi, S.: Intracellular localization of ^{3}H-befunolol by means of electron microscopic radioautography of cryo-fixed ultrathin sections. *J. Clin. Electron Microsc.*16, 737-738, 1983.

Nagata, T., Yoshida, K., Murata, F.: Demonstration of hot and cold mercury in the human thyroid tissues by means of radioautography and chemography. *Acta Pharmacol.Toxicol.*41, 60-61, 1977b.

Nagata, T., Yoshida, K., Ohno, S., Murata, F.: Ultrastructural localization of soluble and insoluble ^{3}H-methyl prednisolone as revealed by electron microscopic dry-mounting radioautography. *Proc. 9th Internat.Congr.Electr.Microsc.*2, 40-41, 1978b.

Nishigaki, T., Momose, Y., Nagata, T.: Light microscopic radioautographic study of the localization of anti-allergic agent, Tranilast, in rat mast cells. *Histochem. J.* 19, 533-536, 1987.

Nishigaki, T., Momose, Y., Nagata, T.: Electron microscopic radioautographic study of the localization of an anti-allergic agent, tranilast, in rat mast cells. *Cell.Mol. Biol.* 36, 65-71, 1990a.

Nishigaki, T., Momose, Y., Nagata, T.: Localization of the anti-allergic agent tranilast in the urinary bladder of rat as demonstrated by light microscopic radioautography. *Drug Res.* 40, 272-275, 1990b.

Oguchi, K., Nagata, T.: A radioautographic study of activated satellite cells in dystrophic chicken muscle. In, Current Research in Muscular Dystrophy Japan.The Proc. Ann. Meet. Muscular Dystrophy Res. 1980, pp. 16-17, Ministry of Welfare of Japan, Tokyo, 1980.

Oguchi, K., Nagata, T.: Electron microscopic radioautographic observation on activated satellite cells in dystrophy chickens. In, Clinical Studies on the Etiology of Muscular Dystrophy.Annual Report on Neurological Diseases 1981, pp. 30-33, Ministry of Welfare of Japan, Tokyo, 1981.

Ohno, S., Fujii, Y., Usuda, N., Endo, T., Hidaka, H., Nagata, T.: Demonstration of intracellular localization of calmodulin antagonist by wet-mounting radioautography. *J. Electron Microsc.* 32, 1-12, 1983.

Ohno, S., Fujii, Y., Usuda, N., Nagata, T., Endo, T., Tanaka, T., Hidaka, H.: Intracellular localization of calmodulin antagonists (W-7). In, Calmodulin and intracellular Ca^{2+} receptors.Kakiuchi, S., Hidaka, H, Means, A. R., Eds., pp. 39-48, Plenum Publishing Co., New York, 1982.

Olea, M. T.: An ultrastructural localization of lysosomal acid phosphatase activity in aging mouse spleen: a quantitative X-ray microanalytical study. *Acta Histochem.Cytochem.*24, 201-208, 1991.

Olea, M. T., Nagata, T.: X-ray microanalysis of cerium in mouse spleen cells demonstrating acid phosphatase activity using high voltage electron microscopy, *Cell. Mol. Biol.* 37, 155-163, 1991.

Olea, M.T., Nagata, T. : Simultaneous localization of ^{3}H-thymidine incorporation and acid phosphatase activity in mouse spleen: EM radioautography and cytochemistry. *Cell.Mol. Biol.* 38, 115-122, 1992a.

Olea, M.T., Nagata, T.: A radioautographic study on RNA synthesis in aging mouse spleen after 3H-uridine labeling in vitro. *Cell.Mol. Biol.* 38, 399-405, 1992b.

Oliveira, S.F., Nagata, T., Abrahamsohn, P.A., Zorn, T.M.T.: Electron microscopic radioautographic study on the incorporation of ^{3}H-proline by mouse decidual cells. *Cell.Mol. Biol.* 37, 315-323, 1991.

Oliveira, S. F., Abrahamsohn, P. A., Nagata, T., Zorn, T. M. T.: Incorporation of ^{3}H-amino acids by endometrial stromal cells during decidualization in the mouse. A radioautographical study.*Cell.Mol. Biol.* 41, 107-116, 1995.

Pearse, A. G. E.: Histochemistry, Theoretical and Applied. 4th Ed. Vol. 1.439 pp., 1980, Vol. 2.1055 pp., 1985, Vol. 3.Ed. with P. Stoward, 728 pp. Churchill Livingstone, Edinburgh, London and New York, 1991.

Sakai, Y., Ikado, S., Nagata, T.: Electron microscopic radioautography of satellite cells in regenerating muscles. *J. Clin. Electr.Microsc.*10, 508-509, 1977.

Sato, A.: Quantitative electron microscopic studies on the kinetics of secretory granules in G-cells. *Cell Tissue Res.* 187, 45-59, 1978.

Sato, A., Iida, F., Furihara, R., Nagata, T.: Electron microscopic raioautography of rat stomach G-cells by means of ^{3}H-amino acids. *J. Clin. Electron Microsc.*10, 358-359, 1977.

Shimizu, T., Usuda, N.,Yamanda, T., Sugenoya, A., Iida, F.: Proliferative activity of human thyroid tumors evaluated by proliferating cell nuclear antigen/cyclin immnohistochemical studies. *Cancer*, 71, 2807-2812, 1993.

Sun, L.: Age related changes of RNA synthesis in the lungs of aging mice by light and electron microscopic radioautography. *Cell.Mol. Biol.* 41, 1061-1072, 1995.

Sun, L., Gao, F., Duan, H., Nagata, T.: Light microscopic radioautography of DNA synthesis in pulmonary cells in aging mice. In, Radioautography in Medicine, Nagata, T. Ed., pp. 201-205, Shinshu University Press, Matsumoto, 1994.

Sun, L., Gao, F., Nagata, T.: Study on the DNA synthesis of pulmonary cells in aging mice by light microscopic radioautography. *Cell.Mol. Biol.* 41, 851-859, 1995a.

Sun, L., Gao, F., Jin, C., Duan, H., Nagata, T.: An electron microscopic radioautographic study on the DNA synthesis of pulmonary tissue cells in aging mice. *Med. Electron. Microsc.*28, 129-131, 1995b.

Sun, L., Gao, F., Jin, C., Nagata, T.: DNA synthesis in the tracheae of aging mice by means of light and electron microscopic radioautography. *Acta Histochem.Cytochem.*30, 211-220, 1997a.

Sun, L., Gao, F., Nagata, T.: A Light Microscopic radioautographic study on protein synthesis in pulmonary cells of aging mice. *Acta Histochem.Cytochem.*30, 463-470, 1997b.

Suzuki, K., Imada, T., Gao, F., Ma, H., Nagata, T.: Radioautographic study of benidipine hydrochloride: localization in the mesenteric artery of spontaneously hypertensive rat. *Drug Res.* 44, 129-133, 1994.

Terauchi, A., Mori, T., Kanda, H., Tsukada, M., Nagata, T.: Radioautographic study of 3H-taurine uptake in mouse skeletal muscle cells. *J. Clin. Electron Microsc.*21, 627-628, 1988.

Terauchi, A., Nagata, T.: Observation on incorporation of ^{3}H-taurine in mouse skeletal muscle cells by light and electron microscopic radioautography. *Cell.Mol. Biol.* 39, 397-404, 1993.

Terauchi, A., Nagata, T.: In corporation of ^{3}H-taurine into the blood capillary cells of mouse skeletal muscle. Radioautography in Medicine, Nagata, T. ed., Shinshu University Press, Matsumoto, 1994.

Toriyama, K.: Study on the aging changes of DNA and protein synthesis of bipolar and photo-receptor cells of mouse retina by light and electron microscopic radioautography. *Cell.Mol. Biol.* 41, 593-601, 1995.

Tsukahara, S., Yoshida, K., Nagata, T.: A radioautographic study on the incorporation of ^{14}C-bupranolol (beta-blocking agent) into the rabbit eye. Histochemistry 68, 237-244, 1980.

Usuda, N., Nagata, T.: Electron microscopic radioautography of acyl-CoA mRNA by in situ hybridization. *J. Clin. Electron Microsc.*25, 332-333, 1992.

Usuda, N., Nagata, T.: The immunohistochemical and in situ hybridization studies on hepatic peroxisomes. *Acta Histochem.Cytochem.*28, 169-172, 1995.

Usuda, N., Hanai, T., Morita, T., Nagata, T.: Radioautographic demonstration of peroxisomal acyl-CoA oxidase mRNA by in situ hybridization. In, Recent advances in cellular and molecular biology, Vol. 6. Molecular biology of nucleus, peroxisomes, organelles and cell movement.Wegmann, R. J., Wegmann, M., Eds, pp.181-184, Peeters Press, Leuven, 1992.

Uwa, H., Nagata, T.: Cell population kinetics of the scleroblast during ethisterone-induced anal-fin process formation in adult females of the Medaka. *Dev. Growth Different.*9, 693-694, 1976.

Watanabe, I., Makiyama, M. C. K., Nagata, T.: Electron microscopic radioautographic observation of the submandibular salivary gland of aging mouse. *Acta Microscopica,* 6. 130-131, 1997.

Yamabayashi, S., Gunarso, W., Tsukahara, S., Nagata, T.: Incorporation of ^{3}H-befunolol (beta blocking agent) into melanin granules of ocular tissues in the pigmented rabbits. I. Light microscopic radioautography. *Histochemistry*, 73, 371-375, 1981.

Yamada, A. T.: Timely and topologically defined protein synthesis in the periimplanting mouse endometrium revealed by light and electron microscopic radioautography. *Cell.Mol. Biol.* 39, 1-12, 1993.

Yamada, A., Nagata, T.: Ribonucleic acid and protein synthesis in the uterus of pregnant mouse during activation of implantation window. *Med. Electron Microsc.*27, 363-365, 1992a.

Yamada, A., Nagata, T.: Light and electron microscopic raioautography of DNA synthesis in the endometria of pregnant-ovariectomized mice during activation of implantation window. *Cell.Mol. Biol.* 38, 763-774, 1992b.

Yamada, A., Nagata, T.: Light and electron microscopic raioautography of RNA synthesis of peri-implanting pregnant mouse during activation of receptivity for blastocyst implantation. *Cell.Mol. Biol.* 38, 211-233, 1993.

Yoshinaga, K.: Uterine receptivity for blastcyst implantation. *Ann. N. Y. Acad. Sci. USA*, 541, 424-431, 1988.

Yoshizawa, S., Nagata, A., Honma, T., Oda, M., Murata, F., Nagata, T.: Study of ethionine pancreatitis by means of electron microscopic radioautography. *J. Clin. Electron Microsc.*7, 349-350, 1974.

Yoshizawa, S., Nagata, A., Honma, T., Oda, M., Murata, F., Nagata, T.: Radioautographic study of protein synthesis in pancreatic exocrine cells of alcoholic rats. *J. Clin. Electron.Microsc.*10, 372-373, 1977.

In: Cell Aging
Editors: Jack W. Perloft and Alexander H. Wong

ISBN: 978-1-61324-369-5
2012 Nova Science Publishers, Inc.

Chapter 2

SARCOPENIA: MOLECULAR MECHANISMS AND CURRENT THERAPEUTIC STRATEGY

Kunihiro Sakuma[1] and Akihiko Yamaguchi[2]*

[1] Research Center for Physical Fitness, Sports and Health, Toyohashi University of Technology, 1-1 Hibarigaoka, Tenpaku-cho, Toyohashi 441-8580, Japan
[2] School of Dentistry, Health Sciences University of Hokkaido, Kanazawa, Ishikari-Tobetsu, Hokkaido, Japan

ABSTRACT

The world's elderly population is expanding rapidly, and we are now faced with the significant challenge of maintaining or improving physical activity, independence, and quality of life in the elderly. Sarcopenia, the age-related loss of skeletal muscle mass, is characterized by a deterioration of muscle quantity and quality leading to a gradual slowing of movement, a decline in strength and power, increased risk of fall-related injury, and often, frailty. Since sarcopenia is largely attributed to various molecular mediators affecting fiber size, mitochondrial homeostatis, and apoptosis, the mechanisms responsible for these deleterious changes present numerous therapeutic targets for drug discovery. Muscle loss has been linked with several proteolytic systems, including the calpain, ubiquitin-proteasome and lysosome-autophagy systems. In addition, reseachers have indicated defects of Akt-mTOR (mammalian target of rapamycin) and RhoA-SRF (Serum response factor) signaling in sarcopenic muscle. In this chapter, we summarize the current understanding of the mechanisms underlying the progressive loss of muscle mass and provide an update on therapeutic strategies (resistance training, myostatin inhibition, amino acids supplementation, calorie restriction, etc) for counteracting sarcopenia. Resistance training combined with amino acid-containing supplements would be the best way to prevent age-related muscle wasting and weakness. Myostatin inhibition seems to be the most interesting strategy for attenuating sarcopenia as well as

* Address correspondence and reprint requests to:
Kunihiro Sakuma, Ph.D.
Research Center for Physical Fitness, Sports and Health, Toyohashi
University of Technology, 1-1 Hibarigaoka, Tenpaku-cho, Toyohashi, 441-8580, Japan
E-mail: ksakuma@las.tut.ac.jp; FAX: 81-532-44-6947

muscular dystrophy in the near future. In contrast, muscle loss with age may not be influenced positively by treatment with a proteasome inhibitor or antioxidant.

Keywords: sarcopenia, SRF, autophagy, myostatin, skeletal muscle.

1. INTRODUCTION

Skeletal muscle contractions power human body movements and are essential for maintaining stability. Skeletal muscle tissue accounts for almost half of the human body mass and, in addition to its power-generating role, is a crucial factor in maintaining homeostasis. Given its central role in human mobility and metabolic function, any deterioration in the contractile, material, and metabolic properties of skeletal muscle has an extremely important effect on human health. Aging is associated with a progressive decline of muscle mass, quality, and strength, a condition known as sarcopenia [1]. The term sarcopenia, coined by I. H. Rosenberg, originates from the Greek words *sarx* (flesh) and *penia* (loss). Although this term is applied clinically to denote loss of muscle mass, it is often used to describe both a set of cellular processes (denervation, mitochondrial dysfunction, inflammatory and hormonal changes) and a set of outcomes such as decreased muscle strength, decreased mobility and function [2], increased fatigue, a greater risk of falls [3], and reduced energy needs [4]. In addition, reduced muscle mass in aged individuals has been associated with decreased survival rates following critical illness [5]. Estimates of the prevalence of sarcopenia range from 13% to 24% in adults over 60 years of age to more than 50% in persons aged 80 and older [2]. The estimated direct healthcare costs attributable to sarcopenia in the United States in 2000 were $18.5 billion ($10.8 billion in men, $7.7 billion in women), which represented about 1.5% of total healthcare expenditures for that year [6]. Therefore, age-related losses in skeletal muscle mass and function present an extremely important current and future public health issue.

Lean muscle mass generally contributes up to ~50% of total body weight in young adults but declines with aging to be 25% at 75-80 yr old [7, 8]. The loss of muscle mass is typically offset by gains in fat mass. The loss of muscle mass is most notable in the lower limb muscle groups, with the cross-sectional area of the vastus lateralis being reduced by as much as 40% between the age of 20 and 80 yrs [9]. On a muscle fiber level, sarcopenia is characterized by specific type II muscle fiber atrophy, fiber necrosis, and fiber-type grouping [9-13]. In elderly men, Verdijk et al. [12] showed a reduction in type II muscle fiber satellite cell content with aging. Although various investigators support such an age-related decrease in the number of satellite cells [12-17], some reports [18-20] indicate no such change. In contrast, most studies point to an age-dependent reduction in muscle regenerative capacity due to reduced satellite cell proliferation and differentiation.

Another morphologic aspect of sarcopenia is the infiltration of muscle tissue components by lipids, because of the increased frequency of adipocyte or lipid deposition [21, 22] within muscle fibers. As with precursor cells in bone marrow, liver, and kidney, muscle satellite cells that can express an adipocytic phenotype increase with age [17, 23], although this process is still relatively poorly understood in terms of its extent and spatial distribution. Lipid deposition, often referred to as intra-myocellular lipid, may result from a net buildup of lipids due to the reduced oxidative capacity of muscle fibers with aging [21, 24].

Several possible mechanisms for age-related muscle atrophy have been described; however the precise contribution of each is unknown. Age-related muscle loss is a result of reductions in the size and number of muscle fibers [25] possibly due to a multi-factoral process that involves physical activity, nutritional intake, oxidative stress, and hormonal changes [3, 26]. The specific contribution of each of these factors is unknown but there is emerging evidence that the disruption of several positive regulators [Akt and SRF (serum response factor)] of muscle hypertrophy with age is an important feature in the progression of sarcopenia [27]. In contrast, many investigators have failed to demonstrate an age-related enhancement in levels of common negative regulators (Atrogin-1, myostatin, and calpain) in senescent mammalian muscles.

A greater understanding of the loss of muscle mass with age could have a dramatic impact on the elderly if such research leads to the maintenance or improvement of functional ability. This review aims to outline the molecular mechanisms of muscle atrophy in sarcopenia, and to address several recent strategies for inhibiting the muscle loss.

2. THE MOLECULAR MECHANISMS OF MUSCLE ATROPHY IN SARCOPENIA

2.1. Positive Regulators

2.1.1. Akt Signaling in Aged Muscle

The mammalian target of rapamycin (mTOR) signaling kinase, which can be activated by Akt/protein kinase B, has emerged as a crucial regulator of skeletal muscle hypertrophy [28-30]. p70S6K is pivotal in the control of translation, as it regulates a subset of mRNA containing 5-terminal polypyrimidine tracts, which encode ribosomal proteins and factors essential to the translational machinery [31]. mTOR also phosphorylates eukaryotic initiation factor 4E (eIF4E) binding protein (4E-BP1) [32]. Several studies using rats and humans indicated differently the levels of Akt/mTOR/p70S6K signaling among fast-type muscles. Compared to those in young Fischer 344×Brown Norway rats, the amounts of phosphorylated mTOR and p70S6K were increased 70-75% in the tibialis anterior (TA) but not plantaris muscles of senescent rats [33]. Kimball et al. [34] showed that, in the gastrocnemius muscle, the level of phosphorylated p70S6K, eIF2B activity, and the amount of eIF4E associated with eIF4G increased between 12 and 27 months of age despite an apparent decrease in Akt activity. In addition, other groups [35, 36] also showed the decreased phosphorylation status of Akt in aged mammalian muscle. Therefore, the activation of Akt-downstream regulators (ex. mTOR, p70S6K) in senescent fast-type muscle may be attributable to the lower level of activated Akt. Akt-downstream regulators in aged muscle would attempt to activate protein synthesis to an extent similar to that in young muscle with Akt in this impaired state.

Old muscles show a marked defect in the contraction-induced activation of these mediators. Parkingdon et al. [33] reported lower levels of phosphorylated p70S6K and mTOR after high-frequency electrical stimulation (HFES) in muscle of senescent rats compared with young rats. The same group [37] also demonstrated that 4E-BP1 was markedly phosphory-lated in the TA muscle of aged but not young rats at 6h after HFES. In addition, they suggested no increase in eIF4E-eIF4G association after HFES in the aged muscle [37].

Furthermore, Thomson and Gordon [38] suggested impaired overload-induced muscle growth in old rats possibly due to diminished phosphorylation of mTOR (Ser[2448]), p70S6K (mTOR-specific Thr[389]), rpS6 (Ser[235/236]) and 4E-BP1. These lines of evidence clearly show that sarcopenic muscle exhibits an impairment of Akt/mTOR/p70S6K signaling after contraction. This defect would explain the limited capacity for hypertrophy after muscle stimulation in aged animals.

2.1.2. The Defect of SRF Signaling with Age

SRF is an ubiquitously expressed member of the MADS box transcription factor family, sharing a highly conserved DNA-binding/dimerization domain, which binds the core sequence of SRE/CArG boxes [CC (A/T)6 GG] as homodimers [39]. Functional CArG boxes have been found in the cis-regulatory regions of various muscle-specific genes, such as the skeletal α-actin [40], muscle creatine kinase, dystrophin [41], tropomyosin, and myosin light chain 1/3 [42] genes. The majority of SRF's targets are genes involved in cell growth, migration, cytoskeletal organization and myogenesis [43]. SRF was first shown to be essential for both skeletal muscle cell growth and differentiation in experiments performed with C2C12 myogenic cells. In this model, SRF inactivation abolished MyoD and myogenin expression, preventing cell fusion in differentiated myotubes [44]. Further experiments demonstrated that MyoD expression was modulated by a RhoA/SRF signaling cascade [45]. Recent results obtained with specific SRF knock-out models by the Cre-LoxP system, emphasize a crucial role for SRF in postnatal skeletal muscle growth and regeneration [46] by direct binding of interleukin-4 (IL-4) and Insulin-like growth factor I (IGF-I) promoters in vivo. SRF also enhances the hypertrophic process in muscle fibers after mechanical overloading [47, 48] as well as muscle differentiation and MyoD gene expression in vitro [49]. For example, we showed that in mechanically overloaded muscles of rats, SRF protein is co-localized with MyoD and myogenin in myoblast-like cells during the active differentiation phase [50]. In contrast, a marked reduction in levels of SRF and SRF-related molecules (RhoA, FAK, paxillin, etc.) was recognized in the muscle of unweighted hindlimbs in mice [48] and in merosin-deficient dystrophic mice [50] along with rapid atrophy. More recently, a decrease of SRF expression achieved using a transgenic approach was found to accelerate the atrophic process in muscle fibers with age [51]. In fact, using crude and fractionated homogenates and immunofluorescence, our recent study clearly demonstrated a blunted expression of SRF protein in the quadriceps and triceps brachii in aged mice [52]. In addition, our data showed a decrease in mRNA and protein levels of MRTF-A, an especially powerful SRF co-activator [53] expressed ubiquitously in cardiac, smooth and skeletal muscles [54, 55], in these muscles of aged mice. The expression and cellular localization of SRF and MRTF-A appear to be regulated by several upstream factors including β1-integrin [56], RhoA, striated muscle activators of Rho signaling (STARS) [53], and MuRF2 [57]. For instance, Lange et al. [57] demonstrated that SRF is blocked and relocalized by the nuclear translocation of MuRF2, which regulates a signaling pathway composed of titin-Nbr1-p62/SQSTM1 at the position of the sarcomere depending on mechanical activity. In addition, the mutation of SRF delineated the translocational action of MRTF-A induced in vitro by STARS, a muscle-specific actin-binding protein [53]. A marked decrease in the amount of STARS mRNA has also been observed in the muscle of senescent mice [52] and of humans subjected to unloading [58]. These lines of evidence clearly show the existence of a defect of SRF-signaling in aged mammalian muscle.

2.2. Negative Regulators

At least four major proteolytic pathways (lysosomal, Ca^{2+}-dependent, caspase-dependent, and ubiquitin-proteasome-dependent) operate in the skeletal muscle and may be altered during aging, thus contributing to sarcopenia. The lysosomal and proteasomal systems lead to an exhaustive degradation of cell proteins into amino acids or small peptides, whereas Ca^{2+}-dependent and caspase systems can perform only limited proteolysis, owing to their restricted specificity.

The endosome-lysosome system is relatively non-selective and mostly involved in the degradation of long-lived proteins [59]. Lysosomal autophagic degradation has been reported to be induced in skeletal muscle by a 24h nutrient starvation [60], whereas autophagins, a class of cysteine proteases putatively involved in the formation of autophagosomes, are particulary abundant in the skeletal muscle [61]. Recently, autophagy-related genes have been shown to be hyperexpressed during muscle atrophy induced by denervation or fasting [62-64].

The ubiquitin-proteasome system, initially described as relevant to the catabolism of regulatory or damaged proteins, is also involved in bulk protein degradation, at least in the skeletal muscle. In particular, the identification of muscle-specific components of the so-called SCF ubiquitin-ligase complexes has improved knowledge of the regulation of the ubiquitin-proteasme system in skeletal muscle [65].

The Ca^{2+}-dependent system comprises several cystein proteases called calpains and a physiological inhibitor named calpastatin. Calpains affect only a limited proteolysis on their substrates (i.e. protein kinase C, calcineurin, and titin), resulting in irreversible modifications that lead to changes in activity or to degradation via other proteolytic pathways [66, 67].

Finally, caspases, a family of cysteine protease, are mostly known for their role in the execution of apoptosis. Whether caspases are relevant to the regulation of skeletal muscle mass still remains to be elucidated [68-70].

2.2.1. The Ubiquitin-Proteasome-Dependent Signaling in Aged Muscle

In a variety of conditions such as cancer, diabetes, denervation, uremia, sepsis, disuse, and fasting, skeletal muscles undergo atrophy through degradation of myofibrillar proteins via the ubiquitin-proteasome pathway [71, 72]. Recent advances assert that muscle atrophy in these conditions shares a common mechanism in the induction of the muscle-specific E3 ubiquitin ligases atrophy gene-1/muscle atrophy F-box (Atrogin-1/MAFbx) and muscle ring-finger protein 1 (MuRF1) [65, 73-76]. Only very indirect measurements (small increases in mRNA levels encoding some components of the ubiquitin-proteasome pathway [77-80] or ubiquitin-conjugate accumulation [81]) in old muscles of rodents or humans suggested a modest activation of this pathway. Atrogin-1/MAFbx and/or MuRF1 mRNA levels in aged muscle are reportedly increased [82] or unchanged [83, 84] in humans and rats, or decreased in rats [81, 85]. When even the mRNA expression of these atrogenes increased in sarcopenic muscles, the induction was very limited (1.5-2.5 fold) as compared with other catabolic situations (10-fold).

Although various findings have been made regarding the mRNA levels of both ubiquitin ligases in aged mammalian muscle, the examination of protein levels revealed another molecular mechanism. For instance, a descriptive analysis using muscle samples (n＝10) by

Edström et al. [85] indicated the marked up-regulation of phosphorylated Akt and Forkhead box O (FOXO) 4 in the gastrocnemius muscle of aged female rats probably contributing to the down-regulation of Atrogin-1/MAFbx and MuRF1 mRNA. This result is further supported by the more recent finding of Léger et al. [35], who, using human subjects 70 years old, demonstrated a decrease in nuclear FOXO1 and FOXO3a, by 73% and 50%, respectively, although they did not recognize an significant age-dependent change in the expression of Atrogin-1/MAFbx and MuRF-1 mRNA. Since Goldberg's group [86] has defined the rapid and transient expression of Atrogin-1/MAFbx and MuRF1, the mRNA levels of these atrogenes would not nesessarily correspond with those of proteins in various conditions. Furthermore, the major peptidase activities of the proteasome (i.e. the chymotrypsin-like, trypsin-like, and caspase-like activities) were always reduced (as reported in other tissues [87]) or unchanged with aging [75, 79, 87, 88]. Altogether, these observations clearly suggest that the activation of the ubiquitin-proteasome system contributed little to the establishment of sarcopenia in accordance with the very slow muscle mass erosion.

2.2.2. Possible Contribution of Autophagy Signaling with Age

Autophagy (derived from Greek and meaning " to eat oneself ") occurs in all eukaryotic cells and is evolutionarily conserved from yeast to humans [89]. Autophagy is a ubiquitous catabolic process that involves the bulk degradation of cytoplasmic components through a lysosomal pathway [90-93]. This process is characterized by the engulfment of part of the cytoplasm inside double-membrane vesicles called autophagosomes. Autophagosomes subsequently fuse with lysosomes to form autophagolysosomes in which the cytoplasmic cargo is degraded and the degradation products are recycled for the synthesis of new molecules [94]. Turnover of most long-lived proteins, macromolecules, biological membranes, and whole organelles, including mitochondria, ribosomes, the endoplasmic reticulum and peroxisomes, is mediated by autophagy [95].

Three major mechanisms of autophagy have been described: (1) microautophagy, in which lysosomes directly take up cytosol, inclusion bodies and organelles for degradation; (2) chaperone-mediated autophagy, in which soluble proteins with a particular pentapeptide motif are recognized and transported across the lysosomal membrane for degradation; and (3) macroautophagy, in which a portion of the cytoplasm, including subcellular organelles, is sequestered within a double membrane-bound vacuole that ultimately fuses with a lysosome [96] [Figure 1].

More recently, Noglaska et al. [98] suggested the prominent accumulation of p62/SQSTM1, an autophagy-regulating protein, in muscle fibers of patients with sporadic myositis. p62/SQSTM1 interacts with the autophagosome membane light chain 3 (LC3) protein, which facilitates delivery of its polyubiquitinated cargo for lysosomal degradation [99, 100]. Interestingly, the ubiquitin-proteasome system and the lysosomal-autophagy system in skeletal muscle are interconnected was suggested by two recent reports by Mammucari et al. [62], and Zhao et al. [101]. Both studies identified FOXO3 as a regulator of the lysosomal and proteasomal pathways in muscle wasting. FOXO3 is a transcriptional regulator of the ubiquitin ligases MuRF1 and Atrogin-1. It has now been linked to the expression of autophagy- related genes in skeletal muscle in vivo and C2C12 myotubes [62, 101].

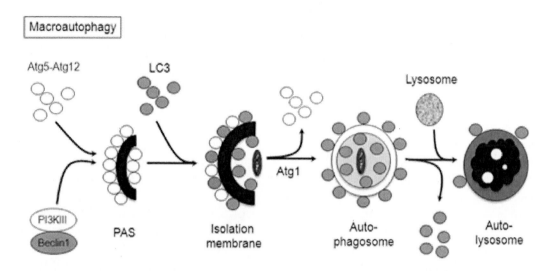

Figure 1. Schematic representation of the major pathway delivering cellular substrates to lysosomes. Macroautophagy engulfs whole portions of cytoplasm together with various organelles. In this process, $PI3K_{III}$ complex and Atg12-Atg5 conjugate are probably required for the generation of pre-autophagosomal structures, and liquid-conjugated LC3 is associated with the elongation of isolation membrane. Atg1 complex likely functions for the completion of the autophagosomes before they fuse with lysosomes [96]. $PI3K_{III}$; Phosphatidylinositol 3-kinase of class III; LC3; microtuble-associated protein light chain 3.

A decline in autophagy during normal aging has been described for invertebrates and higher organisms [102]. Ineffcient autophagy has been attributed to a major role in the apparent age-related accumulation of damaged cellular components, such as undegradable lysosome-bound lipofuscin, protein aggregates, and damaged mitochondria [103]. Interestingly, the degree of age-related changes in autophagy appears to be organ-specific. While the autophagic activity in liver declines with age [104], one group suggests that autophagy is maintained in the heart of aged rats [105]. Importantly, and to the best of our knowledge, only two recent studies have dealt with the changes of autophagy with age in mammalian skeletal muscle [106, 107]. Compared to those in young male Fischer 344 rats, amounts of Beclin-1, important for the formation of pre-autophagosome structures [108, 109], were significantly increased in the plantaris muscles of senescent rats [107]. In contrast, aging did not influence the amounts of Atg7 and Atg9 in rat plantaris muscle, although these autophagy-linking proteins possess a crucial role in the formation and expansion of the autophagosome [110, 111]. Indeed, Western blot analysis by Wohlgemuth et al. [106] clearly showed a marked increase in the amount of microtubule-associated protein light chain (LC3) in muscle during aging. However, they could not demonstrate an aging-related induction of the ratio of LC3-II to LC3-I, a better biochemical marker to assess ongoing autophagy [60, 112]. Therefore, not all contributors to autophagy signaling seem to change similarly in senescent skeletal muscle.

2.2.3. The Adaptive Changes in Myostatin in Aged Muscle

Growth and differentiation factor 8 (GDF8), otherwise known as myostatin, was first discovered during screening for members of a novel transforming growth factor-beta (TGF-β) superfamily, and shown to act as a potent negative regulator of muscle growth [113, 114].

Studies indicate that myostatin inhibits cell cycle progression and levels of myogenic regulatory factors (MRFs), thereby controlling myoblastic proliferation and differentiation during developmental myogenesis [114-118]. Mutations in myostatin can lead to massive hypertrophy and/or hyperplasia in developing animals, as evidenced by knockout experiments in mice and by the phenotype seen in myostatin-null cattle [119] and humans [120]. Moreover, mouse skeletal engineered to over-express myostatin propeptide, the naturally occurring myostatin inhibitor follistatin, or a dominant-negative form of the Activin IIB receptor (ActRIIB-the main myostatin receptor [121, 122]), all display similar, if not greater increases in skeletal muscle size [121]. Myostatin binds to and signals through a combination of Activin IIA/B receptors on the cell membrane, however, it has higher affinity for ActRIIB. On binding ActRIIB, myostatin forms a complex with a second surface type I receptor, either activin receptor-like kinase (ALK4 or ActRIB) or ALK5, to stimulate the phosphorylation of receptor Smad (Rsmad), and Smad2/3 transcription factors in the cytoplasm. Then Smad2/3 translocate and modulate nuclear gene transcription such as MyoD [123] via a TGF-β-like mechanism. In contrast, FOXO1 and Smad2 appear to control the differentiation of C2C12 myoblasts by regulating myostatin mRNA and its promoters [123]. Myostatin levels increased with muscle atrophy due to unloading in mice and humans [124-126], and with severe muscle wasting in HIV patients [127]. For example, our recent study [126], using healthy university students, showed a significantly increased expression of myostatin mRNA in muscle after the unilateral suspension of a leg for two weeks. Together, these studies suggested that increased levels of myostatin lead to muscle wasting. However, studies measuring myostatin levels during aging have yielded conflicting results such as marked increases in humans at the mRNA and protein levels [35], no change in mice at the protein level [128], and a decrease in rats at the mRNA level [36]. The functional role of myostatin in aged mammalian muscle may be revealed by further descriptive analysis using other methods (ex. immunofluorescence) and examining the adaptive changes in downstream modulators (ex. ActRIIB, Smad3) of myostatin signaling.

3. THE MOLECULAR MECHANISMS OF FIBER LOSS IN SARCOPENIA

The mitochondria play an important part in the cell death program (apoptosis) by activating caspases (cystein-dependent, aspartate-specific proteases) [129, 130]. Mitochondrial dysfunction can lead to mitochondrial cytochrome c release [130]. In the cytoplasm, cytochrome c, Apaf-1, caspase-9, and dATP form an apoptosome, which can activate caspase-3, a key cell death protease. Apoptosis is a major contributor to fiber loss in aged skeletal muscle [131]. Proposed mechanisms for these changes include mitochondrial dysfunction and altered apoptotic signaling [132]. Dirks et al. [133] have shown that, in the gastrocnemius muscle of aged rats, the frequency of apoptosis was significantly increased concomitant with a significant rise in the levels of activated (cleaved) caspase-3. Mitochondria are also essential for proper cellular functioning and viability, being the main sites for energy production and playing an essential role in the maintenance of redox homeostatis [134]. Because the mitochondria are the main cellular site of reactive oxygen species (ROS), it follows that mitochondrial components are immediately susceptible to oxidative damage. In particular, mtDNA is especially prone to oxidative damage [135] due to its proximity to the electron transport chain, the lack of protective histones and a less efficient repair system compared to nuclear DNA [136]. Indeed, accumulation of mtDNA deletion mutations has been reported at an advanced age in animal models and humans [137-139]. In addition, some experimental studies using aged mammals have also provided clear evidence of the colocalization of segmental mitochondrial abnormalities and mtDNA deletion mutations [139]. The same group showed that the vastus lateralis muscle, which undergoes a high degree of sarcopenia, exhibited a greater incidence of electron transport system (ETS) abnormalities and associated fiber loss than the soleus and adductor longus muscles, which are more resistant to sarcopenia, suggesting a direct association between ETS abnormalities and fiber loss [137]. In addition, many researchers [140-142] have suggested that the mitochondrial permeability transition pore could induce a release of cytochrome c and eventually initiate apoptosis. Importantly, in the white gastrocnemius and soleus muscles of Fischer 344 rats, it was recently demonstrated that aging significantly increased DNA fragmentation, and the amounts of cleaved caspase-3 and proapoptotic Bax, and decreased anti-apoptotic Bcl-2 protein content [142]. These lines of evidence show the functional role of apoptosis in provoking fiber loss in sarcopenia.

4. THE MOLECULAR MECHANISMS OF AGE-DEPENDENT DEFECTS IN MUSCLE REGENERATION

Satellite cells are muscle-specific stem cells located under the basal lamina of muscle fibers [123], which are responsible for muscle regeneration [144, 145]. Similar to the embryonic stem cells that build organs, adult stem cells that regenerate organs are capable of symmetric and asymmetric division, self-renewal, and differentiation. This precise coordination of complex stem cell responses throughout adult life is regulated by evolutionarily-conserved signaling networks that cooperatively direct and control (1) the breakage of stem cell quiescence, (2) cell proliferation and self-renewal, (3) cell expansion and prevention of premature differentiation and finally, (4) the acquisition of terminal cell fate. This highly regulated process of tissue regeneration recapitulates embryogenic organogenesis with respect to the involvement of interactive signal transduction networks such as HGF, Notch, Wnt, TGF-β, calcineurin, and Ras/MAPK [21, 128, 146].

4.1. The Breaking of Quiescence and Proliferating Process in Satellite Cells

Interestingly, the proliferative potential of satellite cells is nearly constant since satellite cells isolated from both young (9-year-old) and old (> 60-year-old) humans are capable of about 20 to 30 replications *in vitro* under standard conditions [147, 148]. Also, mouse satellite cells isolated from young (2-6 mo), adult (1-10 and 11-13 mo), old (19-25 mo), and senile (29-33 mo) mice show the same proliferative capacity during short-term cultures *in vitro*, provided that the mitogenic milieu of older cells is probably enriched with FGF [18].

In contrast to the *in vitro* observations, satellite cells in aged muscles display an impaired ability to activate and proliferate in response to injury and represents one of the key age-specific defects in muscle repair [21, 149, 150]. Activation of these quiescent cells for proliferation is triggered by injury or attrition of mature muscle fibers. By 24 hours after muscle injury, satellite cells enter the G1/S phase of the cell cycle and robustly proliferate for the next 2-3 days [148]. Nitric oxide (NO) production by increased NO synthase (NOS) activity is very important for satellite cell activation, possibly through the activation of matrix metalloproteinases (MMP), which induce the release of hepatocyte growth factor (HGF), from the extra-cellular matrix [151, 152]. By binding to its receptor, c-Met, which is expressed by satellite cells, HGF is able to stimulate satellite cell activation [153] but at the same time, inhibits muscle differentiation [154]. Inhibition of NO production inhibits HGF release, c-Met/HGF co-localization, and satellite cell activation [152]. NO also induces expression of follistatin [155], a fusigenic secreted molecule, known to antagonize myostatin, a negative regulator of myogenesis expressed by quiescent satellite cells, thus possibly contributing to the exit of satellite cells from quiescence. Several studies have pointed out age-related defects of NO-MMP-HGF signaling during the activation of satellite cells. Although a study with cultured muscle cells demonstrated similar MMP2-9 activity and responsiveness after serum exposure between young and old muscle, MMP2-9 activity *in vivo* was ~2-fold lower in old muscle when measured in crushed muscle extract [156]. Serum exposure induced a lower increase in the expression of the HGF receptor, c-Met, in old satellite cells [156]. In addition, NO supplementation (L-arginine or diethylenetriamine NONOate) rescued the emanation of old satellite cells at 48h of culture following centrifugation (1500g for 30 min). Furthermore, Richmonds and colleagues [157] found a reduction in nNOS activity and the percentage of NOS-containing fibers in muscles from aged rats (24 mo) compared to young rats (8 mo). However, this type of change in NOS content with aging may exhibit muscle and species specificity [158, 159]. Considering these lines of evidence, some age-related defects seems to occur in the NO-MMP-HGF-dependent activation of satellite cells.

The proliferating process in satellite cells appears to be controlled by Notch signaling during muscle regeneration [149]. Within hours to days following muscle injury, there is an increased expression of Notch signaling components (Delta-1, Notch-1 and active Notch) in activated satellite cells and neighboring muscle fibers [19, 149]. Up-regulation of Notch signaling promotes the transition from activated satellite cells to highly proliferative myogenic precursor cells and myoblasts, as well as prevents differentiation to form myotubes [19, 160, 161]. Proliferation was decreased and differentiation was promoted when Notch activity was inhibited in myoblasts with a Notch antagonist, Numb, gamma-secretase inhibitor, or with small-interfering RNA (siRNA) knockdown of presenilin-1 [149, 160, 162]. In addition, mutations in Delta-like 1 or CSL result in excessive premature muscle differentiation and defective muscle growth [163]. Apparent impairment of Notch signaling occurs in aged muscle, because the Notch ligand, Delta, is not up-regulated following injury in this muscle. Forced activation of this pathway with a Notch-activating antibody can restore the regenerative potential by inducing the expression of several positive regulators (PCNA, Cyclin D1) of cell cycle progression [19, 164]. A recent study revealed that levels of TGF-β are higher in aged than young satellite cell niches [128]. Further analysis showed greater activation of the TGF-β pathway in old satellite cells, and physical competition between Notch and pSmad3 at the promoters of multiple cyclin-dependent kinase (CDK) inhibitors [128].

Another stimulus for skeletal muscle proliferation is microenvironment-secreted factors such as fibroblast growth factor. Mitogen activated protein kinase (MAPK)-signaling cascades and p38α/β MAPK are required for satellite cell activation and regulate the quiescent state of satellite cells [165]. Analysis pf p38α mutants showed that p38α abrogation induces delayed cell-cycle exit, and altered expression of cell-cell regulators in cultured myoblasts. As a result of continuous proliferation, and lack of growth arrest feedback, p38α mutants showed increased myoblast proliferation in the neonatal period [166]. Indeed, a more recent study by Jump et al. [167] clearly showed a marked defect of proliferation in aged satellite cells caused by the induction of FGF2 expression.

4.2. The Differentiation of Satellite Cells

Similar to Notch signaling, canonical Wnt signaling is critical for muscle repair [168-171]. The canonical Wnt signaling cascade requires soluble Wnt ligands to interact with Frizzled receptors and low-density lipoprotein receptor-related protein co-receptors. This coordination stimulates phosphorylation of Disheveled and inactivates GSK3β's phosphorylation of β-catenin. In the nucleus, the de-phosphorylated β-catenin binds to TCF/LEF1 transcription factors [172], which may directly activate Myf5 and MyoD or may up-regulate MRF co-activators such as c-Jun N-terminal kinases [173, 174]. It is suggested that Notch activity presides during myoblast proliferation after which there is a temporal switch to Wnt signaling and subsequent myoblast differentiation and fusion into myotubes [168]. Inhibiting Notch (with soluble Jagged ligand or with a γ-secretase inhibitor) or activating Wnt (by inhibiting GSK3β or adding Wnt3a) decreases Myf5 expression and promotes muscle differentiation providing evidence that Notch signaling needs to be turned off and Wnt turned on for differentiation to ensue [168, 175].

This hypothesis was supported by the finding that aberrant activation of the Wnt pathway can lead to fibrogenic conversion of cells in different lineages [176-178]. In fact, Wnt signaling was shown to be enhanced in aged muscle and in myogenic progenitors exposed to aged serum [176]. To directly test the effects of Wnt on cell fate and muscle regeneration, Brack et al. [2007] altered Wnt signaling *in vitro* and *in vivo*. The addition of Wnt3A protein to young serum resulted in increased myogenic-to-fibrogenic conversion of progenitors *in vitro* [176]. Conversely, the myogenic-to-fibrogenic conversion of aged serum was abrogated by Wnt inhibitors [176]. *In vivo*, the injection of Wnt3A into young regenerating muscle 1 day after injury resulted in increased connective tissue deposition and a reduction in satellite cell proliferation [176]. The authors therefore tested whether inhibiting Wnt signaling in aged muscle would reduce fibrosis and enhance muscle regeneration. Indeed, there was reduced fibrosis in aged regenerating muscle injected with Dickkopf-1, a Wnt antagonist [179].

Satellite cell myogenic potential mostly relies on the expression of Pax genes and myogenic regulatory factors (MRFs: MyoD, Myf5, myogenin, and MRF4). Sequential activation and expression of Pax3/7 and MRFs is required for the progression of skeletal myoblasts through myogenesis. Pax7 is expressed by all satellite cells and essential to their postnatal maintenance and self-renewal [180]. Pax7 induces myoblast proliferation and delays their differentiation not by blocking myogenin expression [181] but by regulating MyoD [182], in parallel, myogenin directly down-regulates Pax7 protein expression during differentiation [182]. MyoD is required for the differentiation of skeletal myoblasts [183, 184]. In addition, MyoD null satellite cells showed reduced myogenin and completely absent MRF4 expresion, and displayed a dramatic differentiation deficit [184]. Indeed, muscle regeneration *in vivo* is markedly impaired in MyoD null mice [185]. In contrast, Myf5 regulates the proliferation rate and homeostasis [186, 187]. Interestingly, Myf5 and MyoD can compensate for each other in adults. Myf5 deficiency leading to a lack of myoblast amplification and loss of MyoD induced an increased propensity for self-renewal rather than progression through myogenic differentiation. The differentiation factors myogenin and MRF4 are not involved in satellite cell development or maintenance [186] but the induction of myogenin is necessary and sufficient for the formation of myotubes and fibers. However, senescent mammalian muscle seems not to include a defect of expression in the MyoD family in satellite cells during regeneration. mRNA levels for MyoD, myogenin, and Myf5 have been shown to be elevated in aged compared with young skeletal muscle [187-190]. In addition, young and old mice possess a very similar inducing pattern of these MRF proteins in damaged muscle. However, it is possible that these MRF proteins did not function normally in aged muscle because of the abundant Id family [191, 192], which perturbed the interaction between E protein and MRF proteins [193]. Furthermore, increases in pSmad3 inhibit the normal transcriptional activity of MyoD in senescent muscles [128].

IGF-I positively regulated the proliferation and differentiation of satellite cells/myoblasts *in vitro* via different pathways, the latter possibly through a calcineurin-dependent pathway. Since activated calcineurin promotes the transcription and activation of myocyte enhance factor 2 (MEF2), myogenin, and MyoD [194-196], calcineurin seems to control satellite cell differentiation and myofiber growth and maturation, all of which are involved in muscle regeneration [197, 198]. In fact, our previous study [197] showed a marked increase in the amount of calcineurin protein and the clear colocalization of calcineurin and MyoD or myogenin in many myoblasts and myotubes during muscle regeneration. In addition, we showed that the inhibition of calcineurin by CsA-induced extensive inflammation, marked fiber atrophy, and the appearance of immature myotubes in regenerating muscle compared with placebo-treated mice [197]. Several other studies indicated such defects in skeletal muscle regeneration when calcineurin was inhibited [199, 200], whereas transgenic activation of calcineurin is known to markedly promote the remodeling of muscle fibers after damage [201, 202]. However, the defect in calcineurin signaling may not occur in aged muscle, since several researchers could not detect a marked reduction in calcineurin protein under normal conditions [203, 204].

4.3. Other Factors Modulating Muscle Regeneration

Cell-extrinsic influences on satellite cell function may also change with age [164]. Heterochronic transplantation experiments showed that muscles from old rats grafted into young hosts regenerated greater muscle mass and force than did similar grafts into old hosts; conversely, the regeneration of old muscle in an old host was impaired compared to that in a young host [205]. This result was confirmed by parabiosis experiments in mice, showing that satellite cell proliferation and muscle regeneration is improved and the Notch ligand, Delta, was up-regulated in muscles from an old mouse parabiotically joined to those of a young mouse [19]. This effect is not due to the enlargement of circulating progenitor cells from the young partner but must result from an increase of positive factors in young serum and/or decrease or dilution of inhibitory factors present in old serum. Heterochronic parabiosis also prevented the increase in fibrogenesis during muscle regeneration in aged mice and reversed the age-related myogenic-to-fibrogenic conversion, and these changes were accompanied by decreased Wnt signaling observed in cultured satellite cells exposed to serum from old mice [176]. The systemic factors that affect the response of satellite cells to injury and thus the success of muscle regeneration have not yet been identified.

It was also suggested that age-related decline in muscle regeneration was associated with a concurrent fall in phagocytic macrophage clearance activity during the early stages of muscle repair in SJL/J mice [206]. Indeed, both macrophage infiltration and muscle regeneration after injury occur slower in old than young rats [207, 208]. It is well established that the efficiency of the immune system deteriorates with age [209]; and thus, a role for altered inflammatory cell function in the deficient repair of old muscles cannot be ignored [208].

5. OTHER POSSIBLE CONTRIBUTORS TO SARCOPENIA

5.1. Defects of Spinal Motoneurons and Neuromuscular Junctions

There is a strong association between behavioral motor impairment and degree of muscle wasting [210]. Mounting evidence from both animal and human studies suggests that degeneration of motor neurons [211], followed by changes in structural and functional denervation, and loss of motor units contribute significantly to the progression of skeletal muscle aging. Aging muscle fibers are known to undergo cycles of denervation and re-innervation that lead to remodeling of the motor units. A preferential denervation of the fast-twitch fibers and re-innervation by axonal sprouting from slow motor neurons results in a conversion of fast to slow fibers and fiber type grouping [210].

Age-associated degeneration of the neuromuscular junction (NMJ) is well documented in several animal models as well as in humans [212, 213]. These changes include more axon terminal branches and less sprouting and degeneration per endplate [214], as well as unordered intramuscular branching [214]. Previous studies of aging have shown a decrease in the number of pre-synaptic vesicles and in the number of nerve terminals per NMJ in the thigh muscle of mice [215], and in extensor digitorum longus (EDL) muscles of rats [216], and an increase in the area of motor endplates [216]. The physiological and morphological study of the post-synaptic receptor site in aged Lewis rats, revealed a gradual reduction in the number of acetylcholine receptors (AChRs) per NMJ [217], as well as a shift toward less affinity for AChRs [218]. These abnormalities likely contribute to a reduction in the ability to sustain synaptic transmission, resulting in a "functional denervation" of the aged muscle. This leads to some compensatory changes in the NMJ such as ultraterminal sprouting [216], collateral innervation [219], or remodeling of functioning motor units.

Skeletal muscle produces neurotrophic factors essential to motoneuron survival and muscle fiber innervation during development, such as members of the nerve growth factor (NGF) family of neurotrophins (NTs), glial cell-line derived neurotrophic factor (GDNF), and ciliary neurotrophic factor (CNTF) [220]. Spinal motoneurons express cognate receptors for these neurotrophic factors, and during aging, major changes take place in their expression pattern. Whereas the high -affinity NT receptors (TrkB and Trk C) are down-regulated, the components of the GDNF receptor (GFR-α1 and Ret) are up-regulated [221]. This pattern of regulation mirrors the altered expression of the corresponding neurotrophic factors in the target muscles [220]. Of the NTs, NT-4 is most abundantly expressed in the skeletal muscle of adult mammals [222-224] in spite of negligible levels of brain derived neurotrophic factor (BDNF) and NT-3. In addition, we demonstrated NT-4 to be more abundant in the cytosol of fast-twitch fibers in adult rats [224]. NT-4 is also positively regulated by NMJ impulse activity in adulthood [224, 225] and the down-regulation of NT-4 in senescence probably reflects the decreased motorneuron activity. The consequences of decreased BDNF/NT-3 signaling are not clear but may contribute to the processes underlying the axonal aberrations and slowing of motor axon conduction in senescence [220].

GDNF, one of the most potent neurotrophic factors for motoneurons, is markedly up-regulated in human as well as rat muscle tissue during aging [220]. Our previous study [226] also indicated the up-regulation of GDNF protein in cerebellum, spinal cord motoneurons, and the innervating muscle fibers of merocin-deficient *dy* mice. GDNF induces sprouting and muscle fiber re-innervation [227] and may be involved in the maintenance of cell body size and the cholinergic phenotype of motoneurons [220]. In addition, muscle-derived CNTF receptor a is considered to play an important role in muscle fiber innervation/re-innervation [228]. Edström et al. [229] showed increased levels of CNTF receptorα in sarcopenic muscle compared with normal adult muscle. The up-regulation of GDNF and CNTF receptorα in sarcopenic muscles probably reflects signaling from regenerating/denervated muscle fibers to attract motor axons. Although this is evidence for increased GDNF signaling from muscle to motoneurons during aging, it is evidently not sufficient to restore appropriate innervation of the muscle fibers.

5.2. Metabolic Disorder

Increased central or truncal adiposity is a well-established phenomenon that accompanies aging, irrespective of sex [230] or race [231, 232]. Adipose tissue infiltration of skeletal muscle also increases with age [233, 234]. Sarcopenia may be linked with the increased obesity in the elderly [235]. Indeed, persons who are obese and sarcopenic are reported, independent of age, ethnicity, smoking, and co-morbidity, to have worse outcomes, including functional impairment, disabilities, and falls, than do those who are non-obese and sarcopenic [236]. Recent work has demonstrated that mitochondrial damage occurs in obese individuals due to enhanced ROS and chronic inflammation caused by increased fatty acid load [237-239]. Specifically, in skeletal muscle, the expression of PGC-1α drives not only mitochondrial biogenesis and the establishment of oxidative myofibers, but also vascularization [240, 241]. It was found that a high-fat diet or fatty acid treatment caused a reduction in the expression of PGC-1α and other mitochondrial genes in skeletal muscle [242], which may be a mechanism through which excess caloric intake impairs skeletal muscle function. A recent study also demonstrated that transgenic overexpression of PGC-1α in skeletal muscle improved sarcopenia and obesity associated with aging in mice [243]. MKP-1, a dual-specificity phosphatase and the nuclear-localized product of an immediate-early gene, is responsive to numerous stimuli including ROS, cytokines, growth factors, and fatty acids [237, 244]. Wu et al. clearly showed that in mice lacking MKP-1, mitochondrial oxidative phosphorylation was enhanced in the oxidative portions of skeletal muscle. In addition, they demonstrated that the mice were resistant to age-induced obesity, in the absence of a high-fat diet. Moreover, MKP-1 expression in skeletal muscle of aged patients has been shown to be enhanced [245]. Based on these findings, Roth and Bennett [246] proposed that MKP-1 functions as a common target in the convergence between sarcopenia and overnutrition in a pathophysiological pathway that leads to a loss of skeletal muscle mitochondrial function.

5.3. Chronic Molecular Inflammation

An age-related disruption in the intracellular redox balance appears to be a primary causal factor in producing a chronic state of low-grade inflammation. Chronic molecular inflammation is considered as underlying mechanism of aging and age-related pathological processes [247]. The aging-related redox-sensitive transcription factor NF-κB has been shown to induce inflammation. The age-related up-regulation of key players such as IL-6 and TNF-α is mediated by NF-κB [247]. Moreover, ROS also appear to function as second messengers for TNF-α in skeletal muscle, activating NF-κB either directly or indirectly [248]. In fact, increased oxidative strss and inflammation are known to go hand in hand in many skeletal muscle-associated diseases. Chronic subclinical inflammation may be a marker of functional limitations in older persons across several diseases/health conditions [249]. It has been shown that TNF-α is one of the primary signals inducing apoptosis in muscle. Apoptosis and inflammation closely interact with oxidative damage and are involved in sarcopenic symptoms [250]. It has been suggested that inflammation may negatively influence skeletal muscle through direct catabolic effects or through indirect mechanisms (i.e., decreases in GH and IGF-I concentrations, induction of anorexia etc)[251].

6. Several Strategies For Counteracting Sarcopenia

6.1. Resistance Training

One resistance exercise bout can, within 1h, increase muscle protein synthesis [252], which can last up to 72h after exercise [253]. Resistance training has shown the most promise among interventions aimed to decrease the effects of sarcopenia, as it enhances strength, power, and mobility function and induces varing degrees of skeletal muscle hypertrophy [254-258]. For example, 12 weeks of whole-body resistance training resulted in an increase in type II muscle fiber area in men aged 64-86 yr [259] and 65-72 yr [260, 261]. A 2-year longitudinal trial of resistance training found increases in leg press (32%) and military press (90%) 1 repetition maximum (1 RM) and knee extensor muscle cross-sectional area (9%) in 60-80 yr old men and women [262]. The functional benefits of resistance training have been evaluated in a large-scale trial of 72- to 98-year olds and frail nursing home residents, with resistance training increasing muscle strength (113%), gait velocity (12%), stair-climbing power (28%), spontaneous physical activity, and thigh muscle cross-sectional area

(2.7%)[255]. In addition, 6 weeks of training for elderly persons (68.4±5.4 years) improved their physical activity profile (6-min walk, 30-second chair stand, chair sit & reach and back scratch) as well as muscle strength.

In the elderly, resistance training induces the muscle expression of IGF-I [258], myogenic regulatory factors [263], and IL-6 [264], which contribute to musclular hypertrophy by regulating the activation, proliferation, and differentiation of satellite cells. In addition, resistance training decreases oxidative DNA damage and improves the electron transport chain function [265]. One bout of resistance exercise for elders can enhance the rate of synthesis of muscle protein [266, 267]. However, several studies using humans and rodents indicated a lower degree of activation in mitogen-activated protein kinase (MAPK) and Akt-mTOR pathways after muscle contraction or mechanical overload than occurs in young adults [38, 40, 266]. More recently, Mayhew et al. [268] indicated that one bout of resistance exercise elicited a similar extent of activation in translational signaling (Akt, p70S6K, rpS6, and 4E-BP1) between young and old subjects. Several studies [258, 269] have shown that excess intensive strength training for the elderly impairs the effective gain of muscle strength and mass particularly in women. Therefore, careful attention should be paid when determining the amount and frequency of resistance training for the elderly.

6.2. Myostatin Inhibition

Many researchers have focused on inhibiting myostatin for treating various muscle disorders. The use of neutralizing antibodies to myostatin improved muscle disorders in rodent models of Duchenne muscular dystrophy (*mdx*), limb girdle muscular dystrophy 2F (Sgcg-/-), and amyotrophic lateral sclerosis (SOD1^{G93A} transgenic mouse) [270-272]. Indeed, myostatin inhibition using MYO-029 was tested in a prospective, randomized, placebo-controlled US phase I/II trial in 116 adults with muscular dystrophy such as BMD, FSHD, and LGMD. The participants received MYO-029 or placebo at either 1, 3, 10 or 30 mg/Kg. In addition to the primary objective of assessing safety and adverse events, the biological activity of MYO-029 was determined from muscle biopsies, quantitative muscle testing, and imaging technologies. MYO-029 had good safety and tolerability except for cutaneous hypersensitivity at the 10 and 30 mg/Kg doses, attributed to the need for repeated protein administration [273]. No improvements in muscle function were noted, but dual-energy radiographic absorptiometry and muscle histological investigations revealed that some subjects had increased muscle fiber size. The study concluded that systemic administration of myostatin inhibitors was relatively safe and that more potent myostatin inhibitors for stimulating muscle growth in muscular dystrophy should be considered. Interestingly, Zhou et al. [274] indicated the therapeutic potential of blocking the myostatin receptor ActRIIB, on cancer cachexia, which has been implicated in increased signaling of the ActRIIB pathway [275, 276]. They utilized various models, including colon 26 (C26) tumor-bearing mice and inhibin-deficient mice, as well as nude mice bearing human G361 melanoma and TOV-21G ovarian carcinoma xenografts. Blocking the ActRIIB pathway using a decoy receptor (sActRIIB) completely reversed the pathogenesis and muscle atrophy caused by cancer cachexia. Treatment with sActRIIB abolished activation of the ubiquitin-proteasome system and the induction of atrophy-specific ubiquitin ligases, probably due to the decrease in phospho-Smad2 and increase in inactivated (phospho)-FOXO3 and phospho-Akt proteins in cachexic muscles. Furthermore, this preparation markedly increased numbers of BrdU-, Pax7, and M-cadherin immunoreactive satellite cells (muscle stem cells) in the gastrocnemius muscles of inhibin-α KO mice. The symptoms of cancer cachexia are not the same as those of sarcopenia. Although no researcher has observed similar positive effects after treatment with sActRIIB in patients with cancer cachexia, this would be an intriguing therapy for alleviating various muscular disorders including sarcopenia.

Several trials using pharmacy have been conducted to positively modulate the expression of follistatin, an antagonist of myostatin [277, 278]. For instance, the administration of valproic acid or tricostatin-A, both deacetylase inhibitors, has been shown to increase myogenesis and myofiber size in *mdx* mice, a model of DMD, possibly due to the up-regulation of follistatin [279]. Although a more recent review introduced the potential role of valproic acid or trichostatin-A in inhibiting muscle atrophy in cancer cachexia, neither agent counteracts muscle atrophy or ubiquitin ligase hyperexpression [280]. Inhibiting myostatin to counteract sarcopenia has also been investigated in animals. A lack of myostatin caused by gene manipulation increased the number of satellite cells, and enlarged the cross-sectional area of predominant type IIB/X fibers in TA muscles of mice [281]. In addition, these myostatin-null mice showed prominent regenerative potential including accelerated fiber remodeling after an injection of notexin [281]. Lebrasseur et al. [282] reported several positive effects of 4 weeks of treatment with PF-354 (24 mg/Kg), a drug for myostatin inhibition, in aged mice. They showed that PF-354-treated mice exhibited significantly greater muscle mass (by 12%), and increased performance such as treadmill time, distance to exhaustion, and habitual activity. Furthermore, PF-354-treated mice exhibited decreased levels of phosphorylated Smad3 and MuRF-1 in aged muscle. More recently, Murphy et al. [283] showed, by way of once-a-week injections, that a lower dose of PF-354 (10 mg/Kg) significantly increased the fiber cross-sectional area (by 12%) and in situ force of tibialis anterior muscles (by 35%) of aged mice (21-mo-old). In addition, this form of treatment reduced markers of apoptosis by 56% and reduced caspase3 mRNA levels by 65%. Blocking myostatin enhances muscle protein synthesis [284] by potentially relieving the inhibition normally imposed on the Akt/mTOR signaling pathway by myostatin [285]. The blockade may also attenuate muscle protein degradation by inhibiting the ubiquitin-proteasome system, which is controlled, in part, by Akt [286, 287], although the mechanism involved has not been demonstrated. In contrast, a microarray analysis of the skeletal muscle of $Mstn^{-/-}$ mice showed an increased expression of anti-apoptotic genes compared with that in control mice [287]. These lines of evidence clearly highlight the therapeutic potential of antibody-directed inhibition of myostatin for treating sarcopenia by inhibiting protein degradation and/or apoptosis. Table 1 provides an overview of the effect of myostatin-signaling inhibition on muscle fiber of various muscle wasting including sarcopenia.

Table 1. Effect of myostatin-signaling inhibition on muscle fiber of various muscular dystrophy, ALS, sarcopenia, and cancer cachexia

References	Experimental Design	Modulating method of myostatin activity	Outcomes
Bogdanovich et al. 2 Wagner et al. 200002 [271] 2 [288]	*mdx* mouse (DMD model)	Gene-knock out (complete inactivation of myostatin)	Increase in muscle mass (EDL) ↑ Twitch and tetanus force ↑ Blood CK levels ↓ Utrophin expression →
	mdx mouse (DMD model)	Gene-knock out (complete inactivation of myostatin)	Increase in muscle mass (Pectoralis, Triceps Brachii, Quadriceps, Gastroc) ↑ Forearm grip strength ↑, Muscle fiber diameter ↑
Li et al. 2005 [289]	*dy* mouse (Laminin-deficient congenital muscular dystrophy)	Gene-knock out (complete inactivation of myostatin)	Embryonic MHC-positive fibers ↑ Degeneration and inflammation → Percentage of postnatal lethality ↑
Bogdanovich et al. 2005 [290]	*mdx* mouse (DMD model)	Myostatin propeptide (10 mg/Kg), weekly injection, 3 month	Increase in muscle mass (EDL) ↑ rota-rod performance ↑ Twitch and tetanus force ↑ Blood CK levels ↓ Utrophin expression →
Holzbaur et al. 2006 [272]	SOD1G93A mouse and rat (ALS model)	RF35 (40 mg/Kg → 29/mg/Kg), weeklyinjection, 80 days	Increase in muscle mass (Gastrocnemius, TA, Quadriceps, diaphragm) at early-stage↑ but not end-stage disease Grip strength (Forelimb↑, but Hindlimb →)
Siriett et al. 2006 [281]	Mouse, 24-month-old (Sarcopenia)	Gene-knock out (complete inactivation ofmyostatin)	Inhibition of fiber atrophy (type IIB/X) Percentage of type IIA fibers ↓ Muscle fiber number (TA) ↑
Wagner et al. 2008 [273]	Human, Phase I/II trial for BMD, FSHD, LGMD	Myo-029 (1, 3, 10 or 30 mg/Kg, 6 months	No adverse effects Strength →, Mean body mass →, Muscle fiber diameter →
Lebrasseur et al. 2009 [282]	Mouse, 24-month-old (Sarcopenia)	PF-354 (25 mg/Kg), weekly injection, 4 weeks	Increase in muscle mass (by 12%) ↑, Grip strength → Basal oxygen consumption (VO$_2$) ↑, Basal metabolic rate ↑ Phospho-Smad3 ↓, MuRF1 ↓, PGC-1α↑
Morine et al. 2010 [292]	*mdx* mouse (DMD model)	Inhibition of myostatin receptor (ActRIIB) by adeno-associated virus-mediated gene transfer	Increase in muscle mass (Gastrocnemius: 39%, TA: 46%, Quadriceps: 39%) ↑ Blood CK levels ↓ Muscle fiber size (EDL ↑, Diaphragm →)
Murphy et al. 2010 [283]	Mouse, 21-month-old (Sarcopenia)	PF-354 (10 mg/Kg), weekly injection, 14 weeks	Phospho-Smad3 ↓ MuRF1 ↓ Muscle force (by 35%) ↑, Apoptosis (by 56%) ↓ Cleaved caspase 3 mRNA (by 65%)↓
Zhou et al. 2010 [274]	Several cancer cachexia model (C26 tumor-bearing mice etc)	Inhibition of myostatin receptor (ActRIIB) by Decoy receptor	Ubiquitin-proteasome signaling ↓ Phospho-Smad2 ↓ Inactivated FOXO3 ↑ Phospho-Akt ↑ Inhibition of muscle atrophy

6.3. Amino Acid Supplementation

Many Americans consume more than the Recommended Dietary Allowance (RDA) of protein. However, research shows that a significant number of elderly people do not meet the estimated average requirement let alone the RDA [292]; 32% to 41% of women and 22% to 38% of men aged 50 and older consume less than the RDA of protein [293]. Epidemiological studies show that protein intake is positively associated with preservation of muscle mass. For example, in a recent study, 38 healthy, normal-weight, sedentary women aged 57 to 75 were recruited to determine whether a higher muscle mass index was associated with animal or vegetal protein intake [294].

Many reviews indicate that certain nutritional interventions such as a high protein intake or an increased intake of essential amino acids and the branched chain amino acid (BCAA) leucine with resistance training may help to attenuate fiber atrophy in sarcopenic muscle by enhancing anabolic pathways and inhibiting catabolic pathways [295-297]. In particular, leucine can be considered a regulatory amino acid with unique characteristics. It plays several roles in muscle metabolism regulation, which include translational control of protein synthesis [298] and glucose homeostasis [299]. In addition, leucine has been demonstrated to be a nitrogen donor for the synthesis of muscle alanine and glutamine [298]. Considering these findings, the use of leucine as an anti-atrophic agent is biologically justified.

It has been documented that oral post-exercise amino acid supplementation had a synergistic effect on the contraction-induced escalation of muscle protein synthesis following an acute resistance exercise bout [300, 301]. Treatment with amino acids has been shown to inhibit protein degradation by ubiquitin-proteasome in aged muscle [302], and to induce additive hypertrophy in response to resistance training [254]. Recent studies have shown that amino acids play a role in the phosphorylation of translational factors called eukaryotic initiation factors, specifically eIF4F and p70S6K, through an mTOR-mediated mechanism [301, 303]. On the other hand, a number of studies have not found benefits from protein supplementation [304-306]. Almost all of these latter studies did not conduct a detailed morphological analysis that includes fiber cross-sectional area of muscle biopsy specimens. In addition, these studies utilized a single bout of or short-term (10 days) ingestion to examine the rate of myofibrillar synthesis [306], protein synthesis and mitochondrial function [304]. The administration of many essential amino acids enhanced muscle mass and protein synthesis both under normal conditions [307-309] and with resistance training [303]. Supplementation with essential amino acids not enriched with leucine failed to enhance muscle protein synthesis. Recently, Verhoeven et al. [310] did not find any effect of a 3-month leucine supplementation (7.5 g/kg) on strength and muscle mass in the elderly. In these experiments, lipid profiles and glucose metabolism (assessed using an oral glucose insulin sensitivity index, the homeostasis model assessment of insulin resistance, and glycosylated hemoglobin) remained unchanged as well. The authors observed no significant increase and no apparent decrease in plasma leucine and valine concentrations, respectively. Nicastro et al. [311] proposed the intriguing possibility that the plasma leucine-valine imbalance is attributable to the defect of leucine supplementation for sarcopenia by Verhoeven et al. [310]. In fact, a leucine-induced imbalance in the plasma valine concentration has already been related to body growth deficits [312]. Furthermore, Rieu et al. [313] showed that leucine-enriched meals (0.052 g/kg, twice the post-prandial plasma leucine concentration) increased the rate of fractional myofibrillar protein synthesis when compared to the placebo-treated group. Their diet was also enriched with isoleucine (0.0016 g/kg) and valine (0.0068 g/kg) in order to avoid an imbalance in plasma BCAA levels. Therefore, many essential amino acids including large amounts of leucine are needed to effectively counteract sarcopenia.

6.4. Hormone Supplementation

The idea of using hormone supplementation to treat or prevent sarcopenia is attractive. Estrogen replacement therapy in women has shown a small positive effect on muscle mass [314, 315]. The most promising hormones in this regard are growth hormone and testosterone, although careful attention should be paid to these applications.

6.4.1. Testosterone

Serum levels of testosterone decline with age in males [316]. Testosterone is being evaluated as an anabolic therapy for age-related physical dysfunction in males [317]. Systemic reviews of the literature [318] have concluded that testosterone supplementation increases skeletal muscle mass in hypogonadal men [319], men with chronic illness, and healthy, older men [320, 321]. For instance, a recent study of 6 months of testosterone supplementation in a randomized placebo-controlled trial reported increased lean body mass and leg and arm strength [322]. Testosterone exerts its effects mostly by binding the androgen receptor (AR), which results in a conformational change, allowing for association with the receptor's transcriptional co-activator, β-catenin. The complex of testosterone/SR/β-catenin translocates into the nucleus and binds to AR-binding DNA sequences, located in the promoter of particular genes, usually causing an increase in transcription [323]. Although the mechanisms by which testosterone increases skeletal muscle mass are poorly understood, testosterone has been shown to positively regulate IGF-I [324], Wnt [325], and myostatin [326]. In addition, testosterone treatment at 600 mg in the elderly has been shown to increase the number of proliferating satellite cells possessing proliferating cell nuclear antigen (PCNA) and active Notch-1. The ability of satellite cells to break quiescence is dependent on Notch signaling [149, 164]. A more recent study clearly showed that Notch stimulates satellite cell proliferation by decreasing the expression of the cyclic-dependent kinase inhibitors (CDKs) p15, p16, p21, and p27 by inhibiting the binding of pSmad3 to their promoter regions [128]. Non-canonically, meaning independent of the AR/β-catenin complex, simple binding of testosterone to the AR itself has been shown to modulate the direct activation of Akt, a potent regulator of muscle hypertrophy. The testosterone/AR complex also interacts with glucocorticoid receptors, thus inhibiting the catabolic effects of glucocorticoid [327]. Adverse effects of testosterone include sudden cardiac death, elevated hematocrit levels, increased prostate-specific antigen levels, and worsening lipid profiles [320]. Although substantial gains in muscle mass and strength can be realized in older men with supraphysiological testosterone doses, these high doses are associated with a high frequency of adverse effects [328]. Therefore, it is necessary to monitor the administration of large amounts of testosterone in the elderly.

6.4.2. Growth Hormone

Growth hormone (GH) is a single-chain peptide of 191 amino acids produced and secreted mainly by the somatotrophs of the anterior pituitary gland. GH coordinates the postnatal growth of multiple target tissues, including skeletal muscle [329]. GH secretion occurs in a pulsatile manner with a major surge at the onset of slow-wave sleep and less conspicuous secretory epidsodes a few hours after meals [330], and is controlled by the action of two hypothalamic factors, GHRH, which stimulates GH secretion, and somatostatin, which inhibits GH secretion [331]. The secretion of GH is maximal at puberty accompanied by very high circulating IGF-I levels [332], with a gradual decline during adulthood. Indeed, in aged men, daily GH secretion is 5- to 20-fold lower than that in young adults [333]. The age-dependent decline in GH secretion is secondary to a decrease in GHRH, and to an increase in somatostatin secretion [334].

With respect to the somatomedin hypothesis, the growth-promoting actions of GH are mediated by circulating or locally produced IGF-I [335]. GH-induced muscle growth may be mediated in an endocrine manner by circulating IGF-I derived from liver and/or in an autocrine/ paracrine manner by direct expression of IGF-I from target muscle via GH receptors on muscle membranes. Several studies have shown that GH treatment increases IGF-I mRNA levels in skeletal muscle tissues as well as in the myoblast cell line C2C12 [329, 336]. However, more recent findings by Sotiropoulos et al. [337] clearly showed a direct GH-mediated effect on the growth of muscle cells in primary cultures by facilitating the fusion of myoblasts with nascent myotubes via NFATc2 independent of IGF-I. They demonstrated that GH does not regulate IGF-I expression in myotubes. In addition, they suggested that GH and IGF-I hypertrophic effects are additive and rely on different signaling pathways (GH via NFATc2 vs IGF-I via Akt-mTOR). Taken together, GH may enhance the fusion of muscle cells not via IGF-I in target muscle. The effects of GH administration on muscle mass, strength and physical performance are still under debate [338]. In animal models, GH treatment is very effective at inhibiting sarcopenic symptoms such as muscle atrophy and decreases in protein synthesis, particularly in combination with exercise training [339]. The effect of GH treatment for elderly subjects is controversial. Some groups demonstrated an improvement in strength after long-term administration (3-11 months) of GH [338]. In contrast, many researchers have found that muscle strength or muscle mass did not improve on supplementation with GH [338, 340]. One recent study reported a positive effect for counteracting sarcopenia after the administration of both GH and testosterone [341]. Several reasons may underlie the ineffectiveness of GH treatment in improving muscle mass and strength in the elderly, such as a failure of exogeneous GH to mimic the pulsatile pattern of natural GH secretion or the induction of GH-related insulin resistance. It should also be considered that the majority of the trials conducted on GH supplementation have reported a high incidence of side effects, including soft tissue edema, carpal tunnel syndrome, arthralgias, and gynecomastica, which pose serious concerns especially in older adults. Therefore, one should pay very careful attention when administering GH to the elderly.

64.3. IGF-I

As its name implies, IGF-I is similar to insulin in structure, sharing 50% amino acid identity. IGF-I is perhaps the most important mediator of muscle growth and repair [342] possibly by utilizing Akt-mTOR signaling in the former case [343]. Transgenic overexpression of IGF-I has been shown to elicit marked hypertrophy of skeletal muscle

[344] and enhance muscle regeneration in normal [345] and muscular dystrophic [346] mice. For instance, Musaro et al. [347] demonstrated that upon muscle injury, stem cells expressing c-Kit, Sca-1, and CD45 antigens increased locally and the percentage of recruited cells was conspicuously enhanced by mIGF-I expression. More recently, local expression of IGF-I negatively modulated the inflammatory response, accelerating the regenerative process after muscle injury [348]. In addition, the injection of a recombinant adeno-associated virus directing overexpression of IGF-I in mature muscle fibers has been shown to promote an average increase of 15% in muscle mass and 14% in strength in young adult mice, and remarkably, prevented age-related muscle atrophy in old mice [349]. Therefore, the transgenic approach of up-regulating IGF-I expression in skeletal muscle would be appropriate for inhibiting sarcopenia. In contrast, the systemic administration of IGF-I promoted the regeneration [350], endurance [351], and contractile function of skeletal muscle in a model of DMD in mice. In elderly women (66-82 yr), treatment with recombinant human IGF-I also increased muscle protein synthesis as measured by the incorporation of [1-13C] leucine [352]. However, the administration of IGF-I to the elderly has resulted in controversial findings on muscle strength and function.

This ineffectiveness may be attributable to age-related insulin resistance to amino acid transport and protein synthesis [353] or a marked decrease in IGF-I receptors in muscle with age [353]. Further study is needed to define whether IGF-I treatment has a positive effect on sarcopenia

6.4.4. Selective Androgen Receptor Modulators (SARMs)

SARMs are a class of ligands that bind androgen receptors and display tissue-selective activation of androgenic signaling [354, 355]. Pre-clinical studies in rats had shown some SARMs to have significant anabolic activity, with only moderate to minimal androgenic activity *in vitro* [356]. Much of the pioneering work in developing non-steroidal SARMs was done at Ligand Pharmaceuticals and the University of Tennessee. Gao et al. [357] reported that two SARMS [S-1 and S-4, S-3-(4-fluorophenoxy)- and S-3-(4-acetylamino-phenoxy)-2-hydroxy-2-methyl- N- (4-nitro-3-trifluoromethyl-phenyl)-propionamidesm, respectively] showed strong anabolic effects. They demonstrated that, in orchidectomized (ORX) animals, treatment with S-4 (3 and 10 mg/Kg) for 2 weeks restored soleus muscle mass and strength to that seen in intact animals. More recently, some studies have been conducted to elucidate the molecular mechanisms behind the effectiveness of treatment with SARMS for muscle wasting. Jones et al. [358] showed that SARM blocked the dexamethasone-induced dephosphorylation of Akt and other proteins involved in protein synthesis. In addition, they demonstrated that SARM's administration blocked the up-regulation of ubiquitin ligases after dexamethasone treatment by phosphorylating FOXO.

Several reviews have described the therapeutic potential of SARMs for osteoporosis and for muscle-wasting conditions relevant to sarcopenia [359, 360]. SARMs are under development in many major pharmaceutical firms, but much of the work defining their mechanism of action and structure-activity relationship has not been published. In a December 2006 press release, GTx reported that ostarine had met its primary end point in a Phase II proof-of-concept double-blind, placebo-controlled clinical trial of 60 elderly men and 60 postmenopausal women. Without a prescribed diet or exercise regimen, all subjects treated with ostarine exhibited a dose-dependent increase in total lean body mass after 3 months. Treatment with ostarine also resulted in improvements in functional performance, as determined from increases in speed and power, during a stair climb test. No serious adverse safety events were reported during the trial and a high degree of tissue selectivity was reported, with no changes in serum markers for prostate-specific angtigen, or LH compared with the placebo. SARMs hold considerable promise as a new class of function-promoting anabolic drugs for a variety of clinical conditions, such as frailty, functional limitations associated with aging and chronic illness, cancer cachexia, and osteoporosis. Although the pre-clinical data looks promising, efficacy trials of SARMs are just beginning. Further research is needed to elucidate the molecular basis of tissue selectivity, and to achieve greater potency and tissue selectivity.

6.4.5 β2-agonist

The β-adrenergic signaling pathway represents a novel therapeutic target for the treatment of skeletal muscle wasting and weakness due to critical roles in the mechanisms controlling protein synthesis and degradation. Stimulating the signaling pathway with a β-adrenoceptor agonist (β-agonist) has therapeutic potential for muscle-wasting disorders including sarcopenia [361, 362]. Chronic administration of the β-adrenoceptor agonist formoterol for 4 weeks increased amounts of both slow-type (soleus) and fast-type (EDL) muscle in adult (16 months) and old (27 months) rats.

It should be noted that although β-agonists have demonstrated muscle anabolic properties, they have also been associated with numerous undesirable side-effects including an increased heart rate and muscle tremors that have so far limited their therapeutic potential [362]. In small animal models, new generation β-agonists can elicit an anabolic response in skeletal muscle even at very low doses, with reduced effects on the heart and cardiovascular system compared with older types such as fenoterol and clenbuterol [363]. It should be noted, however, that the potentially deleterious cardiovascular side effects of β-agonists have not been obviated completely: further research is needed to refine their development and therapeutic approach in order to overcome these obstacles, and for their true clinical significance for sarcopenia [and other muscle-wasting disorders] to be realized [361, 362].

6.5. Calorie Restriction (CR)

Although the mechanisms by which calorie restriction (30-40%) delays the aging process remain to be fully elucidated, CR is intricately involved in regulating cellular and systemic redox status and in modulating the expression of genes related to macromolecule and organelle turnover, energy metabolism, and cell death and survival [364-366]. Several studies indicate the protection of age-related functional decline and loss of muscle fibers by CR [364, 366, 367]. These protective effects are likely attributable to the ability of CR to reduce the incidence of mitochondrial abnormalities (mitochondrial proton leak), attenuate oxidative stress (ROS generation), and counteract the age-related increases in proapoptotic signaling in skeletal muscle [364, 367, 368]. In addition, it has been reported that CR reduces the increase of ETS abnormalities and decreases the accumulation of mtDNA deletions in skeletal muscle mitochondria of aged rats [369]. In contrast, more recent long-term CR (40% reduction) for F344BN rats was demonstrated to inhibit the age-related decline of mitochondrial oxidative capacity although not to affect the frequency of mtDNA fragmentation [370]. Noticeably, CR has been shown to modulate the majority of the apoptotic pathways involved in age-associated skeletal muscle loss, such as mitochondria-, cytokine/receptor-, and Ca^{2+}/ER-stress- mediated signaling [364, 367]. For instance, CR markedly inhibits increases in several mediators of the TNF-mediated pathway of apoptosis (TNF-α, TNF-receptor 1, cleaved caspase-3 and -8) possibly by enhancing production of a muscle-derived anabolic cytokine, IL-15, which competes with TNF-mediated signaling [371]. In addition, the combination of calorie restriction with exercise training is proposed to counteract the apoptosis associated with sarcopenia more effectively [372].

In contrast, CR alone or combined with life-long voluntary exercise did not modulate the amount of several autophagy-linked molecules (Beclin-1, Atg9, LC3) at the protein level except Atg7 in sarcopenic muscles of rats [105]. Therefore, the attenuating effect on sarcopenia by CR may be independent of the autophagy signaling. Interestingly, a study using dual-energy X-ray absorptiometry indicated an increase in skeletal muscle among rhesus monkeys on 30% CR for 17 years [373]. More recently, McKiernan et al. [374] showed that CR for rhesus monkeys opposed age-related reductions in the proportion and cross-sectional area of type II fibers in a histochemical analysis of vastus lateralis muscle biopsies. It remains to be determined whether CR is effective in counteracting the age-related loss of muscle in human subjects and to what extent dietary intervention can be applied in human populations. Since excessive CR (over 50%) may have a number of side effects (e.g. weakness, loss of stamina, osteoporosis, depression, anorexia nervosa, etc.)[375], a more mild calorie restriction should be applied in the elderly.

7. OTHER POTENTIAL CANDIDATES

7.1. Antioxidant Supplementation

ROS in skeletal muscle arise most abundantly from the mitochondrial electron transport chain. They account for the production of a significant amount of superoxide anion, generated at complexes I, II, and III of the chain, after reactions with species generated by single-electron transfers, such as those between Fe-S clusters and ubiquinone/ubisemiquinone [376]. ROS production has been shown to increase in skeletal muscle during aging [377, 378], and oxidative stress has been claimed to be relevant to age-related cell damage [379, 380]. The accumulation of ROS can induce higher rates of cellular damage by substances such as deoxyribonucleic acid (DNA), proteins, and lipid-containing structures [381]. During the aging process, increased levels of ROS also lead to the modification of mitochondrial DNA, which can prevent protein synthesis and ATP production, and result in increases in necrosis and apoptosis [372, 382].

In addition to derangements directly occurring at the skeletal muscle level, sarcopenia in aged individuals is also associated with reduced antioxidant dietary intake [85]. In this regard, poor muscle strength and low physical performance have been associated with low levels of carotenoids [383, 384]. The lack of physical activity, which has been proposed to induce an antioxidant response in itself [88], may also contribute to the onset of sarcopenia.

Although a number of studies have investigated the possibility of delaying the aging process by enhancing, endogeneously or exogenously, antioxidative capacity [385, 386], very little is known about the effectiveness of such strategies on the development of sarcopenia.

Rebrin et al. [387] observed significantly enhanced activity of glutathione, an enzymatic antioxidant, after the administration of vitamin E and C as well as a blend of biofravonoids, polyphenols, and carotenoids for 10 months. In contrast, other studies found that the treatment of aged individuals with high-dose beta-carotene or vitamin E does not seem to replace the effects of diets high in fruits, vegetables, and whole grains and low in saturated fats [388, 389]. A recent trial showed that zinc supplementation did not affect the oxidative stress status in the elderly [390]. In addition, supplementation with vitamin C has been shown to significantly lower both training efficiency and the expression of antioxidative enzymes in skeletal muscle [391]. In contrast, the effectiveness of DHEA treatment for correcting/preventing muscle atrophy has been clearly shown in the case of experimental diabetes [391, 392]. The effect on cancer cachexia are partial although significant [393]. On the whole, caution should be used in generalizing, in terms of both mechanisms and treatments, among the various types of muscle atrophies. In this regard, in some cases, a strong contribution of redox imbalance to the pathogenesis of muscle depletion can be recognized and even inferred from the effectiveness of antioxidant therapies. As proposed by Bonetto et al. [391] more recently, oxidative stress probably would behave as an additional factor that certainly amplifies the wasting stimuli, but probably does not play a leading role in many other cases, which did not demonstrate the effectiveness of antioxidant therapy. With this in mind, any approach to treating sarcopenia by antioxidant supplementation should take into account the relative contribution of oxidative stress.

7.2. Creatine Supplementation

Creatine monohydrate (CrM) is among the most widely used and researched ergogenic aids [394]. At the cellular level, CrM may reduce oxidative stress, prevent motor nerve drop-out [395], enhance mitochondrial function [396], and reduce neuronal apoptosis. Creatine is a guanidine compound that is produced endogenously from arginine, glycine, and methionine in liver, kidney, and pancreas [397]. Exogenous creatine is obtained mainly from meat (1 kg of meat contains ~ 5 g of creatine). Phosphocreatine (PCr) plays an important role in supporting metabolism during high-intensity exercise. Impairments in PCr metabolism may therefore hinder muscle performance and reduce muscle mass [398, 399], although it remains unclear whether or not creatine content is altered by the aging process, different to the marked reduction of PCr in muscle of metabolic myopathic patients [400]. Although some studies did not show a beneficial effect from CrM supplementation during resistance training for elderly individuals [401, 402], many studies have reported that CrM supplementation during resistance training increases muscle mass and muscular strength, endurance, and power in older individuals [400, 403-406]. Rawson et al. [402] reported that 1 month of CrM supplementation (20 g/d × 10 d → 4 g/d × 20 d) did not enhance FFM, total body mass, or upper extremity strength gains, yet there was less leg fatigue in the CrM-supplemented group. In addition, a 2-month resistance exercise program (67-80 years) supplemented with CrM (20 g/d × 5 d → 3 g/d) did not influence training-induced increases in total body mass or strength gain [401]. In contrast, studies of longer duration (> 4 months) reported beneficial effects of CrM supplementation in further enhancing the strength and muscle mass gains attained with a resistance-training program [403, 404, 407]. There are several effects of CrM administration that may enhance resistance exercise-induced strength gains in elderly including activation of myogenic determination factors [408], enhancement of satellite cell activation and recruitment [409], and reduction of amino acid oxidation and protein breakdown [410]. There are a number of important mechanistic questions that remain to be answered, including whether the gains are maintained over a longer period (i.e., > 6 months) post-study and if there is a true enhancement of functional capacity, what is the mechanism of action (i.e., more contractions over time vs. activation of satellite cells [411]. CrM supplementation without resistance training in older adults has also been shown to have minimal benefits [402, 411].

7.3. Angiotensin-Converting Enzyme Inhibitors

Angiotensin-converting enzyme (ACE) inhibitors have long been used as a treatment in primary and secondary prevention in cardiovascular disease as well as secondary stroke prevention. It has now been suggested that ACE inhibitors may have a beneficial effect on skeletal muscle. ACE inhibitors may exert their beneficial effects on skeletal muscles through a number of different mechanisms. ACE inhibitors may improve muscle function through improvements in endothelial function, metabolic function, anti-inflammatory effects, and angiogenesis thereby improving skeletal muscle blood flow. ACE inhibitors can increase mitochondrial numbers and IGF-I levels, thereby helping to counter sarcopenia [412-414].

Observational studies have shown that the long-term use of ACE inhibitors was associated with a lower decline in muscle strength and walking speed in older hypertensive people and a greater lower limb lean muscle mass when compared with users of other anti-hypertensive agents [415]. Several studies have shown that ACE inhibitors improved exercise capacity in both younger and older people with heart failure [415, 416], but caused no improvement in grip strength [417]. Although this could be largely attributed to improvements in cardiac function, skeletal muscle atrophy is also associated with chronic heart failure so the evidence of muscle gains should not be discounted. Few interventional studies using ACE inhibitors for physical function have been undertaken. One study looking at functionally-impaired older people without heart failure has shown that ACE inhibitors increase 6-minute walking distance to a degree comparable to that achieved after 6 months of exercise training [418]. Another found that ACE inhibitors increased exercise time in older hypertensive men [419]. However, a study comparing the effects of nifedipine with ACE inhibitors in older people found no difference between treatments in muscle strength, walking distance, or functional performance [420]. It is possible that frailer subjects with slower walking speeds, who have a tendency to more cardiovascular problems, benefit more. Further evidence is required before recommending ACE inhibitors to counter the effects of sarcopenia. However, ACE inhibitors are associated with cardiovascular benefits and as older people frequently have underlying cardiovascular problems, these agents are already commonly prescribed.

7.4. Proteasome Inhibitors

There are several chemical classes of compounds that inhibit proteasomal activity, including peptide analogs of substrates with different C-terminal groups, such as aldehydes, epoxyketones, boronic acids, and vinyl sulfones [421]. A selective boronic acid proteasome inhibitor, Velcade (also known as PS-341 and bortezomib), directly inhibits the proteasome complex without direct effects on ubiquitination. Velcade is well distributed in the body and does not cross the blood-brain barrier.

In addition to being useful research tools for dissecting the roles of the proteasome, proteasome inhibitors have potential applications in biotechnology and medicine. For example, through their ability to block the activation of NF-κB, proteasome inhibitors can dramatically reduce *in vitro* and *in vivo* the production of inflammatory mediators as well as of various leukocyte adhesion molecules, which play a crucial role in many diseases. Indeed, Velcade is orally active and is presently approved by the Food and Drug Administration and the European Medicines Agency and well-tolerated for treating multiple myeloma [422, 423]. Moreover, this compound is currently being tested in different phase II clinical trials as a possible anti-tumor agent in ovarian, lung, prostate, and pancreatic cancers, and melanoma, and glioblastoma [424-427]. Bonuccelli et al. [428] had indicated that Velcade, once injected locally into the gastrocnemius muscles of *mdx* mice, could up-regulate the expression and membrane localization of dystrophin and members of the DGC. Gazzerro et al. [429] suggested that treatment with Velcade (0.8 mg/Kg) over a 2-week period reduced muscle degeneration and necrotic features in *mdx* muscle fibers, as evaluated with Evans blue dye. In addition, they observed many myotubes and/or immature myofibers expressing embryonic MHC in mdx muscle after Velcade administration probably due to the up-regulation of several myogenic differentiating modulators (MyoD and Myf-5). These effects of Velcade on muscle degeneration would differ dependent on muscle-fiber type. Beehler et al. [430] demonstrated selective attenuation on treatment with Velcade (3 mg/Kg, 7 days) for atrophy of denervated slow-type muscle (soleus), but not fast-type muscle (EDL) of rats. In contrast, MG-132 exerts an inhibitory effect on both the proteasome and the calpain system. More recently, Gazzerro et al. [429] clearly demonstrated that MG-132 increased dystrophin, alpha-sarcoglycan and beta-dystroglycan protein levels in explants from BMD patients, whereas it increased the proteins of the dystrophin glycoprotein complex in DMD cases. However, these proteasome inhibitors may not act to attenuate muscle wasting in cases of sarcopenia. As indicated previously [75, 79, 87, 88], almost no studies demonstrated an enhancement of proteasome-linked modulators for protein degradation in sarcopenic mammalian muscles.

7.5 Cyclophillin inhibitor (Debio-025)

Ca2 overload is known to cause cellular necrosis by directly inducing the opening of the mitochondrial permeability transition (MPT) pore [431, 432]. The MPT pore spans the inner and outer membranes of the mitochondria, and when opened for prolonged periods of time, leads to loss of ATP generation, swelling, rupture, and induction of cell death [431, 432]. Cyclophilin D is a mitochondrial matrix prolyl cis-trans isomerase that directly regulates calcium-and reactive oxygen species-dependent MPT and cellular necrosis. Indeed, mice lacking Ppif (the gene encoding cyclophillin D) show protection from necrotic cell death in the brain and heart after ischemic injury, and mitochondria isolated from these mice are resistant to calcium-induced swelling [433, 434]. Additionally, genetic deletion of Ppif attenuated various dystrophic symptoms (fiber atrophy, fiber loss, invasion by inflammatory cells, and swollen mitochondria) of mice lacking δ-sarcoglycan (Scgd$^{-/-}$ mice) and the α2-chain of laminin-2 (Lama2$^{-/-}$ mice) [435]. Millay et al. [435] demonstrated that the subcutaneous injection of Debio-025, a potent inhibitor of the cyclophillin family, improved calcium overload-induced swelling of mitochondria and reduced manifestations of necrotic disease such as fibrosis and central nuclei, in mdx mice, a model of DMD. In addition,

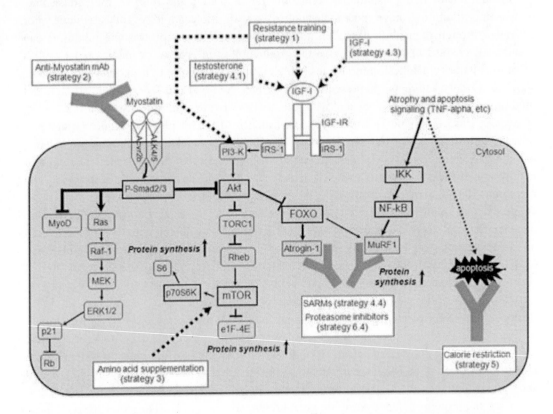

Figure 2. Myostatin signals through the activin receptor IIB (Acvr2b) ALK4/5 heterodimer activate Smad2/3 with blocking of MyoD transactivation in an autoregulatory feedback loop. In addition, Smad3 sequesters MyoD in the cytoplasm to prevent it from entering the nucleus and activating the stem cell population. Activated Smad2/3 also serves to activate the ERK1/2 MAPK pathway. In proliferating myoblasts, this pathway arrests cell proliferation and differentiation through the p21/Rb cascade. Moreover, recent findings [440, 441] suggest that myostatin-Smad pathway inhibit protein synthesis probably due to blocking the functional role of Akt. Resistance training (strategy 1) and the treatment with testosterone (strategy 4.1) and IGF-I (strategy 4.3) stimulate protein synthesis by activating Akt/mTOR/p70S6K pathway. Amino acid supplementation (strategy 3) enhances protein synthesis by stimulating mTOR. Akt blocks the nuclear translocation of FOXO to inhibit the expression of Atrogin-1 and MuRF1 and the consequent protein degradation. SARMs (strategy 4.4) and proteasome inhibitors (6.4) combat the ubiquitin-proteasome signaling activated by these atrogenes. Calorie restriction seems to block the enhanced apoptosis-signaling in sarcopenic muscle fibers. IRS-1; Insulin receptor substrate-1, PI3-K; Phosphatidylinositol 3-kinase, TORC1; a component of TOR signaling complex 1, Rheb; Ras homolog enriched in brain, mTOR; mammalian target of rapamycin, eIF4E; Eukaryotic initiation factor 4E, MEK; Mitogen-activated protein kinase kinase, ERK; Extracellular signal-regulated kinase, FOXO; Forkhead box O, MuRF1; Muscle ring-finger protein 1, NF-κB; Nuclear factor of kappa B, IKK; inhibitor of κB kinase, TNF; Tumor necrosis factor.

treatment with Debio-025 prevented mitochondrial dysfunction and normalized the apoptotic rates and ultrastructural lesions of myopathic Col6a1$^{-/-}$ mice, a model of human Ullrich congenital muscular dystrophy and Bethlem myopathy [436]. More recently, orally administered Debio-025 reduced creatine kinase blood levels and improved grip strength in mdx mice after 6 weeks of treatment [437]. This effect on muscular dystrophy was greater than that of prednisone, currently the standard for treatment of DMD [438, 439]. However, it had not been examined until now whether Debio-025 also has a therapeutic effect on the loss and/or atrophy of muscle fiber with aging in rodents as well as humans. Since there are many symptoms in common between muscular dystrophy and sarcopenia, treatment with Debio-025 may counteract sarcopenic symptoms. Figure 2 provides an overview of the molecular pathways of muscle hypertrophy and several strategies for counteracting sarcopenia.

8. CONCLUSIONS AND PERSPECTIVES

The advances in our understanding of muscle biology that have occurred over the past decade have led to new hopes for pharmacological treatment of muscle wasting. These treatments will be tested in humans in the coming years and offer the possibility of treating sarcopenia/frailty. These treatments should be developed in the setting of appropriate dietary and exercise strategies. Currently, available data shows that resistance training, myostatin inhibition, resistance training plus the ingestion of amino acids, or CR appears to counteract sarcopenia. Resistance training combined with amino acid-containing supplements would be the best way to prevent age-related muscle wasting and weakness. Myostatin inhibition seems to be the most interesting strategy for attenuating sarcopenia in the near future.

ACKNOWLEDGMENTS

This work was supported by a research Grant-in-Aid for Scientific Research C (No. 20500575) from the Ministry of Education, Science, Culture, Sports, Science and Technology of Japan.

REFERENCES

[1]   Candow, D. G.; Chilibeck, P. D. Differences in size, strength, and power of upper and lower body muscle groups in young and older men. *J Gerontol A Biol Sci Med Sci.* 2005, 60, 148-156.

[2]   Melton 3rd, L. J.; Khosla, S.; Crowson, C. S.; O'Fallon, W. M.; Riggs, B. L. Epidemiology of sarcopenia. *J Am Geriat Soc.* 2000, 48, 625-630.

[3]   Baumgartner, R. N.; Waters, D. L.; Gallagher, D.; Morley, J. E.; Garry, P. J. Predictors of skeletal muscle mass in elderly men and women. *Mech Aging Dev.* 1999, 107, 123-136.

[4]   Poehlman, E. T.; Toth, M. J.; Fonong, T. Exercise, substrate utilization and energy requirements in the elderly. *Int J Obesity Relat Metab Disord.* 1995, 19, S93-S96.

[5]   Griffiths, R. D. Muscle mass, survival, and the elderly ICU patient. *Nutrition*, 1996, 12, 456-458.

[6]   Janssen, I.; Shepard, D. S.; Katzmarzyk, P. T.; Roubenoff, R. The healthcare costs of sarcopenia in the United States. *J Am Geriatr Soc.* 2004, 52, 80-85.

[7]   Short, K. R.; Nair, K. S. The effect of age on protein metabolism. *Curr Opin Clin Nutr Metab Care*, 2000, 3, 39-44.

[8]   Short, K. R.; Vittone, J. L.; Bigelow, J. L.; Proctor, D. N.; Nair, K. S. Age and aerobic exercise training effects on whole body and muscle protein metabolism. *Am J Physiol Endocrinol Metab.* 2004, 286, E92-E101.

[9]   Lexell, J. Human aging, muscle mass, and fiber type composition. *J Gerontol A Biol Sci Med Sci.* 1995, 50, 11-16.

[10]   Larsson, L. Morphological and functional characteristics of the aging skeletal muscle in man. A cross-sectional study. *Acta Physiol Scand Suppl.* 1978, 457, 1-36.

[11]   Larsson, L.; Sjodin, B.; Karlsson, J. Histochemical and biochemical changes in human skeletal muscle with age in sedentary males, age 22-65 years. *Acta Physiol Scand.* 1978, 103, 31-39.

[12]   Verdijk, L. B.; Koopman, R.; Schaart, G.; Meijer, K.; Savelberg, H. H.; van Loon, L. J. Satellite cell content is specifically reduced in type II skeletal muscle fibers in the elderly. *Am J Physiol Endocrinol Metab.* 2007, 292, E151-E157.

[13]   Verdijk, L. B.; Gleeson, B. G.; Jonkers, R. A. M.; Meijer, K.; Savelberg, H. H. C. M.; Dendale, P.; van Loon, L. J. C. Skeletal muscle hypertrophy following resistance training is accompanied by a fiber type-specific increase in satellite cell content in elderly men. *J Gerontol A Biol Sci Mde Sci.* 2009, 64, 332-339.

[14]   Brack, A. S.; Bildsoe, H., Hughes, S. M. Evidence that satellite cell decrement contributes to preferential decline in nuclear number from large fibers during murine age-related muscle atrophy. *J Cell Sci.* 2005, 118, 4813-4821.

[15]   Collins, C. A.; Zammit, P. S.; Ruiz, A. P.; Morgan, J. E.; Partridge, T. A. population of myogenic stem cells that survives skeletal muscle aging. *Stem Cells*, 2007, 25, 885-894.

[16]   Day, K.; Shefer, G.; Shearer, A.; Yablonka-Reuveni, Z. The depletion of skeletal muscle satellite cells with age is concomitant with reduced capacity of single progenitors to produce reserve progeny. *Dev Biol.* 2010, 340, 330-343.

[17]   Shefer, G, Van de Mark, D. P.; Richardson, J. B.; Yablonka-Reuveni, Z. Satellite-cell pool size does matter: defining the myogenic potency of aging skeletal muscle. *Dev Biol.* 2006, 294, 50-66.

[18]   Conboy, I. M.; Conboy, M. J.; Smythe, G. M.; Rando, T. A. Notch-mediated restoration of regenerative potential to aged muscle. *Science*, 2003, 302, 1575-1577.

[19]   Roth, S. M.; Martel, G. F.; Ivey, F. M.; Lemmer, J. T.; Metter, E. J.; Hurley, B. F.; Rogers, M. A. Skeletal muscle satellite cell populations in healthy young and older men and women. *Anat Rec.* 2000, 260, 351-358.

[20]   Wagners, A. J.; Conboy, I. M. Cellular and molecular signatures of muscle regeneration: current concepts and controversies in adult myogenesis. *Cell*, 2005, 122, 659-667.

[21]   Dube, J.; Goodpaster, B. H. Assessment of intramuscular triglycerides: contribution to metabolic abnormalities. *Curr Opin Clin Nutr Metab Care*, 2006, 9, 553-559.

[22]   Kraegen, E. W.; Cooney, G. J. Free fatty acids and skeletal muscle insulin resistance. *Curr Opin Lipidol.* 2008, 19, 235-241.

[23] Shefer, G.; Wleklinski-Lee, M.; Yablonka-Reuveni, Z. Skeletal muscle satellite cells can spontaneously enter an alternative mesenchymal pathway. *J Cell Sci.* 2004, 117, 5393-5404.

[24] Hamilton, M. T.; Areiqat, E.; Hamilton, D. G.; Bey, L. Plasma triglyceride metabolism in humans and rats during aging and physical inactivity. *Int J Sport Nutr Exerc Metab.* 2001, 11, S97-S104.

[25] Lexell, J. Aging and human muscle: observations from Sweden. *Can J Appl Physiol.* 1993, 18, 2-18.

[26] Roubenoff, R.; Hughes, V. A. Sarcopenia: current concepts. *J Gerontol A Biol Sci Med Sci.* 2000, 55, M716-M724.

[27] Sakuma, K.; Yamaguchi, A. Molecular mechanisms in aging and current strategies to counteract sarcopenia. *Curr Aging Sci.* 2010, 3, 90-101.

[28] Bodine, S. C.; Stitt, T. N.; Gonzalez, M.; Kline, W. O.; Stover, G. L.; Bauerlein, R.; Zlotchenko, E.; Scrimgeour, A.; Lawrence, J. C.; Glass, D. J.; Yancopoulos, G. D. Akt/mTOR pathway is a crucial regulator of skeletal muscle hypertrophy and can prevent muscle atrophy in vivo. *Nat Cell Biol.* 2001, 3, 1014-1019.

[29] Glass, D. J. Signaling pathways that mediate skeletal muscle hypertrophy and atrophy. *Nat Cell Biol.* 2003, 5, 87-90.

[30] Lai, K. M. V.; Gonzalez, M.; Poueymirou, W. T.; Kline, W. O.; Na, E.; Zlotchenko, E.; Stitt, T. N.; Economides, A. N.; Yancopoulos, G. D.; Glass, D. J. Conditional activation of Akt in adult skeletal muscle induces rapid hypertrophy. *Mol Cell Biol.* 2004, 24, 9295-9304.

[31] Terada, N.; Patel, H. R.; Takase, K.; Kohno, K.; Nairn, A. C.; Gelfand, E. W. Rapamycin selectively inhibits translation of mRNAs encoding elongation factors and ribosomal proteins. *Proc Natl Acad Sci USA*, 1994, 91, 11477-11481.

[32] Gingras, A. C.; Raught, B.; Sonenberg, N. Regulation of translation initiation by FRAP/mTOR. *Genes Dev* 2001, 15, 807-826.

[33] Parkington, J. D.; LeBrasseur, N. K.; Siebert, A. P.; Fielding, R. A. Contraction-mediated mTOR, p70^{S6K}, and ERK1/2 phosphorylation in aged skeletal muscle. *J Appl Physiol.* 2004, 97, 243-248.

[34] Kimball, S. R.; O'Malley, J. P.; Anthony, J. C.; Crozier, S. J.; Jefferson, L. S. Assessment of biomarkers of protein anabolism in skeletal muscle during the life span of the rat: sarcopenia despite elevated protein synthesis. *Am J Physiol Endocrinol Metab.* 2004, 287, E772-E780.

[35] Léger, B.; Derave, W.; De Bock, K.; Hespel, P.; Russell, A. P. Human sarcopenia reveals an increase in SOCS-3 and myostatin and a reduced efficiency of Akt phosphorylation. *Rejuvenation Res.* 2008, 11, 163-175.

[36] Haddad, F.; Adams, G. R. Aging-sensitive cellular and molecular mechanisms associated with skeletal muscle hypertrophy. *J Appl Physiol.* 2006, 100, 1188- 1203.

[37] Funai, K.; Parkington, J. D.; Carambula, S.; Fielding, R. A. Age-associated decrease in contraction-induced activation of downstream targets of Akt/mTOR signaling in skeletal muscle. *Am J Physiol Regul Integr Comp Physiol.* 2006, 290, R1080-R1086.

[38] Thomson, D. M.; Gordon, S. E. Impaired overload-induced muscle growth is associated with diminished translational signaling in aged rat fast-twitch skeletal muscle. *J Physiol.* 2006, 574, 291-305.

[39] Treisman, R. Identification and purification of a polypeptide that binds to the c-fos serum response element. *EMBO J.* 1987, 6, 2711-2717.

[40] Muscat, G. E.; Gustafson, T. A.; Kedes, L. A common factor regulates skeletal and cardiac alpha-actin gene transcription in muscle. *Mol Cell Biol.* 1988, 8, 4120- 4133.

[41] Klamut, H. J.; Gangopadhyay, S. B.; Worton, R. G.; Ray, P. N. Molecular and functional analysis of the muscle-specific promoter region of the Duchenne muscular dystrophy gene. *Mol Cell Biol.* 1990, 10, 193-205.

[42] Ernst, H.; Walsh, K.; Harrison, C. A.; Rosenthal, N. The myosin light chain enhancer and the skeletal actin promoter share a binding site for factors involved in muscle-specific gene expression. *Mol Cell Biol* 1991, 11, 3735-3744.

[43] Pipes, G. C.; Creemers, E. E.; Olson, E. N. The myocardin family of transcriptional co-activators: versatile regulators of cell growth, migration, and myogenesis. *Genes Dev.* 2006, 20: 1545-1556.

[44] Soulez, M.; Rouviere, C. G.; Chafey, P.; Hentzen, D.; Vandromme, M.; Lautredou, N.; Lamb, N.; Kahn, A.; Tuil, D. Growth and differentiation of C2 myogenic cells are dependent on serum response factor. *Mol Cell Biol.* 1996, 16, 6065-6074.

[45] Carnac, G.; Primig, M.; Kitzmann, M.; Chafey, P.; Tuil, D.; Lamb, N.; Fernandez, A. RhoA GTPase and serum response factor control selectively the expression of MyoD without affecting Myf5 in mouse myoblasts. *Mol Biol Cell.* 1998, 9, 1891- 1902.

[46] Charvet, C.; Houbron, C.; Parlakian, A.; Giordani, J.; Lahoute, C.; Bertrand, A.; Sotiropoulos, A.; Renou, L.; Schmitt, A.; Melki, J.; Li, Z.; Daegelen, D.; Tuil, D. New role for serum response factor in postnatal skeletal muscle growth and regeneration via the interleukin 4 and insulin-like growth factor 1 pathways. *Mol Cell Biol.* 2006, 26, 6664-6674.

[47] Sakuma, K.; Nishikawa, J.; Nakao, R.; Nakano, H.; Sano, M.; Yasuhara, M. Serum response factor plays an important role in the mechanically overloaded plantaris muscle of rats. *Histochem Cell Biol.* 2003, 119, 149-160.

[48] Gordon, S. E.; Flück, M.; Booth, F. W. Skeletal muscle focal adhesion kinase, paxillin, and serum response factor are loading dependent. *J Appl Physiol.* 2001, 90, 1174-1183.

[49] Gauthier-Rouviére, C.; Vandromme, M.; Tuil, D.; Lautredou, N.; Morris, M.; Soulez, M.; Kahn, A.; Fernandez, A.; Lamb, N. Expression and activity of serum response factor is required for expression of the muscle-determining factor MyoD in both dividing and differentiating mouse C2C12 myoblasts. *Mol Biol Cell,* 1996, 7, 719-729.

[50] Sakuma, K.; Nakao, R.; Inashima, S.; Hirata, M.; Kubo, T.; Yasuhara, M. Marked reduction of focal adhesion kinase, serum response factor and myocyte enhancer factor 2C, but increase in RhoA and myostatin in the hindlimb dy mouse muscles. *Acta Neuropath* (Berl). 2004, 108, 241-249.

[51] Lahoute, C.; Sotiropoulos, A.; Favier, M.; Guillet-Deniau, I.; Charvet, C.; Ferry, A.; Butler-Browne, G.; Metzger, D.; Tuil, D.; Daegelen, D. Premature aging in skeletal muscle lacking serum response factor. *PLoS One,* 2008, 3, e3910.

[52] Sakuma, K.; Akiho, M.; Nakashima, H.; Akima, H.; Yasuhara, M. Age-related reductions in expression of serum response factor and myocardin-related transcription factor A in mouse skeletal muscles. *Biochim Biophys Acta Mol Basis Dis.* 2008, 1782, 453-461.

[53] Kuwahara, K.; Barrientos, T.; Pipes, G. C.; Li, S.; Olson, E. N. Muscle-specific signaling mechanism that links actin dynamics to serum response factor. *Mol Cell Biol.* 2005, 25, 3173-3181.

[54] Cen, B.; Selvaraj, A.; Burgess, R. C.; Hitzler, J. K.; Ma, Z.; Morris, S. W.; Prywes, R. Megakaryoblastic leukemia 1, a potent transcriptional co-activator for serum response factor (SRF), is required for serum induction of SRF target genes. *Mol Cell Biol.* 2003, 23, 6597-6608.

[55] Wang, D. Z.; Li, S.; Hockemeyer, D.; Sutherland, L.; Wang, Z.; Schratt, G., Richardson, J. A.; Nordheim, A.; Olson, E. N. Potentiation of serum response factor activity by a family of myocardin-related transcription factors. *Proc Natl Acad Sci USA*, 2002, 99, 14855-14860.

[56] Carson, J. A.; Wei, L. Integrin signaling's potential for mediating gene expression in hypertrophying skeletal muscle. *J Appl Physiol.* 2000, 88, 337-343.

[57] Lange, S.; Xiang, F.; Yakovenko, A.; Vihola, A.; Hackman, P.; Rostkova, E.; Kristensen, J.; Brandmeier, B.; Franzen, G.; Hedberg, B.; Gunnarsson, L. G.; Hughes, S. M.; Marchand, S.; Sejersen, T.; Richard, I.; Edström, L.; Ehler, E.; Udd, B.; Gautel, M. The kinase domain of titin controls muscle gene expression and protein turnover. *Science*, 2005, 308, 1599-1603.

[58] Lamon, S.; Wallace, M. A.; Léger, B.; Russell, A. P. Regulation of STARS and its downstream targets suggest a novel pathway involved in human skeletal muscle hypertrophy and atrophy. *J Physiol.* 2009, 587, 1795-1803.

[59] Kadowaki, M.; Kanazawa, T. Amino acids as regulators of proteolysis. *J Nutr.* 2003, 133, 2052S-2056S.

[60] Mizushima, N.; Yamamoto, A.; Matsui, M.; Yoshimori, T.; Ohsumi, Y. In vivo analysis of autophagy in response to nutrient deprivation using transgenic mice expressing a fluorescent autophagosome marker. *Mol Biol Cell.* 2004, 15, 1101-1111

[61] Marino, G.; Uria, J. A.; Puente, X. S.; Quesada, V.; Bordallo, J.; Lopez-Otin, C. Human autophagins, a family of cysteine proteinases potentially implicated in cell degradation by autophagy. *J Biol Chem.* 2003, 278, 3671-3678

[62] Mammucari, C.; Milan, G.; Romanello, V.; Masiero, E.; Rudolf, R.; Del Piccolo, P.; Burden, S. J.; Di Lisi, R.; Sandri, C.; Zhao, J.; Goldberg, A. L.; Schiaffino, S.; Sandri M. FoxO3 controls autophagy in skeletal muscle in vivo. *Cell Metab.* 2007, 6, 458-471.

[63] Ogata, T, Oishi, Y.; Higuchi, M.; Muraoka, I. Fasting-related autophagic response in slow- and fast-twitch skeletal muscles. *Biochem Biophys Res Commun.* 2010, 394, 136-140.

[64] O'Leary, M. F. N.; Hood, D. A. Denervation-induced oxidative stress and autophagy signaling in muscle. *Autophagy*, 2009, 5, 230-231.

[65] Bodine, S. C.; Latres, E.; Baumhueter, S.; Lai, V. K.; Nunez, L.; Clarke, B. A.; Poueymirou, W. T.; Panaro, F. J.; Na, E.; Dharmarajan, K.; Pan, Z. Q.; Valenzuela, D. M.; DeChiara, T. M.; Stitt, T. N.; Yancopoulos, G. D.; Glass, D. J. Identification of ubiquitin ligases required for skeletal muscle atrophy. *Science*, 2001, 294, 1704-1708.

[66] Costelli, P.; Reffo, P.; Penna, F.; Autelli, R.; Bonelli, G.; Baccino, F. M. Ca^{2+}-dependent proteolysis in muscle wasting. *Int J Biochem Cell Biol.* 2005, 37, 2134-2146.

[67] Smith, I. J.; Dodd, S. L. Calpain activation causes a proteasome-dependent increase in protein degradation and inhibits the Akt signaling pathway in rat diaphragm muscle. *Exp Physiol.* 2007, 92, 561-573.

[68] Du, J.; Wang, X.; Miereles, C.; Bailey, J. L.; Debigare, R.; Zheng, B. Activation of caspase-3 is an initial step triggering accelerated muscle proteolysis in catabolic conditions. *J Clin Invest.* 2004, 113, 115-123.

[69] Wei, W.; Fareed, M. U.; Evenson, A.; Menconi, M. J.; Yang, H.; Petkova, V.; Hasselgren, P. O. Sepsis stimulates calpain activity in skeletal muscle by decreasing calpastatin activity but does not activate caspase-3. *Am J Physiol Regul Integr Comp Physiol.* 2005, 288, R580-R590.

[70] Moresi, V.; Presterá, A.; Scicchitano, B. M.; Molinaro, M.; Teodori, L.; Sassoon, D.; Adamo S.; Coletti, D. Tumor necrosis factor-alpha inhibition of skeletal muscle regeneration is mediated by a caspase-dependent stem cell response. *Stem Cells*, 2008, 26, 997-1008.

[71] Attaix, D.; Aurousseau, E.; Combaret, L.; Kee, A.; Larbaud, D.; Ralliére, C.; Souweine, B.; Taillandier, D.; Tilignac, T. Ubiquitin-proteasome-dependent proteolysis in skeletal muscle. *Reprod Nutr Dev.* 1998, 38, 153-165.

[72] Cai, D.; Frantz, J. D.; Tawa, N. E. Jr.; Melendez, P. A.; Oh, B. C.; Lidov, H. G.; Hasselgren, P. O.; Frontera, W. R.; Lee, J.; Glass, D. J.; Shoelson, S. E. IKKbeta/ NF-kappaB activation causes severe muscle wasting in mice. *Cell*, 2004, 119, 285-289.

[73] Lecker, S. H.; Jagoe, R. T.; Gilbert, A.; Gomes, M.; Baracos, V.; Bailey, J.; Price, S. R.; Mitch, W. E.; Goldberg, A. L. Multiple types of skeletal muscle atrophy involve a common program of changes in gene expression. *FASEB J.* 2004, 18, 39-51.

[74] Mitch, W. E.; Goldberg, A. L. Mechanisms of muscle wasting. The role of the ubiquitin-proteasome pathway. *N Engl J Med.* 1996, 335, 1897-1905.

[75] Sandri, M.; Sandri, C.; Gilbert, A.; Skurk, C.; Calabria, E.; Picard, A.; Walsh, K.; Schiaffino, S.; Lecker, S. H.; Goldberg, A. L. Foxo transcription factors induce the atrophy-related ubiquitin ligase atrogin-1 and cause skeletal muscle atrophy. *Cell*, 2004, 117, 399-341.

[76] Stitt, T. N.; Drujan, D.; Clarke, B. A.; Panaro, F.; Timofeyva, Y.; Kline, W. O.; Gonzalez, M.; Yancopoulos, G. D.; Glass, D. J. The IGF-I/PI3K/Akt pathway prevents expression of muscle atrophy-induced ubiquitin ligases by inhibiting FOXO transcription factors. *Mol Cell*, 2004, 14, 395-403.

[77] Bossola, M.; Pacelli, F.; Costelli, P.; Tortorelli, A.; Rosa, F.; Doglietto, G. B. Proteasome activities in the rectus abdominis muscle of young and older individuals. *Biogerontology*, 2008, 9, 261-268.

[78] Cai, D.; Lee, K. K.; Li, M.; Tang, M. K.; Chan, K. M. Ubiquitin expression is up-regulated in human and rat skeletal muscles during aging. *Arch Biochem Biophys.* 2004, 425, 42-50.

[79] Pattison, J. S.; Folk, L. C.; Madsen, R. W.; Childs, T. E.; Booth, F. W. Transcriptional profiling identifies extensive downregulation of extracellular matrix gene expression in sarcopenic rat soleus muscle. *Physiol Genomics*, 2003, 15, 34-43.

[80] Combaret, L.; Dardevet, D.; Béchet, D.; Taillandier, D.; Mosoni, L.; Attaix, D. Skeletal muscle proteolysis in aging. *Curr Opin Clin Nutr Metab Care*, 2009, 12, 37-41.

[81] DeRuisseau, K. C.; Kavazis, A. N.; Powers, S. K. Selective downregulation of ubiquitin conjugation cascade mRNA occurs in the senescent rat soleus muscle. *Exp Gerontol.* 2005, 40, 526-531.

[82] Clavel, S.; Coldefy, A. S.; Kurkdjian, E.; Salles, J.; Margaritis, I.; Derijard, B. Atrophy-related ubiquitin ligases, atrogin-1 and MuRF1 are up-regulated in aged rat tibialis anterior muscle. *Mech Aging Dev.* 2006, 127, 794-801.

[83] Welle, S.; Brooks, A. L.; Delehany, J. M.; Needler, N.; Thornton, C. A. Gene expression profile of aging in human muscle. *Physiol Genomics*, 2003, 14, 149-159.

[84] Whitman, S. A.; Wacker, M. J.; Richmond, S. R.; Godard, M. P. Contributions of the ubiquitin-proteasome pathway and apoptosis to human skeletal muscle wasting with age. *Pflügers Arch.* 2005, 450, 437-446.

[85] Edström, E.; Altun, M.; Hägglund, M.; Ulfhake, B. Atrogin-1/MAFbx and MuRF1 are downregulated in aging-related loss of skeletal muscle. *J Gerontol A Biol Sci Med Sci.* 2006, 61, 663-674.

[86] Sacheck, J. M.; Hyatt, J. P.; Raffaello, A.; Jagoe, R. T.; Roy, R. R.; Edgerton, V. R.; Lecker, S. H.; Goldberg, A. L. Rapid disuse and denervation atrophy involve transcriptional changes similar to those of muscle wasting during systemic diseases. *FASEB J.* 2007, 21, 140-155.

[87] Attaix, D.; Mosoni, L.; Dardevet, D.; Combaret, L.; Mirand, P. P.; Grizard, J. Altered responses in skeletal muscle protein turnover during aging in anabolic and catabolic periods. *Int J Biochem Cell Biol.* 2005, 37, 1962-1973.

[88] Husom, A. D.; Peters, E. A.; Kolling, E. A.; Fugere, N. A.; Thompson, L. V.; Ferrington, D. A. Altered proteasome function and subunit composition in aged muscle. *Arch Biochem Biophys.* 2004, 421, 67-76.

[89] Levine, B.; Klionsky, D. J. Development by self-digestion: molecular mechanisms and biological functions of autophagy. *Dev Cell,* 2004, 6, 463-477.

[90] Meijer, A. J.; Codogno, P. Regulation and role of autophagy in mammalian cell. *Int J Biochem Cell Biol.* 2004, 36, 2445-2462.

[91] Xie, Z.; Klionsky, D. J. Autophagosome formation: core machinery and adaptations. *Nat Cell Biol.* 2007, 9, 1102-1109.

[92] Sandri, M. Autophagy in health and disease. 3. Involvement of autophagy in muscle atrophy. *Am J Physiol Cell Physiol.* 2010, 298, C1291-C1297.

[93] Sandri, M. Autophagy in skeletal muscle. *FEBS Lett.* 2010, 584, 1411-1416.

[94] Lee, H. K.; Marzella, L. Regulation of intracellular protein degradation with special reference to lysosoms: role in cell physiology and pathology. *Int Rev Exp Pathol.* 1994, 35, 39-147.

[95] Cuervo, A. M. Autophagy: many paths to the same end. *Mol Cell Biochem.* 2004, 263, 55-72.

[96] Wang, C. W.; Klionsky, D. J. The molecular mechanism of autophagy. *Mol Med.* 2003, 9, 65-76.

[97] Bechet, D.; Tassa, A.; Taillandier, D.; Combare, L.; Attaix, D. Lysosomal proteolysis in skeletal muscle. *Int J Biochem Cell Biol.* 2005, 37, 2098-2114.

[98] Noglaska, A.; Terracciano, C.; D'Agostino, C.; Engel, W. K.; Askanas, V. p62/SQSTM1 is overexpressed and prominently accumulated in inclusion of sporadic inclusion-body myositis muscle fibers, and can help differentiating it from polymyositis and dermatomyositis. *Acta Neuropath(Berl).* 2009, 118, 407-413.

[99] Bjorkoy, G.; Lamak, T.; Brech, A.; Outzen, H.; Perander, M.; Overvatn, A.; Stenmark, H.; Johansen, T. p62/SQSTM1 forms protein aggregates degraded by autophagy and

has a protective effect on huntingtin-induced cell death. *J Cell Biol*. 2005, 171, 603-614.

[100] Pankiv, S.; Clausen, T. H.; Lamak, T.; Brech, A.; Bruun, J. A.; Outzen, H.; Overvatn, A.; Bjorkoy, G.; Johansen, T. p62/SQSTM1 binds directly to Atg8/LC3 to facilitate degradation of ubiquitinated protein aggregates by autophagy. *J Biol Chem*. 2007, 282, 24131-24145.

[101] Zhao, J.; Brault, J. J.; Schild, A.; Cao, P.; Sandri, M.; Schiaffino, S.; Lecker, S. H.; Goldberg, A. L. FoxO3 coordinately activates protein degradation by the autophagic/lysosomal and proteasomal pathways in atrophying muscle cells. *Cell Metab*. 2007, 6, 472-483.

[102] Cuervo, A. M.; Bergamini, E.; Brunk, U. T.; Droge, W.; Ffrench, M.; Terman, A. Autophagy and aging: the importance of maintaining "clean" cells. *Autophagy*, 2005, 1, 131-140.

[103] Terman, A.; Brunk, U. T. Oxidative stress, accumulation of biological 'garbage', and aging. *Antioxid Redox Signal*, 2006, 8, 197-204.

[104] Donati, A.; Cavallini, G.; Paradiso, C.; Vittorini, S.; Pollera, M.; Gori, Z.; Bergamini, E. Age-related changes in the regulation of autophagic proteolysis in rat isolated hepatocytes. *J Gerontol A Biol Sci Med Sci*. 2001, 56, B288-B293.

[105] Wohlgemuth, S. E.; Julian, D.; Akin, D. E.; Fried, J.; Toscano, K.; Leeuwenburgh, C, Dunn Jr, W. A. Autophagy in the heart and liver during normal aging and calorie restriction. *Rejuvenat Res*. 2007, 10, 281-292.

[106] Wohlgemuth, S. E.; Seo, A. Y.; Marzetti, E.; Lees, H. A.; Leeuwenburgh, C. Skeletal muscle autophagy and apoptosis during aging: effects of calorie restriction and life-long exercise. *Exp Gerontol*. 2010, 45, 138-148.

[107] McMullen, C. A.; Ferry, A. L.; Gamboa, J. L.; Andrade, F. H.; Dupont-Versteegden, E. E. Age-related changes of cell death pathways in rat extraocular muscle. *Exp Gerontol*. 2009, 44, 420-425.

[108] Kihara, A.; Kabeya, Y, Ohsumi, Y.; Yoshimori, T. Beclin-phosphatidylinositol 3-kinase complex functions at the trans-Golgi network. *EMBO Rep*. 2001, 2, 330-335.

[109] Kihara, A.; Noda, T.; Ishihara, N, Ohsumi, Y. Two distinct Vps34 phosphatidylinositol 3-kinase complexes function in autophagy and carboxypeptidase Y sorting in *Saccharomyces cerevisiae*. *J Cell Biol*. 2001, 152, 519-530.

[110] Komatsu, M.; Waguri, S.; Ueno, T.; Iwata, J.; Murata, S.; Tanida, I.; Ezaki, J.; Mizushima, N.; Ohsumi, Y.; Uchiyama, Y.; Kominami, E.; Tanaka, K.; Chiba, T. Impairment of starvation-induced and constitutive autophagy in Atg7-deficient mice. *J Cell Biol*. 2005, 169, 425-434.

[111] Noda, T.; Kim, J.; Huang, W. P.; Baba, M.; Tokunaga, C.; Ohsumi, Y.; Klionky, D. J. Apg9p/Cvt7p is an integral membrane protein required to transport vesicle formation in the Cvt and autophagy pathways. *J Cell Biol*. 2000, 148, 465-480.

[112] Yan, L.; Vatner, D. E.; Kim, S. J.; Ge, H.; Masurekar, M.; Massover, W. H.; Yang, G.; Matsui, Y.; Sadoshima, J.; Vatner, S. F. Autophagy in chronically ischemic myocardium. *Proc Natl Acad Sci USA*, 2005, 102, 13807-13812.

[113] McPherron, A. C.; Lawler, A. M.; Lee, S. J. Regulation of skeletal muscle mass in mice by a new member TGF-beta superfamily member. *Nature*, 1997, 387, 83-90.

[114] Lee, S. J. Regulation of muscle mass by myostatin. *Annu Rev Cell Dev Biol*. 2004, 20, 61-86.

[115] Amthor, H.; Huang, R.; McKinnell, I.; Christ, B.; Kambadur, R.; Sharma, M.; Patel, K. The regulation and action of myostatin as a negative regulator of muscle development during avian embryogenesis. *Dev Biol.* 2002, 251, 241-257.

[116] Langley, B.; Thomas, M.; Bishop, A.; Sharma, M.; Gilmour, S.; Kambadur, R. Myostatin inhibits myoblast differentiation by down-regulating MyoD expression. *J Biol Chem.* 2002, 277, 49831-49840.

[117] Thomas, M.; Langley, B.; Berry, C.; Sharma, M.; Kirk, S.; Bass, J.; Kambadur, R. Myostatin, a negative regulator of muscle growth, functions by inhibiting myoblast differentiation. *J Biol Chem.* 2000, 275, 40235-40243.

[118] Yang, W.; Zhang, Y.; Li, Y.; Wu, Z.; Zhu, D. Myostatin induces cyclin D1 degradation to cause cell cycle arrest through a phosphatidylinositol 3-kinase/AKT/GSK-3β pathway and is antagonized by insulin-like growth factor 1. *J Biol Chem.* 2007, 282, 3799-3808.

[119] McPherron, A. C.; Lee, S. J. Double muscling in cattle due to mutations in the myostatin gene. *Proc Natl Acad Sci USA,* 1997, 94, 12457–12461.

[120] Schuelke, M.; Wagner, K. R.; Stolz, L. E.; Hübner, C.; Riebel, T.; Kömen, W.; Braun, T.; Tobin, J. F.; Lee, S. J. Myostatin mutation associated with gross muscle hypertrophy in a child. *N Engl J Med.* 2004, 350, 2682-2688.

[121] Lee, S. J.; McPherron, A. M.; Regulation of myostatin activity and muscle growth. *Proc Natl Acad Sci USA,* 2001, 98, 9306-9311.

[122] Rebbapragada, A.; Benchabane, H.; Wrana, J. L.; Celeste, A. J.; Attisano, L. Myostatin signals through a transforming growth factor beta-like signaling pathway to block adipogenesis. *Mol Cell Biol.* 2003, 23, 7230-7242.

[123] Allen, D. L.; Unterman, T. G. Regulation of myostatin expression and myoblast differentiation by FoxO and SMAD transcription factors. *Am J Physiol Cell Physiol.* 2007, 292, C188-C199.

[124] Wehling, M.; Cai, B.; Tidball, J. G. Modulation of myostatin expression during modified muscle use. *FASEB J.* 2000, 14, 103-110.

[125] Sakuma, K.; Watanabe, K.; Sano, M.; Uramoto, I.; Totsuka, T. Differential adaptation of growth and differentiation factor 8/myostatin, fibroblast growth factor 6 and leukemia inhibitory factor in overloaded, regenerating, and denervated rat muscles. *Biochim Biophys Acta Mol Cell Res.* 2000, 1497, 77-88.

[126] Sakuma, K.; Watanabe, K.; Hotta, N.; Koike, T.; Ishida, K.; Katayama, K.; Akima, H. The adaptive responses in several mediators linked with hypertrophy and atrophy of skeletal muscle after lower limb unloading in humans. *Acta Physiol(Oxf).* 2009, 197, 151-159.

[127] Gonzalez-Cadavid, N. F.; Taylor, W. E.; Yarasheski, K.; Sinha-Hikim, I.; Ma, K.; Ezzat, S.; Shen, R.; Lalani, R.; Asa, S.; Mamita, M.; Nair, G.; Arver, S.; Bhasin, S. Organization of the human myostatin gene and expression in health men and HIV-infected men with muscle wasting. *Proc Natl Acad Sci USA,* 1998, 95, 14938-14943.

[128] Carlson, M. E.; Hsu, M.; Conboy, I. M. Imbalance between pSmad3 and Notch induces CDK inhibitors is old muscle stem cells. *Nature,* 2008, 454, 528-532.

[129] Marzetti, E.; Lawler, J. M.; Hiona, A.; Manini, T.; Seo, A. Y.; Leeuwenburgh, C. Modulation of age-induced apoptotic signaling and cellular remodeling by exercise and calorie restriction in skeletal muscle. *Free Radic Biol Med.* 2008, 44, 160-168.

[130] Wang, C.; Youle, R. J. The role of mitochondria in apoptosis. *Ann Rev Genet.* 2009, 43, 95-118.

[131] Adhihetty, P. J.; O'Leary, M. F.; Hood, D. A. Mitochondria in skeletal muscle: adaptable rheostats of apoptotic susceptibility. *Exerc Sport Sci Rev.* 2008, 36, 116-121.

[132] Dirks, A.; Leeuwenburgh, C. Apoptosis in skeletal muscle with aging. *Am J Physiol Regul Integr Comp Physiol.* 2002, 282, R519-R527.

[133] Dirks, A. J.; Leeuwenburgh, C. Aging and lifelong calorie restriction result in adaptations of skeletal muscle apoptosis repressor, apoptosis-inducing factor, X-linked inhibitor of apoptosis, caspase-3, and caspase-12. *Free Radic Biol Med.* 2004, 36, 27-39.

[134] Jezek, P.; Hlavata, L. Mitochondria in homeostasis of reactive oxygen species in cell, tissues, and organism. *Int J Biochem Cell Biol.* 2005, 37, 2478-2503.

[135] Yakes, F. M.; Van, H. B. Mitochondrial DNA damage is more extensive and persists longer than nuclear DNA damage in human cells following oxidative stress. *Proc Natl Acad Sci USA*, 1997, 94, 514-519.

[136] Wei, Y. H.; Lee, H. C. Oxidative stress, mitochondrial DNA mutation, and impairment of antioxidant enzymes in aging. *Exp Biol Med*(Maywood). 2002, 227, 671-682.

[137] Bua, E. A.; McKiernan, S. H.; Eanagat, J.; McKenzie, D.; Aiken, J. M. Mitochondrial abnormalities are more frequent in muscles undergoing sarcopenia. *J Appl Physiol.* 2002, 92, 2617-2624.

[138] Coral-Debrinski, M.; Horton, T.; Lott, M. T.; Shoffner, J. M.; Beal, M. F.; Wallace, D. C. Mitochondrial DNA deletions in the human brain: regional variability and increase with advanced age. *Nat Genet.* 1992, 2, 324-329.

[139] Wanagat, J.; Cao, Z.; Pathare, P.; Aiken, J. M. Mitochondrial DNA deletion mutations co-localize with segmental electron transport system abnormalities, muscle fiber atrophy, fiber splitting, and oxidative damage in sarcopenia. *FASEB J.* 2001, 15, 322-332.

[140] Cortopassi, G. A.; and Wong, A. Mitochondria in organismal aging and degeneration. *Biochim Biophys Acta*, 1999, 1410, 183-193.

[141] Pollack, M.; Leeuwenburgh, C. Apoptosis and aging: role of the mitochondria. *J Gerontol A Biol Sci Med Sci.* 2001, 56, B475-B482.

[142] Song, W.; Kwak, H. B.; Lawler, J. M. Exercise training attenuates age-induced changes in apoptotic signaling in rat skeletal muscle. *Antioxid Redox Signal*, 2006, 8, 517-518.

[143] Mauro, A. Satellite cell of skeletal muscle fibers. *J Biophys Biochem Cytol.* 1961, 9, 493-495.

[144] Bischoff, R. The satellite cell and muscle regeneration. In: Myology. 2nd ed. Engel, A. G.; Franzini-Armstrong, C. Eds. New York: McGraw-Hill 1994; pp. 97-118.

[145] Zammit, P. S. The muscle satellite cell: the story of a cell on the edge! In: Skeletal muscle repair and regeneration. Schiaffino, S.; Partridge, T. Eds. Dordevet: Springer 2008; pp. 45-64.

[146] Sakuma, K.; Yamaguchi, A. The functional role of calcineurin in hypertrophy, regeneration, and disorders of skeletal muscle. *J Biomed Biotechnol.* 2010, 2010, Article ID 721219, 8 pages.

[147] Renault, V.; Thornell, L. E.; Eriksson, P. O.; Butler-Browne, G.; Mouly, V. Regenerative potential of human skeletal muscle during aging. *Aging Cell*, 2002, 1, 132-139.

[148] Hawke, T. J.; Garry, D. J. Myogenic satellite cells: physiology and molecular biology. *J Appl Physiol*. 2001, 91, 534-551.

[149] Conboy, I. M.; Rando, T. A. The regulation of Notch signaling controls satellite cell activation and cell fate determination in postnatal myogenesis. *Dev Cell*, 2002, 3, 397-409.

[150] Katsanos, C. S.; Kobayashi, H.; Sheffield-Moore, M.; Aarsland, A.; Wolfe, R. R. A high proportion of leucine is required for optimal stimulation of the rate of muscle protein synthesis by essential amino acids in the elderly. *Am J Physiol Endocrinol Metab*. 2006, 291, E381-E387.

[151] Tatsumi, R.; Liu, X.; Pulido, A.; Morales, M.; Sakata, T.; Dial, S.; Hattori, A.; Ikeuchi, Y.; Allen, R. E. Satellite cell activation in stretched skeletal muscle and the role of nitric oxide and hepatocyte growth factor. *Am J Physiol Cell Physiol*. 2006, 290, C1487-C1494.

[152] Anderson, J. E. A role for nitric oxide in muscle repair: NO-mediated satellite cell activation. *Mol Biol Cell*, 2000, 11, 1859-1874.

[153] Tatsumi, R.; Anderson, J. E.; Nevoret, C. J.; Halevy, O.; Allen, R.E. HGF/SF is present in normal adult skeletal muscle and is capable of activating satellite cells. *Dev Biol*. 1998, 194, 114-128.

[154] Miller, K. J.; Thaloor, D.; Matteson, S.; Pavlath, G. K. Hepatocyte growth factor affects satellite cell activation and differentiation in regenerating skeletal muscle. *Am J Physiol Cell Physiol*. 2000, 278, C174-C181.

[155] Pisconti, A.; Brunelli, S.; Di Padova, M.; De Palma, C.; Deponti, D.; Baesso, S.; Sartorelli, V.; Cossu, G.; Clementi, E. Follistatin induction by nitric oxide through cyclic GMP: a tightly regulated signaling pathway that controls myoblast fusion. *J Cell Biol*. 2006, 172, 233-244.

[156] Barani, A. E.; Durieux, A.; Sabido, O.; Freyssenet, D. Age-related changes in the mitotic and metabolic characteristics of muscle-derived cells. *J Appl Physiol*. 2003, 95, 2089-2098.

[157] Richmonds, C. R.; Boonyapisit, K.; Kusner, L. L.; Kaminski, H. J. Nitric oxide synthase in aging rat skeletal muscle. *Mech Aging Dev*. 1999, 109, 177-189.

[158] Betters, J. L.; Lira, V. A.; Soltow, Q. A.; Drenning, J. A.; Criswell, D. S. Supplemental nitric oxide augments satellite cell activity on cultured myofibers from aged mice. *Exp Gerontol*. 2008, 43, 1094-1101.

[159] Capanni, C.; Squarzoni, S.; Petrini, S.; Villanova, M.; Muscari, C.; Maraldi, N. M.; Guarnieri, C.; Caldarera, C. M. Increase of neuronal nitric oxide synthase in rat skeletal muscle during aging. *Biochem Biophys Res Commun*. 1998, 245, 216-219.

[160] Kitzmann, M.; Bonnieu, A.; Duret, C.; Vernus, B.; Barro, M.; Laoudj-Chenivesse, D.; Verdi, J. M.; Carnac, G. Inhibition of Notch signaling induces myotube hypertrophy by recruiting a subpopulation of reserve cells. *J Cell Physiol*. 2006, 208, 538-548.

[161] Buas, M.; Kabak, S.; Kadesch, T. Inhibition of myogenesis by Notch: evidence for multiple pathways. *J Cell Physiol*. 2009, 218, 84-93.

[162] Ono, Y.; Gnocchi, V. F.; Zammit, P. S.; Nagatomi, R. Presenilin-1 acts via Id1 to regulate the function of muscle satellite cells in a gamma-secretase-independent manner. *J Cell Sci*. 2009, 122, 4427-4438.

[163] Vasyutina, E.; Lenhard, D. C.; Wende, H.; Erdmann, B.; Epstein, J. A.; Birchmeier, C. RBP-J (Rbpsuh) is essential to maintain muscle progenitor cells and to generate satellite cells. *Proc Natl Acad Sci USA*, 2007, 104, 4443-4448.

[164] Conboy, I. M.; Conboy, M. J.; Wagners, A. J.; Girma, E. R.; Weissman, I. L.; Rando, T. A. Rejuvenation of aged progenitor cells by exposure to a young systemic environment. *Nature*, 2005, 433, 760-764.

[165] Jones, N. C.; Tyner, K. J.; Nibarger, L.; Stanley, H. M.; Cornelison, D. D.; Fedorov, Y. V.; Olwin, B. B. The p38alpha/beta MAPK functions as a molecular switch to activate the quiescent satellite cell. *J Cell Biol*. 2005, 169, 105-116.

[166] Perdiguero, E.; Ruiz-Bonilla, V.; Gresh, L.; Hui, L.; Ballestar, E.; Sousa-Victor, P.; Baeza-Raja, B.; Jardi, M.; Bosch-Comas, A.; Esteller, M.; Caelles, C.; Serrano, A. L.; Wagner, E. F.; Muñoz-Cánoves, P. Genetic analysis of p38 MAP kinases in myogenesis: fundamental role of p38alpha in abrogating myoblast proliferation. *EMBO J*. 2007, 26, 1245-1256.

[167] Jump, S. S.; Childs, T. E.; Zwetsloot, K. A.; Booth, F. W.; Lees, S. J. Fibroblast growth factor 2-stimulated proliferation is lower in muscle precursor cells from old rats. *Exp Physiol*. 2009, 94, 739-748.

[168] Brack, A. S.; Conboy, I. M.; Conboy, M. J.; Shen, J.; Rando, T. A. A temporal switch from Notch to Wnt signaling in muscle stem cells is necessary for normal adult myogenesis. *Cell Stem Cell*, 2008, 2, 50-59.

[169] Brack, A. S.; Rando, T. A. Age-dependent changes in skeletal muscle regeneration. In: Skeletal muscle repair and rgeneration. Schiaffino, S.; Partridge, T. Eds. Dordrecht: Springer 2008; pp. 359-374.

[170] Brack, A.; Murphy-Seiler, F.; Hanifi, J.; Deka, J.; Eyckerman, S.; Keller, C.; Aguet, M.; Rando, T. A. BCL9 is an essential component of canonical Wnt signaling that mediates the differentiation of myogenic progenitors during muscle regeneration. *Dev Biol*. 2009, 335, 93-105.

[171] Le Grand, F.; Rudnicki, M. A. Skeletal muscle satellite cells and adult myogenesis. *Curr Opin Cell Biol*. 2007, 19, 628-633.

[172] Church, V.; Francis-West, P. Wnt signaling during limb development. *Int J Dev Biol*. 2002, 46, 927-936.

[173] Cossu, G.; Borello, U. Wnt signaling and the activation of myogenesis in mammals. *EMBO J*. 1999, 18, 6867-6872.

[174] Ridgeway, A.; Petropoulos, H.; Wilton, S.; Skerjanc, I. S. Wnt signaling regulates the function of MyoD and myogenin. *J Biol Chem*. 2000, 275, 32398-32405.

[175] van der Velden, J.; Langen, R. C.; Kelders, M. C.; Wouters, E. F.; Janssen-Heininger, Y. M.; Schols, A. M. Inhibition of glycogen synthase kinase-3beta activity is sufficient to stimulate myogenic differentiation. *Am J Physiol Cell Physiol*. 2006, 290, C453-C462.

[176] Brack, A. S.; Conboy, M. J.; Roy, S.; Lee, M.; Kuo, C. J.; Keller, C.; et al. Increased Wnt signaling during aging alters muscle stem cell fate and inceased fibrosis. *Science*, 2007, 317, 807-810.

[177] Chilosi, M.; Poletti, V.; Zamò, A.; Lestani, M.; Montagna, L.; Piccoli, P.; Pedron, S.; Bertaso, M.; Scarpa, A.; Murer, B.; Cancellieri, A.; Maestro, R.; Semenzato, G.; Doglioni, C. Aberrant Wnt/beta-catenin pathway activation in idiopathic pulmonary fibrosis. *Am J Pathol*. 2003, 162, 1495-1502.

[178] Jiang, F.; Parsons, C. J.; Stefanovic, B. Gene expression profile of quiescent and activated rat hepatic stellate cells implicates Wnt signaling pathway in activation. *J Hepathol.* 2006, 45, 401-409.

[179] Kuhnert, F.; Davis, C. R.; Wang, H. T.; Chu, P.; Lee, M.; Yuan, J.; Njusse, R.; Kuo, C. J. Essential requirement for Wnt signaling in proliferation of adult small intestine and colon revealed by adenoviral expression of Dickkopf-1. *Proc Natl Acad Sci USA*, 2004, 101, 266-271.

[180] Kuang, S.; Charge, S. B.; Seale, O.; Huh, M.; Rudnicki, M. A. Distinct roles for Pax7 and Pax3 in adult regenerative myogenesis. *J Cell Biol.* 2006, 172, 103- 113.

[181] Zammit, P. S.; Relaix, F.; Nagata, Y.; Ruiz, A. P.; Collins, C. A.; Partridge, T. A.; Beauchamp, J. R. Pax7 and myogenic progression in skeletal muscle satellite cells. *J Cell Sci.* 2006, 119, 1824-1832.

[182] Olguin, H. C.; Yang, Z.; Tapscott, S. J.; Olwin, B. B. Reciprocal inhibition between Pax7 and muscle regulatory factors modulates myogenic cell fate determination. *J Cell Biol.* 2007, 177, 769-779.

[183] Sabourin, L. A.; Girgis-Gabardo, A.; Seale, P.; Asakura, A.; Rudnicki, M. A. Reduced differentiation potential of primary MyoD-/- myogenic cells derived from adult skeletal muscle. *J Cell Biol.* 1999, 144, 631-643.

[184] Cornelison, D. D.; Olwin, B. B.; Rudnicki, M. A.; Wold, B. J. MyoD(-/-) satellite cells in single-fiber culture are differentiation defective and MRF4 deficient. *Dev Biol.* 2000, 224, 122-137.

[185] Megeney, L. A.; Kablar, B.; Garrett, K.; Anderson, J. E.; Rudnicki, M. A. MyoD is required for myogenic stem cell function in adult skeletal muscle. *Genes Dev.* 1996, 10, 1173-1183.

[186] Gayraud-Morel, B.; Chretien, F.; Flamant, P.; Gomes, D.; Zammit, P. S.; Tajbakhsh, A. A role for the myogenic determination gene Myf5 in adult regenerative myogenesis. *Dev Biol.* 2007, 312, 13-28.

[187] Ustanina, S.; Carvajal, J.; Rigby, P.; Braun, T. The myogenic factor Myf5 supports efficient skeletal muscle regeneration by enabling transient myoblast amplification. *Stem Cells*, 2007, 25, 2006-2016.

[188] Marsh, D. R.; Criswell, D. S.; Carson, J. A.; Booth, F. W. Myogenic regulatory factors during regeneration of skeletal muscle in young, adult, and old rats. *J Appl Physiol.* 1997, 83, 1270-1275.

[189] Lowe, D. A.; Lund, T.; Always, S. E. Hypertrophy-stimulated myogenic regulatory factor mRNA increases are attenuated in fast muscle of aged quails. *Am J Physiol Cell Physiol.* 1998, 275, C155-C162.

[190] Musaró, A.; Cusella De Angelis, M. G.; Germani, A.; Ciccarelli, C.; Molinaro, M; Zani, B. M. Enhanced expression of myogenic regulatory genes in aging skeletal muscle. *Exp Cell Res.* 1995, 221, 241-248.

[191] Always, S. E.; Degens, H.; Lowe, D. A.; Krishnamurthy, G. Increased myogenic repressor Id mRNA and protein levels in hindlimb muscles of aged rats. *Am J Physiol Regul Integr Comp Physiol.* 2002, 282, R411-R422.

[192] Always, S. E.; Degens, H.; Krishnamurthy, G.; Smith, C. A. Potential role for Id myogenic repressors in apoptosis and attenuation of hypertrophy in muscles of aged rats. *Am J Physiol Cell Physiol.* 2002, 283, C66-C76.

[193] Benezra, R.; Davis, R. L.; Lockshon, D.; Turner, D. L.; Weintraub, H. The protein Id: a negative regulator of helix-loop-helix DNA binding proteins. *Cell*, 1990, 61, 49-59.

[194] Delling, U,; Tureckova, J.; Lim, H. W.; De Windt, L. J.; Rotwein, P.; Molkentin, J. D. A calcineurin-NFATc3-dependent pathway regulates skeletal muscle differentiation and slow myosin heavy-chain expression. *Mol Cell Biol*. 2000, 20, 6600–6611.

[195] Friday, B. B.; Horsley, V.; Pavlath, G. K. Calcineurin activity is required for the initiation of skeletal muscle differentiation. *J Cell Biol*. 2000, 149, 657–666.

[196] Friday, B. B.; Mitchell, P. O.; Kegley, K. M.; Pavlath, G. K. Calcineurin initiates skeletal muscle differentiation by activating MEF2 and MyoD. *Differentiation*, 2003, 71, 217–227.

[197] Sakuma, K.; Nishikawa, J.; Nakao, R.; Watanabe, K.; Totsuka, T.; Nakano, H.; Sano, M.; Yasuhara, M. Calcineurin is a potent regulator for skeletal muscle regeneration by association with NFATc1 and GATA-2. *Acta Neuropath(Berl)*. 2003, 105, 271-80.

[198] Sakuma, K.; Nakao, R.; Aoi, W.; Inashima, S.; Fujikawa, T.; Hirata, M.; Sano, M.; Yasuhara, M. Cyclosporin A treatment upregulates Id1 and Smad3 expression and delays skeletal muscle regeneration. *Acta Neuropath(Berl)*. 2005, 110, 269-280.

[199] Abbott, K. L.; Friday, B. B.; Thaloor, D.; Murphy, T. J.; Pavlath, G. K. Activation and cellular localization of the cyclosporine A-sensitive transcription factor NF-AT in skeletal muscle cells. *Mol Biol Cell*, 1998, 9, 2905–2916.

[200] Koulmann, N.; Sanchez, B.; N'Guessan, Chapot, R.; Serrurier, B.; Peinnequin, A.; Ventura-Clapier, R.; Bigard, X. The responsiveness of regenerated soleus muscle to pharmacological calcineurin inhibition. *J Cell Physiol*. 2006, 208, 116–122.

[201] Stupka, N.; Schertzer, J. D.; Bassel-Duby, R.; Olson, E. N.; Lynch, G. S. Calcineurin-Aα activation enhances the structure and function of regenerating muscles after myotoxic injury. *Am J Physiol Regul Integr Comp Physiol*. 2007, 293, R686–R694.

[202] Lara-Pezzi, E.; Winn, N.; Paul, A.; et al. A naturally occurring calcineurin variant inhibits FoxO activity and enhances skeletal muscle regeneration. *J Cell Biol*. 2007, 179, 1205–1218.

[203] Paturi, S.; Gutta, A. K.; Katta, A.; Kakarla, S. K.; Arvapalli, R. K.; Gadde, M. K.; Nalabotu, S. K.; Rice, K. M.; Wu, M.; Blough, E. Effects of aging and gender on muscle mass and regulation of Akt-mTOR-p70S6K related signaling in the F344BN rat model. *Mech Aging Dev*. 2010, 131, 202-209.

[204] Kinnard, R. S.; Mylabathula, D. B.; Uddemarri, S.; Rice, K. M.; Wright, G. L.; Blough, E. R. Regulation of p70^{S6K}, GSK-3b, and calcineurin in rat striated muscle during aging. *Biogerontology*, 2005, 6, 173-184.

[205] Carlson, B. M.; Faulkner, J. A. Muscle transplantation between young and old rats: age of host determines recovery. *Am J Physiol*. 1989, 256, C1262-C1266.

[206] Zacks, S. I.; Sheff, M. F. Age-related impeded regeneration of mouse minced anterior tibial muscle. *Muscle Nerve*, 1982, 5, 152-161.

[207] Grounds, M. D. Age-associated changes in the response of skeletal muscle cells exercise and regeneration. *Ann N Y Acad Sci*. 1998, 854, 78-91. 1998,

[208] Sadeh, M. Effects of aging on skeletal muscle regeneration. *J Neurol Sci*. 1988, 67-74. 1988

[209] Warren, L. A.; Rossi, D. J. Stem cells and aging in the hematopoietic system. *Mech Aging Dev*. 2009, 130, 46-53.

[210] Edström, E.; Ulfhake, B. Sarcopenia is not due to lack of regenerative drive in senescent skeletal muscle. *Aging Cell*, 2005, 4, 65-77.

[211] Larsson, L. Motor units: remodeling in aged animals. *J Gerontol A Biol Sci Med Sci.* 1995, 50, 91-95.

[212] Deschenes, M. R.; Roby, M. A.; Eason, M. K.; Harris, M. B. Remodeling of the neuromuscular junction precedes sarcopenia-related alterations in myofibers. *Exp Gerontol.* 2010, 45, 389-393.

[213] Luff, A. R. Age-associated changes in the innervation of muscle fibers and changes in the mechanical properties of motor units. *Ann NY Acad Sci.* 1998, 854, 92-101.

[214] Kawabuchi, M.; Tan, H.; Wang, S. Age affects reciprocal cellular interactions in neuromuscular synapses following peripheral nerve injury. *Aging Res Rev.* 2011, 10, 43-53.

[215] Fahim, M. A.; Robbins, N. Ultrastructural studies of young and old mouse neuromuscular junctions. *J Neurocytol.* 1982, 11, 641-656.

[216] Rosenheimer, J. L. Factors affecting denervation-like changes at the neuromuscular junction during aging. *Int Dev Neurosci.* 1990, 8, 643-654.

[217] Courtney, J.; Steinbach, J. H. Age changes in neuromuscular junction morphology and acetylcholine receptor distribution on rat skeletal muscle fibers. *J Physiol.* 1981, 320, 435-447.

[218] Smith, D. O.; Chapman, M. R. Acetylcholine receptor binding properties at the rat neuromuscular junction during aging. *J Neurochem.* 1987, 48, 1834-1841.

[219] Kawabuchi, M.; Zhou, C.; Nakamura, K.; Hirata, K. Morphological features of collateral innervation and supernumerary innervation in the skeletal muscles of presenile rats. *Ann Anat.* 1995, 177, 251-265.

[220] Ulfhake, B.; Bergman, E.; Edström, E.; Fundin, B. T.; Johnson, H.; Kullberg, S.; Ming, Y. Regulation of neurotrophin signaling in aging sensory and motoneurons: dissipation of target support? *Mol Neurobiol.* 2000, 21, 109-135.

[221] Johnson, H.; Mossberg, K.; Arvidsson, U.; Piehl, F.; Hökfelt, T.; Ulfhake, B. Increase in alpha-CGRP and GAP-43 in aged motoneurons: a study of peptides, growth factors, and ChAT mRNA in the lumbar spinal cord of senescent rats with symptoms of hindlimb incapacities. *J Comp Neurol.* 1995, 359, 69-89.

[222] Funakoshi, H.; Frisén, J.; Barbany, G.; Timmusk, T.; Zachrisson, O.; Verge, V. M.; Persson, H. Differential expression of mRNAs for neurotrophins and their receptors after axotomy of the sciatic nerve. *J Cell Biol.* 1993, 123, 455-465.

[223] Küst, B. M.; Copray, J. C.; Brouwer, N.; Troost, D.; Boddeke, H. W. Elevated levels of neurotrophins in human biceps brachii tisuue of amyotrophic lateral sclerosis. *Exp Neurol.* 2002, 177, 419-427.

[224] Sakuma, K.; Watanabe, K.; Sano, M.; Uramoto, I.; Nakano, H.; Li, Y. J.; Kaneda, S.; Sorimachi, Y.; Yoshimoto, K.; Yasuhara, M.; Totsuka, T. A possible role for BDNF, NT-4 and TrkB in the spinal cord and muscle of rat subjected to mechanical overload, bupivacaine injection and axotomy. *Brain Res.* 2001, 907, 1-19.

[225] Funakoshi, H.; Belluardo, N.; Arenas, E.; Yamamoto, Y.; Casabona, A.; Persson, H.; Ibáñez, C. F. Muscle-derived neurotrophin-4 as an activity-dependent trophic signal for adult motoneuron. *Science*, 1995, 268, 1495-1499.

[226] Sakuma, K.; Watanabe, K.; Totsuka, T.; Sano, M.; Nakano, H.; Nakao, R.; Nishikawa, J.; Sorimachi, Y.; Yoshimoto, K.; Yasuhara, M. The reciprocal change of neurotrophin-

4 and glial cell line-derived neurotrophic factor protein in the muscles, spinal cord and cerebellum of the dy mouse. *Acta Neuropath(Berl)*. 2002, 104, 482-492.

[227] Nguyen, Q. T.; Parsadanian, A. S.; Snider, W. D.; Lichtman, J. W. Hyperinnervation of neuromuscular junctions caused by GDNF overexpression in muscle. *Science*, 1998, 279, 1725-1729.

[228] Dechiara, T. M.; Vejsada, R.; Poueymirou, W. T.; Acheson, A.; Suri, C.; Conover, J. C.; Friedman, B.; McClain, J.; Pan, L.; Stahl, N.; Ip, N. Y.; Yancopoulos, G. D. Mice lacking the CNTF receptor, unlike mice lacking CNTF, exhibit profound motor neuron deficits at birth. *Cell*, 1995, 83, 313-322.

[229] Edström E.; Altun, M.; Bergman, E.; Johnson, H.; Kullberg, S.; Ramírez-León, V.; Ulfhake, B. Factors contributing to neuromuscular impairment and sarcopenia during aging. *Physiol Behav*. 2007, 92, 129-135.

[230] Horber, F. F.; Gruber, B.; Thomi, F.; Jensen, E. X.; Jaeger, P. Effect of sex and age on bone mass, body composition and fuel metabolism in humans. *Nutrition*, 1997, 13, 524-534.

[231] Ito, H.; Ohshima, A.; Ohto, N.; Ogasawara, M.; Tsuzuki, M.; Takao, K.; Hijii, C.; Tanaka, H.; Nishioka, K. Relation between body composition and age in health Japanese subjects. *Eur J Clin Nutr*. 2001, 55, 462-470.

[232] Lara-Castro, C.; Weinsier, R. L.; Hunter, G. R.; Desmond, R. Visceral adipose tissue in women: longitudinal study of effects of fat gain, time, and race. *Obes Res*. 2002, 10, 868-874.

[233] Goodpaster, B. H.; Carlson, C. L.; Visser, M.; Kelley, D. E.; Scherzinger, A.; Harris, T. B.; Stamm, E.; Newman, A. B. Attenuation of skeletal muscle and strength in the elderly: The Health ABC study. *J Appl Physiol*. 2001, 90, 2157-2165.

[234] Song, M. -Y.; Ruts, E.; Kim, J.; Janumala, I.; Heymsfield, S.; Gallagher, D. Sarcopenia and increased adipose tissue infiltration of muscle in elderly African American women. *Am J Clin Nutr*. 2004, 79, 874-880.

[235] Zamboni, M.; Mazzali, G.; Fantin, F.; Rossi, A.; Di Francesco, V. Sarcopenic obesity: a new category of obesity in the elderly. *Nutr Metab Cardiovasc Dis*. 2008, 1, 388-395.

[236] Morley, J. E.; Baumgartner, R. N.; Roubenoff, R.; Mayer, J.; Nair, K. S. Sarcopenia. *J Lab Clin Med*. 2001, 137, 231-243.

[237] Roth, R. J.; Le, A. M.; Zhang, L.; Kahn, M.; Samuel, V. T.; Shulman, G. I.; Bennett, A. M. MAPK phosphatase-1 facilitates the loss of oxidative myofibers associated with obesity in mice. *J Clin Invest*. 2009, 119, 3817-3829.

[238] Anderson, E. J.; Lustig, M. E.; Boyle, K. E.; Woodlief, T. L.; Kane, D. A.; Lin, C. T.; Price, J. W. 3rd.; Kang, L.; Rabinovitch, P. S.; Szeto, H. H.; Houmard, J. A.; Cortright, R. N.; Wasserman, D. H.; Neufer, P. D. Mitochondrial H2O2 emission and cellular redox state link excess fat intake to insulin resistance in both rodents and humans. *J Clin Invest*. 2009, 10, 1-9.

[239] Bonnard, C.; Durand, A.; Peyrol, S.; Chanseaume, F.; Chauvin, M. A.; Morio, B.; Vidal, H.; Rieusset, J. Mitochondrial dysfunction results from oxidative stress in the skeletal muscle of diet-induced insulin-resistant mice. *J Clin Invest*. 2008, 118, 789-800.

[240] Arany, Z.; Foo, S. Y.; Ma, Y.; Ruas, J. L.; Bommi-Reddy, A.; Girnun, G.; Cooper, M.; Laznik, D.; Chinsomboon, J.; Rangwala, S. M.; Baek, K. H.; Rosenzweig, A.;

Spiegelman, B. M. HIF-independent regulation of VEGF and angiogenesis by the transcriptional co-activator PGC-1 alpha. *Nature*, 2008, 451, 1008-1012.

[241] Handschin, C.; Chin, S.; Li, P.; Liu, F.; Maratos-Flier, E.; Lebrasseur, N. K.; Yan, Z.; Spiegelman, B. M. Skeletal muscle fiber-type switching, exercise intolerance, and myopathy in PGC-1alpha muscle-specific knockout animals. *J Biol Chem*. 2007, 282, 30014-30021.

[242] Crunkhorn, S.; Dearie, F.; Mantzoros, C.; Gami, H.; da Silva, W. S.; Espinoza, D.; Faucette, R.; Barry, K.; Bianco, A. C.; Patti, M. E. Peroxisome proliferator activator receptor gamma co-activator-1 expression is reduced in obesity: potential pathogenic role of saturated fatty acids and p38 mitogen-activated protein kinase activation. *J Biol Chem*. 2007, 282, 15439-15450.

[243] Wenz, T.; Rossi, S. G.; Rotundo, R. L.; Spiegelman, B. M.; Moraes, C. T. Increased muscle PGC-1alpha expression protects from sarcopenia and metabolic disease during aging. *Proc Natl Acad Sci USA*, 2009, 106, 20405-20410.

[244] Boutros, T.; Chevet, E.; Metrakos, P. Mitogen-activated protein (MAP) kinase/ MAP kinase phosphatase regulation: roles in cell growth, death, and cancer. *Pharmacol Rev*. 2008, 60, 261-310.

[245] Williamson, D.; Gallagher, P.; Harber, M.; Hollon, C.; Trappe, S. Mitogen-activated protein kinase (MAPK) pathway activation: effects of age and acute exercise on human skeletal muscle. *J Physiol*. 2003, 547, 977-987.

[246] Roth, R. J.; Bennett, A. M. MAP kinase phosphotase-1- a new player at the nexus between sarcopenia and metabolic disease. *Aging*, 2010, 2, 170-176.

[247] Chung, H. Y.; Cesari, M.; Anton, S.; Marzetti, E.; Giovannini, S.; Seo, A. Y.; Carter, C.; Yu, B. P.; Leeuwenburgh, C. Molecular inflammation: Underpinnings of aging and age-related diseases. *Aging Res Rev*. 2009, 8, 18-30.

[248] Reid, M. B.; Li, Y. P. Tumor necrosis factor-α and muscle wasting: A cellular perspective. *Respir Res*. 2001, 2, 269-272.

[249] Brinkley, T. E.; Leng, X.; Miller, M. E.; Kitzman, D. W.; Pahor, M.; Berry, M. J.; Marsh, A. P.; Kritchevsky, S. B.; Nicklas, B. J. Chronic inflammation is associated with low physical function in older adults across multiple comorbidities. *J Gerontol A Biol Sci Med Sci*. 2009, 64, 455-461.

[250] Carter, C. S.; Hofer, T.; Seo, A. T. Leeuwenburgh, C. Molecular mechanisms if life- and health span extension: Role of calorie restriction and exercise intervention. *Appl Physiol Nutr Metab*. 2007, 32, 954-966.

[251] Roubenoff R. Catabolism of aging: Is it an inflammatory process? *Curr Opin Clin Nutr Metab Care*, 2003, 6, 295-299.

[252] Dreyer, H. C.; Fujita, S.; Cadenas, J. G.; Chinkes, D. L.; Volpi, E.; Rasmussen, B. B. Resistance exercise increases AMPK activity and reduces 4E-BP1 phosphory-lation and protein synthesis in human skeletal muscle. *J Physiol*. 2006, 576, 613-624.

[253] Miller, B. F.; Olesen, J. L.; Hansen, M.; Dossing, S.; Crameri, R. M.; Welling, R. J.; Langberg, H.; Flyvbjerg, A.; Kjaer, M.; Babraj, J. A.; Smith, K.; Rennie, M. J. Coordinated collagen and muscle protein synthesis in human patella tendon and quadriceps muscle after exercise. *J Physiol*. 2005, 567, 1021-1033.

[254] Esmarck, B.; Andersen, J. L.; Olsen, S.; Richter, E. A.; Mizuno, M.; Kjaer, M. Timing of postexercise protein intake is important for muscle hypertrophy with resistance training in elderly humans. *J Physiol*. 2001, 535, 301-311.

[255] Fiatarone, M. A.; O'Neil, E. F.; Ryan, N. D.; Clements, K. M.; Solares, G. R.; Nelson, M. E.; Roberts, S. B.; Kehayias, J. J.; Lipsitz, L. A.; Evans, W. J. Exercise training and nutritional supplementation for physical frailty in very elderly people. *N Engl J Med.* 1994, 330, 1769-1775.

[256] Hunter, G. R.; McCarthy, J. P.; Bamman, M. M. Effects of resistance training in older adults. *Sports Med.* 2004, 34: 329-348.

[257] Lenk, K.; Schuler, G.; Adams, V. Skeletal muscle wasting in cachexia and sarcopenia: molecular pathophysiology and impact of exercise training. *J Cachexia Sarcopenia Muscle*, 2010, 1, 9-21.

[258] Singh, M. A.; Ding, W.; Manfredi, T. J.; Solares, G. S.; O'Neil, E. F.; Clements, K. M.; Ryan, N. D.; Kehayias, J. J.; Fielding, R. A.; Evans, W. J. Insulin-like growth factor I in skeletal muscle after weight-lifting exercise in frail elders. *Am J Physiol Endocrinol Metab.* 1999, 277, E135-E143.

[259] Charette, S. L.; McEvoy, L.; Pyka, G.; Snow-Harter, C.; Guido, D.; Wiswell, R. A.; Marcus, R. Muscle hypertrophy response to resistance training in older women. *J Appl Physiol.* 1991, 70, 1912-1916.

[260] Frontera, W. R.; Meredith, C. N.; O'Reilly, K. P.; Knuttgen, H. G.; Evans, W. J. Strength conditioning in older men: skeletal muscle hypertrophy and improved function. *J Appl Physiol.* 1988, 64, 1038-1044.

[261] Campbell, W. W.; Joseph, L. J.; Davey, S. L.; Cyr-Campbell, D.; Anderson, R. A.; Evans, W. J. Effects of resistance training and chromium picolinate on body composition and skeletal muscle in older men. *J Appl Physiol.* 1999, 86, 29-39.

[262] McCartney, N.; Hicks, A. L.; Martin, J.; Webber, C. E. A longituidinal trial of weight training in the elderly: continued improvements in year 2. *J Gerontol A Biol Sci Med Sci.* 1996, 51, B425-B433.

[263] Raue, U.; Slivka, D.; Jemiolo, B.; Hollon, C.; Trappe, S. Myogenic gene expression at rest and after a bout of resistance exercise in young (18-30 yr) and old (80-89 yr) women. *J Appl Physiol.* 2006, 101, 53-59.

[264] Pedersen, B. K. The anti-inflammatory effect of exercise: its role in diabetes and cardiovascular disease control. *Essay Biochem.* 2006, 42, 105-117.

[265] Parise, G.; Brose, A. N.; Tarnopolsky, M. A. Resistance exercise training decreases oxidative damage to DNA and increases cytochrome oxidase activity in older adults. *Exp Gerontol.* 2005, 40, 173-180.

[266] Hasten, D. L.; Pak-Loduca, J.; Obert, K. A.; Yarasheski, K. E. Resistance exercise acutely increases MHC and mixed muscle protein synthesis rates in 78-84 and 23-32 yr olds. *Am J Physiol Endocrinol Metab.* 2000, 278, E620-E626.

[267] Yarasheski, K. E.; Pak-Loduca, J.; Hasten, D. L.; Obert, K. A.; Brown, M. B.; Sinacore, D. R. Resistance exercise training increases mixed muscle protein synthesis rate in frail women and men > 76 yr old. *Am J Physiol Endocrinol Metab.* 1999, 277, E118-E125.

[268] Mayhew, D. L.; Kim, J. S.; Cross, J. M.; Bamman, M. M. Translational signaling responses preceding resistance training-mediated myofibers hypertrophy in young and old humans. *J Appl Physiol.* 2009, 107, 1655-1662.

[269] Kosek, D. J.; Kim, J. S.; Petrella, J. K.; Cross, J. M.; Bamman, M. M. Efficacy of 3 day/wk resistance training on myofiber hypertrophy and myogenic mechanisms in young vs. older adults. *J Appl Physiol.* 2006, 101, 531-544.

[270] Bradley, L.; Yaworsky, P. J.; Walsh, F. S. Myostatin as a therapeutic target for musculoskeletal disease. *Cell Mol Life Sci.* 2008, 65, 2119-2124.

[271] Bogdanovich, S.; Krag, T. O.; Barton, E. R.; Morris, L. D.; Whittemore, L. A.; Ahima, R. S.; Khurana, T. S. Functional improvement of dystrophic muscle by myostatin blockade. *Nature,* 2002, 420, 418-421.

[272] Holzbaur, E. L.; Howland, D. S.; Weber, N.; Wallace, K.; She, Y.; Kwak, S.; Tchistiakova, L. A.; Murphy, E.; Hinson, J.; Karim, R.; Tan, X. Y.; Kelley, P.; McGill, K. C.; Williams G.; Hobbs, C.; Doherty, P.; Zaleska, M. M.; Pangalos, M. N.; Walsh, F. S. Myostatin inhibition slows muscle atrophy in rodent models of amyotrophic lateral sclerosis. *Neurobiol Dis.* 2006, 23, 697-707.

[273] Wagner, K. R.; Fleckenstein, J. L.; Amato, A. A.; Barohn, R. J.; Bushby, K.; Escolar, D. M.; Flanigan, K. M.; Pestronk, A.; Tawil, R.; Wolfe, G. I.; Eagle, M.; Florence, J. M.; King, W. M.; Pandya, S.; Straub, V.; Juneau, P.; Meyers, K.; Csimma, C.; Araujo, T.; Allen, R.; Parsons, S. A.; Wozney, J. M.; Lavallie, E. R.; Mendell, J. R. A phase I/II trial of MYO-029 in adult subjects with muscular dystrophy. *Ann Neurol.* 2008, 63, 561-571.

[274] Zhou, X.; Wang, J. L.; Lu, J.; Song, Y.; Kwak, K. S.; Jiao, Q.; Rosenfeld, R.; Chen, Q.; Boone, T.; Simonet, W. S.; Lacey, D. L.; Goldberg, A. L.; Han, H. Q. Reversal of cancer cachexia and muscle wasting by ActRIIB antagonism leads to prolonged survival. *Cell,* 2010, 142, 531-543.

[275] Petraglia, F.; Florio, P.; Luisi, S.; Gallo, R.; Gadducci, A.; Viganó, P.; Di Blasio, A. M.; Genazzani, A. R.; Vale, W. Expression and secretion of inhibin and activin A in normal and neoplastic uterine tissues. High levels of serum activin A in women with endometrial and cervical carcinoma. *J Clin Endocrinol Metab.* 1998, 83, 1194-1200.

[276] Thomas, T. Z.; Wang, H.; Niclasen, P.; Obryan, M. K.; Evans, L. W.; Groome, N. P.; Pedersen, J.; Risbridger, G. P. Expression and localization of activin subunits and follistatins in tissues from men with high grade prostate cancer. *J Clin Endocrinol Metab.* 1997, 82, 3851-3858.

[277] Amthor, H.; Nicholas, G.; McKinnell, I.; Kemp, C. F.; Sharma, R.; Kambadur, R.; Patel, K. Follistatin complexes myostatin and antagonizes myostatin-mediated inhibition of myogenesis. *Dev Biol.* 2004, 270, 19-30.

[278] Gilson, H; Schakman, O.; Kalista, S.; Lause, P.; Tsuchida, K.; Thissen, J. P. Follistatin induces muscle hypertrophy through satellite cell proliferation and inhibiton of both myostatin and activin. *Am J Physiol Endocrinol Metab.* 2009, 297, E157-E164.

[279] Minetti, G. C.; Colussi, C.; Adami, R.; Serral, C.; Mozzetta, C.; Parente, V.; Fortuni, S.; Straino, S.; Sampaolesi, M.; Di Padova, M.; Illi, B.; Gallinari, P.; Steinkühler, C.; Capogrossi, M. C.; Sartorelli, V.; Bottinelli, R.; Gaetano, C.; Puri, P. L. Functional and morphological recovery of dystrophic muscles in mice treated with deacetylase inhibitors. *Nature Med.* 2006, 12, 1147-1150.

[280] Bonetto, A.; Penna, F.; Minero, V. G.; Reffo, P.; Bonelli, G.; Baccino, F. M.; Costelli, P. Deacetylase inhibitors modulate the myostatin/follistatin axis without improving cachexia in tumor-bearing mice. *Curr Cancer Drug Targets,* 2009, 9, 608-616.

[281] Siriett, V.; Platt, L.; Salerno, M. S.; Ling, N.; Kambadur, R.; Sharma, M. Prolonged absence of myostatin reduces sarcopenia. *J Cell Physiol.* 2006, 209, 866-873.

[282] Lebrasseur, N. K.; Schelhorn, T. M.; Bernardo, B. L.; Cosgrove, P. G.; Loria, P. M.; Brown, T. A. Myostatin inhibition enhances the effects on performance and metabolic outcomes in aged mice. *J Gerontol A Biol Sci Med Sci*. 2009, 64, 940- 948.

[283] Murphy, K. T.; Koopman, R.; Naim, T.; Léger, B.; Trieu, J.; Ibebunjo, C.; Lynch, G. S. Antibody-directed myostatin inhibition in 21-mo-old mice reveals novel roles for myostatin signaling in skeletal muscle structure and function. *FASEB J*. 2010, 24, 4433-4442.

[284] Welle, S.; Burgess, K.; Meht, S. Stimulation of skeletal muscle myofibrillar protein synthesis, p70 S6 kinase phosphorylation, and ribosomal protein S6 phosphory-lation by inhibition of myostatin in mature mice. *Am J Physiol Endocrinol Metab*. 2009, 296, E567-E572.

[285] Amirouche, A.; Durieux, A. C.; Banzet, S.; Koulmann, N.; Bonnefoy, R.; Mouret, C.; Bigard, X.; Peinnequin, A.; Freyssent, D. Down-regulation of Akt/mammalian target of rapamycin signaling pathway in response to myostatin overexpression in skeletal muscle. *Endocrinology,* 2009, 150, 286-294.

[286] McFarlane, C.; Plummer, E.; Thomas, M.; Hennebry, A.; Ashby, M.; Ling, N.; Smith, H.; Sharma, M.; Kambadur, R. Myostatin induces cachexia by activating the ubiquitin proteolytic system through an NF-κB-independent, FoxO1- dependent mechanism. *J Cell Physiol*. 2006, 209, 501-514.

[287] Chelh, I.; Meunier, B.; Picard, B.; Reecy, M. J.; Chevalier, C.; Hocquette, J. -F.; Cassar-Malek, I. Molecular profiles of quadriceps muscle in myostatin-null mice reveals PI3K and apoptotic pathways as myostatin targets. *BMC Genomics*, 2009, 10, 196. 13 pages.

[288] Wagner, K. R.; McPherron, A. C.; Winik, N.; Lee, S. J. Loss of myostatin attenuates severity of muscular dystrophy in mdx mice. *Ann Neurol*. 2002, 52, 832-836.

[289] Li, Z. F.; Shelton, G. D.; Engvall, E. Elimination of myostatin does not combat muscular dystrophy in dy mice but increases postnatal lethality. *Am J Pathol*. 2005, 166, 491-497.

[290] Bogdanovich, S.; Perkins, K. J.; Krang, T. O.; Whittemore, L. A.; Khurana, T. S. Myostatin propeptide-mediated amelioration of dystrophic pathophysiology. *FASEB J*. 2005, 19, 543-549.

[291] Morine, K. J.; Bish, L. T.; Selsby, J. T.; Gazzara, J. A.; Pendrak, K.; Sleeper, M. M.; Barton, E. R.; Lee, S. J.; Sweeney, H. L. Activin IIB receptor blockade attenuates dystrophic pathology in a mouse model of Duchenne muscular dystrophy. *Muscle Nerve*, 2010, 42, 722-730.

[292] Fulgoni, V. L. III. Current protein intake in America. Analysis of the National Health and Nutrition Examination Survey, 2003-2004. *Am J Clin Nutr*. 2008, 87, 1554S-1557S.

[293] Kerstetter, J. E.; O'Brien, K. O.; Insogna, K. L. Low protein intake. The impact on calcium and bone homeostasis in humans. *J Nutr*. 2003, 133, 855S-861S.

[294] Lord, C.; Chaput, J. P. Aubertin-Leheure, M.; Labonté, M.; Dionne, I. J. Dietary animal protein intake: Association with muscle mass index in older women. *J Nutr Health Aging*, 2007, 11, 383-387.

[295] Timmerman, K. L.; Volpi, E. Amino acid metabolism and regulatory effects in aging. *Curr Opin Clin Nutr Metab Care*, 2008, 11, 45-49.

[296] Henderson, G. C.; Irving, B. A.; Nair, K. S. Potential application of essential amino acid supplementation to treat sarcopenia in elderly people. *J Clin Endocrinol Metab.* 2009, 94, 1524-1526.

[297] Paddon-Jones, D.; Rasmussen, B. B. Dietary protein recommendations and the prevention of sarcopenia. *Curr Opin Nutr Metab Care,* 2009, 12, 86-90.

[298] Norton, L. E.; Layman, D. K. Leucine regulates translation initiation of protein synthesis in skeletal muscle after exercise. *J Nutr.* 2006, 136, 533S-537S.

[299] Nair, K. S.; Woolf, P. D.; Welle, S. L.; Matthews, D. E. Leucine, glucose, and energy metabolism after 3 days of fasting in health human subjects. *Am J Clin Nutr.* 1987, 46, 557-562.

[300] Tipton, K. D.; Ferrando, A. A.; Phillips, S. M.; Doyle, D. Jr.; Wolfe, R. R. Post-exercise net protein synthesis in human muscle from orally administered amino acids. *Am J Physiol Endocr Metab.* 1999, 276, E628-E634.

[301] Dreyer, H. C.; Drummond, M. J.; Pennings, B.; Fujita, S.; Glynn, E. L.; Chinkes, D. L. Leucine-enriched essential amino acid and carbohydrate ingestion following resistance exercise enhances mTOR signaling and protein synthesis in human muscle. *Am J Physiol Endocr Metab.* 2008, 294, E392-E400.

[302] Combaret, L.; Dardevet, D.; Rieu, I.; Pouch, M. -N.; Béchet, D.; Taillandier, D.; Grizard, J.; Attaix, D. A. leucine-supplemented diet restores the defective post-prandial inhibition of proteasome-dependent proteolysis in aged rat skeletal muscle. *J Physiol.* 2005, 569, 489-499.

[303] Drummond, M. J.; Dreyer, H. C.; Pennings, B.; Fry, C. S.; Dhanani, S.; Dillon, E. L.; Sheffield-Moore, M.; Volpi, E.; Rasmussen, B. B. Skeletal muscle protein anabolic response to resistance exercise and essential amino acids is delayed with aging. *J Appl Physiol.* 2008, 104, 1452-1461.

[304] Walrand, S.; Short, K. R.; Bigelow, M. L.; Sweatt, A. J.; Hutson, S. M., Nair, K. S. Functional impact of high protein intake on healthy elderly people. *Am J Physiol Endocrinol Metab.* 2008, 295, E921-E928.

[305] Godard, M. P.; Williamson, D. L.; Trappe, S. W. Oral amino-acid provision does not affect muscle strength or size gains in older men. *Med Sci Sports Exerc.* 2002, 34, 1126-1131.

[306] Welle, S.; Thornton, C. A. High-protein meals do not enhance myofibrillar synthesis after resistance exercise in 62- to 75-yr-old men and women. *Am J Physiol Endocrinol Metab.* 1998, 274, E677-E683.

[307] Chen, Z.; Bassford, T.; Green, S. B.; Cauley, J. A.; Jackson, R. D.; LaCroix, A. Z.; Leboff, M.; Stefanick, M. L.; Margolis, K. L. Postmenopausal hormone therapy and body composition-a substudy of the estrogen plus progestin trial of the Women's Health Initiative. *Am J Clin Nutr.* 2005, 82, 651-656.

[308] Solerte, S. B.; Gazzaruso, C.; Bonacasa, R.; Rondanelli, M., Zamboni, M.; Basso, C.; Locatelli, E.; Schifino, N.; Giustina, A.; Fioravanti, M. Nutritional supplements with oral amino acid mixtures increases whole-body lean mass and insulin sensitivity in elderly subjects with sarcopenia. *Am J Cardiol.* 2008, 101, 69E-77E.

[309] Dillon, E. L.; Sheffield-Moore, M.; Paddon-Jones, D.; Gilkison, C.; Sanford, A. P.; Casperson, S. L.; Jiang, J.; Chinkes, D. L.; Urban, R. J. Amino acid supplementation increases lean body mass, basal muscle protein synthesis, and insulin-like growth factor-I expression in older women. *J Clin Endocrinol Metab.* 2009, 94, 1630-1637.

[310] Verhoeven, S.; Vanschoonbeek, K.; Verdijk, L. B.; Koopman, R.; Wodzig, W. K.; Dendale, P.; van Loon, L. J. Long-term leucine supplementation does not increase muscle mass or strength in healthy elderly men. *Am J Clin Nutr.* 2009, 89, 1468-1475.

[311] Nicastro, H.; Artioli, G. G.; dos Santos Costa, A.; Solis, M. Y.; da Luz, C. R.; Blachier, F.; Lancha, A. H. Jr. An overview of the therapeutic effects of leucine supplementation on skeletal muscle under atrophic conditions. *Amino Acids*, 2010, in press, DOI10.1007

[312] Harper, A. E.; Miller, R. H.; Block, K. P. Branched-chain amino acid metabolism. *Annu Rev Nutr.* 1984, 4, 409-454.

[313] Rieu, I.; Barage, M.; Sornet, C.; Giraudet, C.; Pujos, E.; Grizard, J.; Mosoni, L.; Dardevet, D. Leucine supplementation improves muscle protein synthesis in elderly mem independently of hyperaminoacidaemia. *J Physiol.* 2006, 575, 305-315.

[314] Dieli-Conwright, C. M.; Spektor, T. M.; Rice, J. C.; Schroeder, E. T. Oestradiol and SERM treatments influence oestrogen receptor coregulator gene expression in human skeletal muscle cells. *Acta Physiol(Oxf).* 2009, 197, 187-196.

[315] Pöllänen, E.; Ronkainen, P. H.; Suominen, H.; Takala, T.; Koskinen, S.; Puolakka, J.; Sipilä, S.; Kovanen, V. Muscular transcriptome in postmenopausal women with or without hormone replacement. *Rejuvenation Res.* 2007, 10, 485-500.

[316] Tenover, J. L. Testosterone and the aging male. *J Androl.* 1997, 18, 103-106.

[317] Bhasin, S.; Storer, T. W.; Javenbakht, M.; Berman, N.; Yarasheski, K. E.; Phillips, J.; Dike, M.; Sinha-Hikim, I.; Shen, R.; Hays, R. D.; Beall, G. Testosterone replacement and resistance exercise in HIV-infected men with weight loss and low testosterone levels. *JAMA,* 2005, 283, 763-770.

[318] Bhasin, S.; Calof, O.; Storer, T. W.; Lee, M.; Mazer, N.; Jasuja, R.; Montori, V. M.; Gao, W.; Dalton, J. T. Drug insight: testosterone and selective androgen receptor modulators as anabolic therapies for physical dysfunction in chronic illness and aging. *Nature Clin Pract Endocrinol Metab.* 2006, 2, 146-159.

[319] Wang, C.; Swedloff, R. S.; Iranmanesh, A.; Dobs, A.; Snyder, P. J.; Cunningham, G.; Matsumoto, A. M.; Weber, T.; Berman, N. Transdermal testosterone gel improves sexual function, mood, muscle strength, and body composition parameters in hypogonadal men. Testosterone gel study group. *J Clin Endocrinol Metab.* 2000, 85, 2839-2853.

[320] Morley, J. E.; Perry, H. M. 3rd. Androgen deficiency in aging men: role of testosterone replacement therapy. *J Lab Clin Med.* 2000, 135, 370-378.

[321] Ferrando, A. A.; Sheffield-Moore, M.; Yeckel, C. W.; Gilkison, C.; Jiang, J.; Achacosa, A.; Lieberman, S. A.; Tipton, K.; Wolfe, R. R.; Urban, R. J. Testosterone administration to older men improves muscle function: molecular and physiological mechanism. *Am J Physiol Endocrinol Metab.* 2002, 282, E601-E607.

[322] Sinha-Hikim, I.; Cornford, M.; Gaytan, H.; Lee, M. L.; Bhasin, S. Effects of testosterone supplementation on skeletal muscle fiber hypertrophy and satellite cells in community-dwelling older men. *J Clin Endocrinol Metab.* 2006, 91, 3024-3033.

[323] Singh, R.; Artaza, J. N.; Taylor, W. E.; Braga, M.; Yuan, X.; Gonzalez-Cadavid, N. F.; Bhasin, S. Testosterone inhibits adipogenic differentiation in 3T3-L1 cells: nuclear translocation of androgen receptor complex with beta-catenin and T-cell factor 4 may bypass canonical Wnt signaling to down-regulate adipogenic transcription factors. *Endocrinology*, 2006, 147, 141-154.

[324] Ferrando, A. A.; Sheffield-Moore, M., Yeckel, C. W.; Gilkison, C.; Jiang, J.; Achacosa, A.; Lieberman, S. A.; Tipton, K.; Wolfe, R. R.; Urban, R. J. Testosterone administration to older men improves muscle function: molecular and physiological mechanisms. *Am J Physiol Endocrinol Metab.* 2002, 282, E601-E607.

[325] Singh, R.; Bhasin, S.; Braga, M.; Artaza, J. N.; Pervin, S.; Taylor, W. E.; Krishnan, V.; Sinha, S. K.; Rajavashisth, T. B.; Jasuja, R. Regulation of myogenic differentiation by androgens: cross talk between androgen receptor/beta-catenin and follistatin/transforming gowth factor-beta signaling pathways. *Endocrinology*, 2009, 150, 1259-1268.

[326] Mendler, L.; Baka, Z.; Kovács-Simon, A.; Dux, L. Androgens negatively regulate myostatin expression in an androgen-dependent skeletal muscle. *Biochem Biphys Res Commun.* 2007, 361, 237-242.

[327] Zhao, W.; Pan, J.; Zhao, Z.; Wu, Y.; Bauman, W. A.; Cardozo, C. P. Testosterone protects against dexamethasone-induced muscle atrophy, protein degradation and MAFbx upregulation. *J Steroid Biochem Mol Biol.* 2008, 110, 125-129.

[328] Bhasin, S.; Woodhouse, L.; Casaburi, R.; Singh, A. B.; Mac, R. P.; Lee, M.; Yarasheski, K. E.; Sinha-Hikim, I.; Dzekov, C.; Dzekov, J.; Magliano, L.; Storer, T. W. Older men are as responsive as young men to the anabolic effects of graded doses of testosterone on the skeletal muscle. *J Clin Endocrinol Metab.* 2005, 90, 678-688.

[329] Florini, J. R. Ewton, D. Z.; Coolican, S. A. Growth hormone and the insulin-like growth factor system in myogenesis. *Endocr Rev.* 1996, 17, 481-517.

[330] Ho, K. Y.; Veldhuis, J. D.; Johnson, M. L.; Furlanetto, R.; Evans, W, S.; Alberti, K. G.; Thorner, M. O. Fasting enhances growth hormone secretion and amplifies the complex rhythm of growth hormone secretion in man. *J Clin Invest.* 1988, 81, 968-975.

[331] Giustina, A.; Mazziotti, G.; Canalis, E. Growth hormone, insulin-like growth factors, and the skeleton. *Endocr Rev.* 2008, 29, 535-559.

[332] Moran, A.; Jacobs, D. R. J.; Steinberger, J.; Cohen, P.; Hong, C.; Prineas, R.; Sinaiko, A. R. Association between the insulin resistance of puberty and the insulin-like growth factor-I/growth hormone axis. *J Clin Endocrinol Metab.* 2002, 87, 4817-4820.

[333] Ryall, J. G.; Schertzer, J. D.; Lynch, G. S. Cellular and molecular mechanisms underlying age-related skeletal muscle wasting and weakness. *Biogerontology*, 2008, 9, 213-228.

[334] Veldhuis, J. D.; Iranmanesh, A. Physiological regulation of the human growth hormone (GH)-insulin-like growth factor type I (IGF-I) axis: predominant impact of age, obesity, gonadal function, and sleep. *Sleep*, 1996, 19, S221-S224.

[335] Le Roith, D.; Bondy, C.; Yakar, S.; Liu, J. L.; Butler, A. The somatomedin hypothesis: 2001. *Endocr Rev.* 2001, 22, 53-74.

[336] Frost, R. A.; Nystrom, G. J.; Lang, C. H. Regulation of IGF-I mRNA and signal transducers and activators of transcription-3 and -5 (Stat-3 and -5) by GH in C2C12 myoblasts. *Endocrinology*, 2002, 143, 492-503.

[337] Sotiropoulos, A.; Ohanna, M.; Kedzia, C.; Menon, R. K.; Kopchick, J. J.; Kelly, P. A.; Pende, M. Growth hormone promotes skeletal muscle cell fusion independent of insulin-like growth factor 1 up-regulation. *Proc Natl Acad Sci USA*, 2006, 103, 7315-7320.

[338] Giovannini, S.; Marzetti, E.; Borst, S. E.; Leeuwenburgh, C. Modulation of GH/ IGF-I axis: Potential strategies to counteract sarcopenia in older adults. *Mech Aging Dev.* 2008, 129, 593-601.

[339] Andersen, N. B.; Andreassen, T. T.; Orskov, H.; Oxlund, H. Growth hormone and mild exercise in combination increases markedly muscle mass and tetanic tension in old rats. *Eur J Endocrinol.* 2000, 143, 409-418.

[340] Blackman, M. R.; Sorkin, J. D.; Munzer, T.; Bellantoni, M. F.; Busby-Whitehead, J.; Stevens, T. E.; Jayme, J.; O'Connor, K. G.; Christmas, C.; Tobin, J. D.; Stewart, K. J.; Cottrell, E.; St Clair, C.; Pabst, K. M.; Harman, S. M. Growth hormone and sex steroid administration in healthy aged women and men: a randomized controlled trial. *JAMA,* 2002, 288, 2282-2292.

[341] Giannoulis, M. G.; Sonksen, P. H.; Umpleby, M.; Breen, L.; Pentecost, C.; Whyte, M.; McMillan, C. V.; Bradley, C.; Martin, F. C. The effects of growth hormone and/or testosterone in health elderly men: a randomized controlled trial. *J Clin Endoctinol Metab.* 2006, 91, 477-484.

[342] Philippou, A.; Maridaki, M.; Halapas, A.; Koutsilieris, M. The role of the insulin-like growth factor 1 (IGF-1) in skeletal muscle physiology. *In Vivo,* 2007, 21, 45-54.

[343] Clemmons, D. R. Role of IGF-I in skeletal muscle mass maintenance. *Trends Endocrinol Metab.* 2009, 20, 349-356.

[344] Coleman, M. E.; DeMayo, F.; Yin, K. C.; Lee, H. M.; Geske, R.; Montgomery, C.; Schwartz, R. J. Myogenic vector expression of insulin-like growth factor I stimulates muscle cell differentiation and myofibers hypertrophy in transgenic mice. *J Biol Chem.* 1995, 270, 12109-12116.

[345] Musaró, A.; McCullagh, K.; Paul, A.; Houghton, L.; Dobrowolny, G.; Molinaro, M.; Barton, E. R.; Sweeney, H. L.; Rosenthal, N. Localized Igf-I transgene expression sustains hypertrophy and regeneration in senescent skeletal muscle. *Nat Genet.* 2001, 27, 195-200.

[346] Barton, E. R.; Morris, L.; Musaró A.; Rosenthal, N.; Sweeney, H. L. Muscle-specific expression of insulin-like growth factor I counters muscle decline in mdx mice. *J Cell Biol.* 2002, 157, 137-148.

[347] Musaró A.; Giacinti, C.; Nardis, C.; Borsellino, G.; Dobrowolny, G.; Pelosi, L.; Cairns, L.; Ottolenghi, S.; Bernardi, G.; Cossu, G.; Battistini, L.; Molinaro, M.; Rosenthal, N. Muscle restricted expression of mIGF-I enhances the recruitment of stem cells during muscle regeneration. *Proc Natl Acad Sci USA,* 2004, 101, 1206-1210.

[348] Pelosi, L.; Giacinti, C.; Nardis, C.; Borsellino, G.; Rizzuto, E.; Nicoletti, C.; Wannenes, F.; Battistini, L.; Rosenthal, N.; Molinaro, M.; Musaró, A. Local expression of IGF-I accelerates muscle regeneration by rapidly modulating inflammatory cytokines and chemokines. *FASEB J.* 2007, 21, 1393-1402.

[349] Barton-Davis, E. R.; Shoturma, D. I.; Musaró, A.; Rosenthal, N.; Sweeney, H. L. Viral mediated expression of insulin-like growth factor I blocks the aging- related loss of skeletal muscle function. *Proc Natl Acad Sci USA,* 1998, 95, 15603-15607.

[350] Schertzer, J. D.; Ryall, J. G.; Lynch, G. S. Systemic administration of IGF-I enhances oxidative status and reduces contraction-induced injury in skeletal muscles of mdx dystrophic mice. *Am J Physiol Endocrinol Metab.* 2006, 291, E499-E505.

[351] Gregorevic, P.; Plant, D. R.; Lynch, G. S. Administration of insulin-like growth factor-I improves fatigue resistance of skeletal muscles from dystrophic mdx mice. *Muscle Nerve*, 2004, 30, 295-304.

[352] Butterfield, G. E.; Thompson, J.; Rennie, M. J.; Marcus, R.; Hintz, R. L.; Hoffman, A. R. Effect of rhGH and rhIGF-I treatment on protein utilization in elderly women. *Am J Physiol Endocrinol Metab*. 1997, 272, E94-E99.

[353] Dardevet, D.; Sornet, C.; Attaix, D.; Baracos, V. E.; Grizard, J. Insulin-like growth factor-1 and insulin resistance in skeletal muscles of adult and old rats. *Endocrinology*, 1994, 134, 1475-1484.

[354] Bhasin, S.; Calof, O. M.; Storer, T. W.; Lee, M. L.; Mazer, N. A.; Jasuja, R.; Montori, V. M.; Gao, W.; Dalton, J. T. Drug insights: Testosterone and selective androgen receptor modulators as anabolic therapies for chronic illness and aging. *Nat Clin Pract Endocrinol Metab*. 2006, 2, 146-159.

[355] Narayanan R, Mohler, M. L.; Bohl, C. E.; Miller, D. D.; Dalton, J. T. Selective androgen receptor modulators in pre-clinical and clinical development. *Nuclear Receptor Signaling*, 2008, 6, e010.

[356] Lynch, G. S. Emerging drugs for sarcopenia: age-related muscle wasting. *Expert Opin Emerg Drugs*, 2004, 9, 345-361.

[357] Gao, W.; Reiser, P. J.; Coss, C. C.; Phelps, M. A.; Kearbey, J. D.; Miller, D. D.; Dalton, J. T. Selective androgen receptor modulator treatment improves muscle strength and body composition and prevents bone loss in orchidectomized rats. *Endocrinology*, 2005, 146, 4887-4897.

[358] Jones, A.; Hwang, D. J.; Narayanan, R.; Miller, D. D.; Dalton, J. T. Effects of a novel selective androgen receptor modulator on dexamethasone-induced and hypogonadism-induced muscle atrophy. *Endocrinology*, 2010, 151, 3706-3719.

[359] Cadilla, R.; Turnbull, P. Selective androgen receptor modulators in drug discovery: medicinal chemistry and therapeutic potential. *Curr Top Med Chem*. 2006, 6, 245-270.

[360] Lynch G. S. Update on emerging drugs for sarcopenia-age-related muscle wasting. *Expert Opin Emerging Drugs*, 2008, 13, 655-673.

[361] Lynch, G. S.; Ryall, J. G. Role of β-adrenoceptor signaling in skeletal muscle: implications for muscle wasting and disease. *Physiol Rev*. 2008, 88, 729-767.

[362] Ryall, J. G.; Lynch, G. S. The potential and the pitfalls of beta-adrenoceptor agonists for the management of skeletal muscle wasting. *Pharm Ther*. 2008, 120, 219-232.

[363] Ryall, J. G.; Sillence, M. N.; Lynch, G. S. Systemic administration of β2-adrenoceptor agonists, formoterol and salmeterol, elicit skeletal muscle hypertrophy in rats at micromolar doses. *Br J Pharmacol*. 2006, 147, 587-595.

[364] Dirks, A. J.; Leeuwenburgh, C. Aging and lifelong calorie restriction result in adaptations of skeletal muscle apoptosis repressor, apoptosis-inducing factor, X-linked inhibitor of apoptosis, caspase-3, and caspase-12. *Free Radic Biol Med*. 2004, 36, 27-39.

[365] López-Lluch, G.; Hunt, N.; Jones, B.; Zhu, M.; Jamieson, H.; Hilmer, S.; Cascajo, M. V.; Allard, J.; Ingram, D. K.; Navas, P.; de Cabo, R. Calorie restriction induces mitochondrial biogenesis and bioenergetic efficiency. *Proc Natl Acad Sci USA*, 2006, 103, 1768-1773.

[366] Payne, A. M.; Dodds, S. L.; Leeuwenburgh, C. Life-long calorie restriction in Fischer 344 rats attenuates age-related loss in skeletal muscle-specific force and reduces extracellular space. *J Appl Physiol.* 2003, 95, 2554-2562.

[367] Phillips, T.; Leeuwenburgh, C. Muscle fiber specific apoptosis and TNF-alpha signaling in sarcopenia are attenuated by life-long calorie restriction. *FASEB J.* 2005, 95, 668-670.

[368] Bevilacqua, L.; Ramsey, J. J.; Hagopian, K.; Weindruch, R.; Harper, M. E. Effects of short- and medium-term calorie restriction on muscle mitochondrial proton leak and reactive oxygen species production. *Am J Physiol Endocrinol Metab.* 2004, 286, E852-E861.

[369] Aspnes, L. E.; Lee, C. M.; Weindruch, R.; Chung, S. S.; Roecker, E. B.; Aiken, J. M. Caloric restriction reduces fiber loss and mitochondrial abnormalities in aged rat muscle. *FASEB J.* 1997, 11, 573-581.

[370] Baker, D. J.; Betik, A. C.; Krause, D. J.; Hepple, R. T. No decline in skeletal muscle oxidative capacity with aging in long-term calorically restricted rats: Effects are independent of mitochondrial DNA integrity. *J Gerontol A Biol Sci Med Sci.* 2006, 61, 675-684.

[371] Marzetti, E.; Carter, C. S.; Wohlgemuth, S. E.; Lees, H. A.; Giovannini, S.; Anderson, B.; Quinn, L. S.; Leeuwenburgh, C. Changes in IL-15 expression and death-receptor apoptotic signaling in rat gastrocnemius muscle with aging and life-long calorie restriction. *Mech Aging Dev.* 2009, 130, 272-280.

[372] Marzetti, E.; Hwang, J. C.; Lees, H. A.; Wohlgemuth, S. E.; Dupont-Versteegden, E. E.; Carter, C. S.; Bernabei, R.; Leeuwenburgh, C. Mitochondrial death effectors: relevance to sarcopenia and disuse muscle atrophy. *Biochim Biophys Acta Gen Subj.* 2010, 1800, 235-244.

[373] Colman, R. J.; Beasley, T. M.; Allison, D. B.; Weindruch, R. Attenuation of sarcopenia by dietary restriction in rhesus monkeys. *J Gerontol Biol Sci.* 2008, 63A, 556-559.

[374] McKiernan, S. H.; Colman, R. J.; Lopez, M.; Beasley, T. M.; Aiken, J. M.; Anderson, R. M.; Weindruch, R. Caloric restrictin delays aging-induced cellular phenotypes in rhesus monkey skeletal muscle. *Exp Gerontol.* 2011, 46, 23-29.

[375] Dirks, A. J.; Leeuwenburgh, C. Caloric restriction in humans: Potential pitfalls and health concerns. *Mech Aging Dev.* 2006, 127, 1-7.

[376] Zhang, Y.; Eriksson, M.; Dallner, G.; Appelkvist, E. L. Analysis of ubiquinone and tocophenol levels in normal and hyperlipidemic human plasma. *Lipids*, 1998, 33, 811-815.

[377] Vasilaki, A.; McArdle, F.; Iwanejko, L. M.; McArdle, A. Adaptive responses of mouse skeletal muscle to contractile activity: the effect of age. *Mech Aging Dev.* 2006, 127, 830-839.

[378] Shigenaga, M. K.; Hagen, T. M.; Ames, B. N. Oxidative damage and mitochondrial decay in aging. *Proc Natl Acad Sci USA*, 1994, 91, 10771-10778.

[379] Cadenas, E.; Davies, K. J. Mitochondrial free radical generation, oxidative stress, and aging. *Free Radic Biol Med.* 2000, 29, 222-230.

[380] Mosoni, L.; Breuille, D.; Buffiere, C.; Obled, C.; Mirand, P. P. Age-related changes in glutathione availability and skeletal muscle carbonyl content in healthy rats. *Exp Gerontol.* 2004, 39, 203-210.

[381] Fulle, S.; Protasi, F.; Di Tano, G.; Pietrangelo, T.; Beltramin, A.; Boncompagni, S.; Vecchiet, L.; Fanò, G. The contribution of reactive oxygen species to sarcopenia and muscle aging. *Exp Gerontol.* 2004, 39, 17-24.

[382] Ishii N. Role of oxidative stress from mitochondria on aging and cancer. *Cornea*, 2008, 26, S3-S9.

[383] Semba, R. D.; Blaum, C.; Guralnik, J. M.; Moncrief, D. T.; Ricks, M. O.; Fried, L. P. Carotenoid and vitamin E status are associated with indicators of sarcopenia among older women living in the community. *Aging Clin Exp Res.* 2003, 15, 482-487.

[384] Cesari, M.; Pahor, M.; Bartali, B.; Cherubini, A.; Penninx, B. W.; Williams, G. R.; Atkinson, H.; Martin, A.; Guralnik, J. M.; Ferrucci, L. Antioxidants and physical performance in elderly persons: the Invecchiare in Chianti (InCHIANTI) study. *Am J Clin Nutr.* 2004, 79, 289-294.

[385] Boots, A. W.; Haenen, G. R.; Bast, A. Health effects of quercetin: from antioxidant to nutraceutical. *Eur J Pharmacol.* 2008, 44, 126-131.

[386] Harikumar, K. B.; Aggarwal, B. B. Resveratrol: a multi-targeted agent for age-associated chronic diseases. *Cell Cycle,* 2008, 7, 1020-1035.

[387] Rebrin, I.; Zicker, S.; Wedekind, K. J.; Paetau-Robinson, I.; Packer, L.; Sohal, R. S. Effect of antioxidant-enriched diets on glutathione redox status in tissue homogenates and mitochondria of the senescence accerelated mouse. *Free Radic Biol Med.* 2005, 39, 549-557.

[388] Semba, R. D.; Varadhan, R.; Bartali, B, Ferrucci, L.; Ricks, M. O.; Blaum, C.; Fried, L. P. Low serum carotenoids and development of severe walking disability among older women living in the community: the women's health and aging study I. *Age Ageing,* 2007, 36, 62-67.

[389] Roman, B.; Carta, L.; Martínez-González, M. A.; Serra-Majem, L. Effectiveness of the Mediterranean diet in the elderly. *Clin Interv Aging,* 2008, 3, 97-109.

[390] Andriollo-Sanchez, M.; Hininger-Favier, I.; Meunier, N.; Venneria, E.; O'Connor, J. M.; Maiani, G.; Polito, A.; Bord, S.; Ferry, M.; Courday, C.; Roussel, A. M.; No antioxidant beneficial effect of zinc supplementation on oxidative stress markers and antioxidant defenses in middle-aged and elderly subjects: the Zenith study. *J Am Coll Nutr.* 2008, 27, 463-469.

[391] Bonetto, A.; Penna, F.; Muscaritoli, M.; Minero, V. G.; Fanelli, F. R.; Baccino, F. M.; Costelli, P. Are antioxidants useful for treating skeletal muscle atrophy? *Free Radic Biol Med.* 2009, 47, 906-916.

[392] Leoncini, S.; Rossi, V.; Signorini, C.; Tanganelli, I.; Comporti, M.; Ciccoli, L. Oxidative stress, erythrocyte aging and plasma non-protein-bound iron in diabetic patients. *Free Radic Res.* 2008, 42, 716-724.

[393] Mastrocola, R.; Reffo, P.; Penna, F.; Tomasinelli, C. E.; Boccuzzi, G.; Baccino, F. M.; Aragano, M.; Costelli, P. Muscle wasting in diabetic and tumor-bearing rats: role of oxidative stress. *Free Radic Biol Med.* 2008, 44, 584-593.

[394] Jacobs, I. Dietary creatine monohydrate supplementation. *Can J Appl Physiol.* 1999, 24, 503-514.

[395] Klivenyi, P.; Ferrante, R. J.; Matthews, R. T.; Bogdanov, M. B.; Klein, A. M.; Andreassen, O. A.; Mueller, G.; Wermer, M.; Kaddurah-Daouk, R.; Beal, M. F. Neuroprotective effects of creatine in a transgenic animal model of amyotrophic lateral sclerosis. *Nat Med.* 1999, 5, 347-350.

[396] Passaquin, A. C.; Renard, M.; Kay, L.; Challet, C.; Mokhtarian, A.; Wallimann, T.; Ruegg, U. T. Creatine supplementation reduces skeletal muscle degeneration and enhances mitochondrial function in mdx mice. *Neuromuscul Disord.* 2002, 12, 174-182.

[397] Guthmiller, P.; Van Pilsum, J. F.; Boen, J. R.; McGuire, D. M. Cloning and sequencing of rat kidney L-arginine:glycine amidinotransferase. Studies on the mechanism of regulation by growth hormone and creatine. *J Biol Chem.* 1994, 269, 17556-17560.

[398] Heinänen, K.; Näntö-Salonen, K.; Komu, M.; Erkintalo, M.; Heinonen, O. J.; Pulkki, K.; Nikoskelainen, E.; Sipilä, I.; Simell, O. Muscle creatine phosphate in gyrate atrophy of the choroids and retina with hyperornithinaemia-clues to pathogenesis. *Eur J Clin Invest.* 1999, 29, 426-431.

[399] Levine, S.; Tikunov, B.; Henson, D.; LaManca, J.; Sweeney, H. L. Creatine depletion elicits structural, biochemical, and physiological adaptations in rat costal diaphragm. *Am J Physiol.* 1996, 271, C1480-C1486.

[400] Tarnopolsky, M.; Martin, J. Creatine monohydrate increases strength in patients with neuromuscular disease. *Neurology,* 1999, 52, 854-857.

[401] Bermon, S.; Venembre, P.; Sachet, C.; Valour, S.; Dolisi, C. Effects of creatine monohydrate ingestion in sedentary and weight-trained older adults. *Acta Physiol Scand.* 1998, 164, 147-155.

[402] Rawson, E. S.; Wehnert, M. L.; Clarkson, P. M. Effects of 30 days of creatine ingestion in older men. *Eur J Appl Physiol Occup Physiol.* 1999, 80, 139-144.

[403] Brose, A.; Parise, G.; Tarnopolsky, M. A. Creatine supplementation enhances isometric strength and body composition improvements following strength exercise training in older adults. *J Gerontol A Biol Sci Med Sci.* 2003, 58, 11-19.

[404] Chrusch, M. J.; Chilibeck, P.; Chad, K. E.; Davison, K. S.; Burke, D. G. Creatine supplementation combined with resistance training in older men. *Med Sci Sports Exerc.* 2001, 33, 2111-2117.

[405] Gotshalk, L. A.; Volek, J. S.; Staron, R. S.; Denegar, C. R.; Hagerman, F. C.; Kraemer, W. J. Creatine supplementation improves muscular performance in older men. *Med Sci Sports Exerc.* 2002, 34, 537-543.

[406] Tarnopolsky, M. A.; Parise, G.; Yardley, N. J.; Ballantyne, C. S.; Olatinji, S.; Phillips, S. M. Creatine-dextrose and protein-dextrose induce similar strength gains during training. *Med Sci Sports Exerc.* 2001, 33, 2044-2052.

[407] Chilibeck, P. D.; Chrusch, M. J.; Chad, K. E.; Shawn Davison, K.; Burke, D. G. Creatine monohydrate and resistance training increase bone mineral content and density in older men. *J Nutr Health Aging,* 2005, 9, 352-353.

[408] Hespel, P.; OP't Eijnde, B.; Van Leemputte, M.; Ursø, B.; Greenhalf, P. L.; Labarque, V.; Dymarkowski, S.; Van Hecke, P.; Richter, E. A. Oral creatine supplementation facilitates the rehabilitation of disuse atrophy and alters the expression of muscle myogenic factors in humans. *J Physiol.* 2001, 536, 625-633.

[409] Olsen, S.; Aagaard, P.; Kadi, F.; Tufekovic, G.; Verney, J.; Olsen, J. L.; Suetta, C.; Kjaer, M. Creatine supplementation augments the increase in satellite cell and myonuclei number in human skeletal muscle induced by strength training. *J Physiol.* 2006, 573, 525-534.

[410] Parise, G.; Mihic, S.; MacLennan, D.; Yarasheski, K. E.; Tarnopolsky, M. A. Effects of acute creatine monohydrate supplementation on leucine kinetics and mixed-muscle protein synthesis. *J Appl Physiol.* 2001. 91, 1041-1047.

[411] Rawson, E. S.; Clarkson, P. M. Acute creatine supplementation in older men. *Int J Sports Med*. 2000, 21, 71-75.

[412] Fabre, J. E.; Rivard, A.; Magner, M.; Silver, M.; Isner, J. M. Tissue inhibition of angiotensin-converting enzyme activity stimulates angiogenesis in vivo. *Circulation*, 1999, 99, 3043-3049.

[413] de Cavanagh, E. M. V.; Piotrkowski, B.; Basso, N.; Stella, I.; Inserra, F.; Ferder, L.; Fraga, C. G. Enalapril and losartan attenuate mitochondrial dysfunction in aged rats. *FASEB J*. 2003, 17, 1096-1098.

[414] Maggio, M.; Ceda, G. P.; Lauretani, F.; Pahor, M.; Bandinelli, S.; Najjar, S. S.; Ling, S. M.; Basaria, S.; Ruggiero, C.; Valenti, G.; Ferrucci, L. Relation of angiotensin converting enzyme inhibitor treatment to insulin-like growth factor-1 serum levels in subjects. 65 years of age (the InCHIANTI study). *Am J Cardiol*. 2006, 97, 1525-1529.

[415] Onder, G.; Penninx, B. W. J. H.; Balkrishnan, R.; Fried, L. P.; Chaves, P. H.; Williamson, J.; Carter, C.; Di Bari, M.; Guralnik, J. M.; Pahor, M. Relation between use of angiotensin-converting enzyme inhibitors and muscle strength and physical function in older women: an observational study. *Lancet*, 2002, 359, 926-930.

[416] Dössegger, L.; Aldor, E.; Baird, M. G.; Braun, S.; Cleland, J. G.; Donaldson, R.; Jansen, L. J.; Joy, M. D.; Marin-Neto, J. A.; Nogueira, E. Influence of angiotensin converting enzyme-inhibition on exercise performance and clinical symptoms in chronic heart-failure - a multi-center, double-blind, placebo-controlled trial. *Eur Heart J*. 1993, 14, 18-23.

[417] Schellenbaum, G. D.; Smith, N. L.; Heckbert, S. R.; Lumley, T.; Rea, T. D.; Furberg, C. D.; Lyles, M. F.; Psaty, B. M. Weight loss, muscle strength, and angiotensin-converting enzyme inhibitors in older adults with congestive heart failure or hypertension. *J Am Geriatr Soc*. 2005, 53, 1996-2000.

[418] Sumukadas, D.; Witham, M. D.; Struthers, A. D.; Mcmurdo, M. E. T. Effect of perindopril on physical function in elderly people with functional impairment: a randomized controlled trial. *Can Med Assoc J*. 2007, 177, 867-874.

[419] Leonetti, G.; Mazzola, C.; Pasotti, C.; Angioni, L.; Vaccarella, A.; Capra, A.; Botta, G.; Zanchetti, A. Treatment of hypertension in the elderly - effects on blood-pressure, heart-rate, and physical-fitness. *Am J Med*. 1991, 90, S12-S13.

[420] Bunout, D.; Barrera, G.; de la Maza, M. P.; Leiva, L.; Backhouse, C.; Hirsch, S. Effects of enalapril or nifedipine on muscle strength or functional capacity in elderly subjects. a double blind trial. *J Renin Angiotensin Aldosterone Syst*. 2009, 10, 77-84.

[421] Lee, D. H.; Goldberg, A. L. Proteasome inhibitors: valuable new tools for cell biologists. *Trends Cell Biol*. 1998, 8, 397-403.

[422] Adams, J.; Palombella, V. J.; Sausville, E. A. Proteasome inhibitors: a novel class of potent and effective anti-tumor agents. *Cancer Res*. 1999, 59, 2615-2622.

[423] Orlowski, R. Z. Proteasome inhibitors in cancer therapy. *Methods Mol Biol*. 1997, 100, 197-203.

[424] Davis, N. B.; Taber, D. A.; Ansari, R. H.; Ryan, C. W.; George, C.; Vokes, E. E.; Vogelzang, N. J.; Stadler, W. M. Phase II trial of PS-341 in patients with renal cell cancer: a university of Chicago phase II consortium study. *J Clin Oncol*. 2004, 22, 115-119.

[425] Faderi, S.; Rai, K.; Gribben, J.; Byrd, J. C.; Flinn, I. W.; O'Brien, S.; Sheng, S.; Esseltine, D. L.; Keating, M. J. Phase II study of single-agent bortezomib for the

treatment of patients with fludarabine-refractory B-cell chronic lymphocytic leukemia. *Cancer*, 2006, 107, 916-924.

[426] Armand, J. P.; Burnett, A. K.; Drach, J.; Harousseau, J. L.; Lowenberg, B.; San Miguel, J. The emerging role of targeted therapy for hematologic malignancies: update on bortezomib and tipifarnib. *Oncologist,* 2007, 12, 281-290.

[427] Nencioni, A.; Grunebach, F.; Patrone, F.; Ballestrero, A.; Brossart, P. Proteasome inhibitors: anti-tumor effects and beyond. *Leukemia,* 2007, 21, 30-36.

[428] Bonuccelli, G.; Sotgia, F.; Capozza, E.; Gazzerro, E.; Minetti, C.; Lisanti, M. P. Localized treatment with a novel FDA-approved proteasome inhibitor blocks the degradation of dystrophin and dystrophin-associated proteins in mdx mice. *Cell Cycle*, 2007, 6, 1242-1248.

[429] Gazzerro, E.; Assereto, S.; Bonetto, A.; Sotgia, F.; Scarfi, S.; Pistorio, A.; Bonuccelli, G.; Cilli, M.; Bruno, C.; Zara, F.; Lisanti, M. P.; Minetti, C. Therapeutic potential of proteasome inhibition in Duchenne and Becker muscular dystrophies. *Am J Pathol.* 2010, 176, 1863-1877.

[430] Beehler, B. C.; Sleph, P. G.; Benmassaoud, L.; Grover, G. J. Reduction of skeletal muscle atrophy by a proteasome inhibitor in a rat model of denervation. *Exp Biol Med.* 2006, 231, 335-341.

[431] Bernardi, P. Mitochondrial transport of cations: channels, exchangers, and permeability transition. *Physiol Rev.* 1999, 79, 1127-1155.

[432] Zamzami, N.; Kroemer, G. The mitochondrion in apoptosis: how Pandora's box opens. *Nat Rev Mol Cell Biol.* 2001, 2, 67-71.

[433] Nakagawa, T.; Shimizu, S.; Watanabe, T.; Yamaguchi, O.; Otsu, K.; Yamagata, H.; Inohara, H.; Kubo, T.; Tsujimoto, Y. Cyclophilin D-dependent mitochondrial permeability transition regulates some necrotic but not apoptotic cell death. *Nature*, 2005, 434, 652-658.

[434] Schinzel, A. C.; Takeuchi, O.; Huang, Z.; Fisher, J. K.; Zhou, Z.; Rubens, J.; Hetz, C.; Danial, N. N.; Moskowitz, M. A.; Korsmeyer, S. J. Cyclophilin D is a component of mitochondrial permeability transition and mediates neuronal cell death after focal cerebral ischemia. *Proc Natl Acad Sci USA*, 2005, 102, 12005- 12010.

[435] Millay, D. P.; Sargent, M. A.; Osinska, H.; Baines, C. P.; Barton, E. R.; Vuagniaux G.; Sweeney, H. L.; Robbins, J.; Molkentin, J. D. Genetic and pharmacologic inhibiton of mitochondrial-dependent necrosis attenuates muscular dystrophy. *Nat Med.* 2008, 14, 442-447.

[436] Wissing, E. R.; Millay, D. P.; Vuagniaux, G.; Molkentin, J. D. Debio-025 is more effective than prednisone in reducing muscular pathology in mdx mice. *Neuromusc Disord.* 2010, 20, 753-760.

[437] Tiepolo, T.; Angelin, A.; Palma, E.; Sabatelli, P.; Merlini, L.; Nicolosi, L.; Finetti, F.; Braghetta, P.; Vuagniaux, G.; Dumont, J. -M.; Baldari, C. T.; Bernardi, P. The cyclophilin inhibitor Debio 025 normalizes mitochondrial function, muscle apoptosis and ultrastructural defects in Col6a1-/- myopathic mice. *Brit J Pharmacol.* 2009, 157, 1045-1052.

[438] Angelini, C. The role of corticosteroids in muscular dystrophy: a critical appraisal. *Muscle Nerve*, 2007, 36, 424-435.

[439] Balaban, B.; Matthews, D. J.; Clayton, G. H.; Carry, T. Corticosteroid treatment and functional improvement in Duchenne muscular dystrophy: long-term effect. *Am J Phys Med Rehabil.* 2005, 84, 843-850.

[440] Morissette, M. R.; Cook, S. A.; Buranasombati, C.; Rosenberg, M. A.; Rosenzweig, A. Myostatin inhibits IGF-I-induced myotube hypertrophy through Akt. *Am J Physiol Cell Physiol.* 2009, 297, C1124-C1132.

[441] Trendelenburg, A. U.; Meyer, A.; Rohner, D.; Boyle, J.; Hatakeyama, S.; Glass, D. J. Myostatin reduces Akt/TORC1/p70S6K signaling, inhibiting myoblast differenti-ation and myotube size. *Am J Physiol Cell Physiol.* 2009, 296, C1258-C1270.

In: Cell Aging
Editors: Jack W. Perloft and Alexander H. Wong

ISBN: 978-1-61324-369-5
2012 Nova Science Publishers, Inc.

Chapter 3

MOLECULAR AND CELLULAR MECHANISMS LINKING AGING TO COGNITIVE DECLINE AND ALZHEIMER'S DISEASE

*Mariana Pehar[1] and Luigi Puglielli[1,2]**

[1]Department of Medicine, University of Wisconsin-Madison, Madison, WI, U. S.
[2]Geriatric Research Education Clinical Center, VA Medical Center, Madison, WI, U. S.

ABSTRACT

Because of the increased lifespan of our population, problems linked to age-associated disabilities are becoming more important and are absorbing a growing fraction of the costs associated with health care. In particular, the disability for cognitive loss and dementia combined is currently the second most expensive among medical conditions affecting the aging population and is expected to become the most expensive in the next decade. In fact, aging is accompanied by cognitive loss and reduced high-order brain functions in a large segment of the population. Aging is also the most important risk factor for sporadic Alzheimer's disease, which is one of the most common forms of dementia in the world. Despite this important role in disease pathogenesis and morbidity, we have yet to fully dissect the aging process of the brain at a molecular level. However, the powerful tools of molecular biology applied to model organisms have recently revealed that the natural process of aging is driven by specific signaling pathways, which in turn regulate molecular events that are essential for the general physiology of the cell. Careful dissection of these events is now providing new information as well as possible ways of intervention for age-associated diseases. In this chapter we seek to integrate recent advances in understanding basic molecular mechanisms that underlie aging with findings that are shedding light on the pathogenesis and development of age-associated cognitive loss and Alzheimer's disease neuropathology.

* Mailing address:
Luigi Puglielli, MD, PhD
VAH-GRECC 11G, 2500 Overlook Terrace, Madison, WI, U. S. 53705
Phone: (608) 2561901 (ext. 11569); Fax: (608) 2807291
E-mail: lp1@medicine.wisc.edu

INTRODUCTION

"The increase in the number of old people in the world will be one of the most profound forces affecting health and social services in the next century"

Dr. Hiroshi Nakajima, Director General,
World Health Organization, May 1996

"The very successes of the past few decades will generate a transition to societies with rapidly increasing numbers of middle aged and elderly. A new set of diseases will rise to prominence"

Dr. Gro Harlem Brundtland, Director General,
World Health Organization, May 1999

During the last century we have witnessed a significant increase in average lifespan of the world population, from about 30-45 years in the early 20th century to the current ~65 years (or more than 80 years in the richest countries). As a result, problems linked to age-associated disabilities are becoming more important and are absorbing a larger fraction of our wealth. The aging of the population is also forcing medicine to face a growing number of patients with chronic debilitating, rather than acute, diseases. Finally, the major causes of death and disabilities are shifting from diseases that affect childhood to diseases that strike during adulthood. The above comments from two different Director General of the World Health Organization clearly emphasize the need to understand diseases that have remained at the margin of medicine and society for a long time. In most cases these are also diseases that can be halted, but not cured and for which prevention will remain our best hope.

"Aging has to be fought against; its faults need vigilant resistance. We must combat them as we should fight a disease ... However, the mind and spirit need even more attention than the body, for old age easily extinguishes them, like lamps when they are not given oil ... Aging will only be respected if it fights for itself, maintains its own rights, avoids dependence, and asserts control over its own sphere as long as life lasts."

Marcus Tullius Cicero, 106-43 BC

The above sentence is from a letter written by Marcus Tullius Cicero, a philosopher, statesman, constitutionalist, Quaestor and Consul of Rome. He was regarded by many as one of the greatest orators of his time, famous for his memory and his ability to talk without notes for long period of times. While approaching the last season of his life he was concerned with the progressive loss of memory that he was experiencing and for the loss of independence that this implied. Like Cicero, a large segment of the aging population experiences a progressive decline of working memory, short-term recall and spatial memory. For many, this decline leads to dependence and reduced activities; for some, it ultimately leads to dementia. The disability for cognitive loss and dementia combined is currently the second most expensive among medical conditions affecting the aging population and is expected to become the most expensive in the next decade (Ferri et al., 2005). Particularly alarming is the rise of late-onset Alzheimer's disease (AD), one of the most common forms of dementia in the world

(Puglielli, 2008; Yankner et al., 2008). Because of the ongoing increase in life expectancy, late-onset AD is projected to affect as many as 115 million individuals worldwide by the year 2050 (World Alzheimer Report, 2009 Executive Summary). The worldwide cost of AD in 2005, when 29 million individuals were affected by the disease, was US$315 billion (Wimo et al., 2007). This number is expected to rise in the upcoming years.

Although considerable progress has been made with other diseases and causes of disability associated with aging, such as cancer, cardiovascular disease and diabetes, we have not been able to fully understand the pathogenesis of AD and its connection with aging or significantly halt the progression of the dementia associated with AD pathology. This is largely due to the complexity of the brain and its diseases, as well as the lack of appropriate models of aging and neurodegenerative diseases. However, the last two decades have surprisingly revealed that the lifespan of an organism is regulated by specific signaling pathways. These findings support the view that aging is not a collection of different and unrelated processes affecting multiple organs at the same time, but rather a general and unified process that "changes" a young adult into an old adult. This has spurred active interest toward the understanding of the common molecular and biochemical events associated with aging. In fact, careful dissection of these events could offer possible ways of intervention for the prevention of the many age-associated diseases and disabilities that affect the growing aging population. At the same time, some of the biochemical and cellular events that characterize AD have also been clarified. As a result, improved (albeit not perfect) models of the disease as well as aging have been generated.

More recent work aimed at bridging the gap between aging research and neuroscience has revealed that one of the signaling pathways involved in the regulation of lifespan in the organism is also implicated in the cognitive decline that accompanies aging and in the pathogenesis of AD dementia. This pathway, which acts down-stream of the insulin-like growth factor 1 receptor (IGF-1R), has been dissected with a combination of *in vitro*, *ex vivo* and *in vivo* approaches. In this chapter we seek to integrate these advances offering a new platform for the study of the biochemical and molecular basis of the cognitive loss associated with age and the AD form of neuropathology.

AGING AND COGNITION

Aging is accompanied by progressive cognitive decline in a large segment of the population. Both cross-sectional and longitudinal human studies indicate that the decline most typically affects working memory, short-term and delayed memory recall, although speed of processing information and spatial memory are also affected (Grady, 2008; Hedden and Gabrieli, 2004; Hedden and Gabrieli, 2005; Yankner et al., 2008). Functional magnetic resonance imaging and positron emission tomography studies in both humans and non-human primates suggest a functional association between memory decline and a progressive and age-associated reduction in the volume of cerebral regions that play an important role in the formation and retrieval of memory. They include the hippocampal formation and entorhinal cortex as well as the prefrontal cortex and the anterior corpus callosum (Logan et al., 2002; Persson et al., 2004; Small et al., 2004; Alexander et al., 2008; Dumitriu et al., 2010; Luebke et al., 2010). Cognitive decline/memory loss as well as reduction in the volume of the hippocampal formation, entorhinal cortex, and corpus callosum are also observed in aged dogs, rats, and mice suggesting intrinsic age-associated events (Grady, 2008; Hedden and Gabrieli, 2004; Yankner et al., 2008).

Stereological quantification and histological assessment of post-mortem brain tissue in humans, non-human primates, and rodents show a progressive and age-associated loss of neurons and synaptic density in the hippocampus, dentate gyrus, and prefrontal cortex, as well as loss of the association fibers of the entorhinal cortex-hippocampal system. These features are also accompanied by reduced white matter density in the corpus callosum and prefrontal cortex (Grady, 2008; Hedden and Gabrieli, 2004; Hedden and Gabrieli, 2005; Yankner et al., 2008). At the functional level, it has been suggested that negative changes in long-term potentiation (LTP) of hippocampal brain slices represent the electrophysiological correlate of the learning decline that characterizes aging in mice and rats (Barnes et al., 2000a; Barnes et al., 2000b; Landfield et al., 1981; Landfield and Lynch, 1977). LTP is a direct measure of pre- and post-synaptic functions (Box 1).

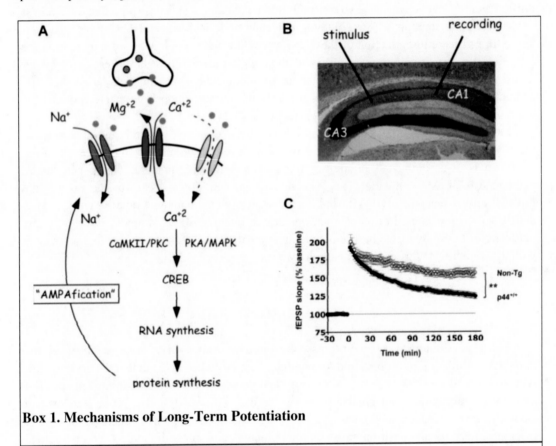

Box 1. Mechanisms of Long-Term Potentiation

Long-term potentiation (LTP) can be defined as a persistent (or long-lasting) increase in synaptic strength induced by brief high-frequency electrical stimulation of afferent neuronal fibers or coincident activation of pre- and post-synaptic neurons. The release of neurotransmitters from the pre-synaptic element induced by electrical stimuli causes early (immediate) or late (lasting) changes in potential across the membrane of the post-synaptic element (see panel A). Although the postsynaptic potential is also a function of the quantity of neurotransmitter released into the synapse, the postsynaptic element can adapt to recurrent identical stimuli by increasing the efficiency of the response (adaptation). This is achieved by regulating both the ionic currents across the ion channels (i.e., by affecting the number of available receptors/channels or channel biochemical properties) and the signaling cascade that is initiated as a response to the ionic current itself. These changes define the strength of a synapse and can be short-lived (seconds to few minutes), without permanent structural changes of the synapse itself, or long-lived (several minutes to hours) when accompanied by structural changes of the synapse. These long-lasting changes require *de novo* protein synthesis and are caused by activation of second messengers and/or transcriptional/ translational machineries.

LTP shares several features with long-term memory, which is the ability to recollect memory days or decades after the event. As a result, LTP is often considered as the electrophysiological correlate of long-term memory and the cellular events underlying LTP are deemed to be essential for the consolidation of short-term memories into long-term memories. LTP, as it relates to the ability to generate long-lasting memories in the hippocampal formation, is often assessed on isolated hippocampal slices. LTP is typically induced by stimulation of the Schaffer collaterals at the border of CA3 and CA1 while the recordings (field excitatory postsynaptic potential, fEPSP) are made in the stratum radiatum of CA1 pyramidal cells (see panel B).

Typically, LTP involves glutamate (panel A; gray circles) release from the pre-synaptic element and Ca^{+2} influx into the post-synaptic element through N-metyl D-aspartate (NMDA) receptors (NMDAR; shown in blue in panel A) and/or L-type voltage-gated calcium channels (VGCC; shown in yellow in panel A). The increase in Ca^{+2} levels leads to activation of a complex -and only partially characterized- signaling cascade that results in the activation of transcription factors such as the cAMP response element-binding (CREB) protein. The RNA and protein synthesis induced by CREB and similar transcriptional activators is at the basis of the structural and functional changes of the synapse that are thought to cause the long-term synaptic plasticity and memory formation. One important feature of these long-lasting changes is the increased expression of the α-amino-3-hydroxy-5-methyl-4-isoxazolepropionic (AMPA) receptors (AMPAR; in red in panel A). This event is often described as the "AMPAfication" of the synapse and is considered essential for the increase in synaptic strength. However, the reader should also be aware that the signaling molecules involved in the long-lasting LTP events are not simply limited to those delineated in this figure and that additional non-NMDAR forms of LTP have been described. Finally, the molecular mechanisms that govern both the early and late phase of LTP may vary in different brain regions and/or animal species.

Panel C shows the LTP generated by hippocampal slices of wild-type (non-transgenic; Non-Tg) and $p44^{+/+}$ transgenic mice, which develop a progeroid phenotype that is reminiscent of an accelerated form of aging ((Maier et al., 2004); see also later in this chapter). $p44^{+/+}$ mice display a clear defect in the late-phase of LTP. This occurs in the absence of evident defects in the presynaptic component or in the early phase of LTP, supporting the view that negative changes in LTP correlate with the learning decline that characterizes aging. The electrophysiological as well as cognitive assessment of $p44^{+/+}$ mice is described in (Pehar et al., 2010). Panel B and panel C are reproduced with permission from M. Pehar *et al.*, Altered longevity-assurance activity of p53:p44 in the mouse causes memory loss, neurodegeneration and premature death. Aging Cell 2010; 9: 174-190.

Additional information on the mechanisms that govern LTP can be found in (Crozier et al., 2004).

During the last decade several microarray studies have been performed in mice, rats, non-human primates, and humans to analyze the genetic profile of the aging brain. They all indicate that normal aging is not accompanied by a genome-wide dysregulation of transcription but rather by specific changes that affect only a small subset of genes, which accounts for less than 5% of all the genes expressed in the brain (Blalock et al., 2003; Erraji-Benchekroun et al., 2005; Fraser et al., 2005; Jiang et al., 2001; Lee et al., 2000; Lu et al., 2004). Interestingly, the genetic profile of patients affected by late-onset AD appears to overlap with the changes that occur during normal aging, although an increased number of genes and more significant changes appear to distinguish AD from normal aging (Blalock et al., 2004; Butler et al., 2004a; Butler et al., 2004b; Yankner et al., 2008). The overlapping of aging and AD has suggested that a continuum exists between the normal cognitive decline that accompanies aging and the more severe memory loss that characterizes the AD form of neuropathology (Butler et al., 2004a; Butler et al., 2004b; Yankner et al., 2008). Interestingly, in normal individuals the synaptic density of memory-forming and -retrieving areas of the brain declines steadily between the age of 20 and 100. It is predicted that normal old humans would reach the same reduced density that is seen in AD patients by the age of 130 (Terry and Katzman, 2001). A similar prediction is reached by analyzing the amyloid β-peptide (Aβ) and tau phases in the brain of normal individuals (Hebert et al., 2003; Thal et al., 2004). Aβ and tau are two important hallmarks of AD neuropathology and are analyzed in more detail in the section below.

ALZHEIMER'S DISEASE NEUROPATHOLOGY

AD represents one of the most common causes of dementia in the World. It is characterized by progressive memory deficits, cognitive impairments and personality changes accompanied by diffuse structural abnormalities in the brain. The symptoms of the disease include memory loss, confusion, impaired judgment, disorientation, and loss of language skills. Based on the onset of the symptoms, AD is normally divided into two groups: early- (<60 years) and late- (>60 years) onset. Early-onset, also called familial AD (FAD), accounts for ~3% of all AD cases and has so far been linked to mutations in the genes encoding the amyloid precursor protein (APP), presenilin 1 (PS1), and presenilin 2 (PS2). Late-onset AD, also called sporadic AD, accounts for ~ 97% of all AD cases and represents one of the most common age-associated diseases. Aging is the single most important risk-factor for late-onset AD. In fact, the prevalence of AD increases sharply after the age of 60 and doubles with every decade of life reaching ~50% in individuals that are 85 years of age or older (Hebert et al., 2003; Thal et al., 2004). In addition, with increasing age, the frequency of higher Aβ and tau/neurofibrillary stages in normal individuals increases as well, underscoring a strong relationship between aging and AD pathology (Thal et al., 2004). Because of the ongoing increase in life expectancy, late-onset AD is projected to affect as many as 115 million individuals worldwide by the year 2050 (World Alzheimer Report. 2009 Executive Summary). In addition to aging, late-onset AD has been associated with different environmental and genetic risk factors (Carlsson et al., 2009).

The three main pathological hallmarks of AD include (i) amyloid plaques (also called senile plaques), (ii) neurofibrillary tangles (NFT) and (iii) diffuse loss of neurons and synapses (Figure 1). Although it can differ from one patient to another, the distribution of the above AD features follows a disease-specific regional pattern with the amyloid plaques more prevalent in the neocortex and the neuronal/synaptic loss more prevalent in the hippocampus, posterior cingulate, corpus callosum, and other areas involved with formation and retrieval of memory as well as high-order activities that require information transfer between the two hemispheres. The distribution of the NFT overlaps with both the amyloid plaques and the neuronal loss; however, it seems to mirror the neuronal loss more closely than the amyloid plaques. In addition to the above pathological hallmarks, AD patients often display amyloid angiopathy, which is characterized by diffuse accumulation of Aβ in the vascular wall. Finally, AD brains are also characterized by a diffuse and widespread presence of reactive astrocytes and microglia, mostly concentrated in areas of neuronal loss but also surrounding plaques. It should be noted that the pathology can surprisingly reveal great variability among different patients, even in the face of similar clinical manifestations.

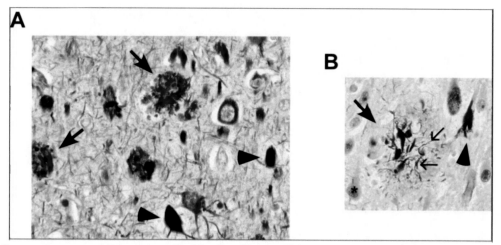

Micrographs were generously provided by Dr. Shahriar Salamat, University of Wisconsin-Madison, Madison, WI, USA.

Figure 1. Pathological alterations in late-onset (sporadic) Alzheimer's disease brains.

Two microphotographs showing amyloid plaques (arrows) and intraneuronal NFT (arrow heads). Dystrophic neurites surround both types of plaques (indicated with small arrows in B). Asterisk (*) in (B) points to gliosis, which often accompanies AD alterations.

Amyloid plaques are extracellular protein deposits (Figure 1). The typical plaque has a dense core of protein aggregates surrounded by dystrophic dendrites and axons, activated microglia, and reactive astrocytes (Puglielli, 2008; Selkoe, 1999; Selkoe, 2002). The core is mostly made of a small hydrophobic peptide of 38-43 amino acids called Aβ. Aβ is organized in fibrils of approximately 7-10 nm that are intermixed with non-fibrillar forms of the peptide. Aβ originates from APP following proteolysis at β and γ sites (Figure 2). The β cleavage is carried out by the β-site APP cleaving enzyme (BACE1) whereas the γ cleavage is carried out by γ-secretase, a multimeric protein complex containing presenilin, nicastrin, anterior pharynx defective 1 and presenilin enhancer 2 (De Strooper, 2003). The β cleavage of APP is the rate-limiting step for the generation of Aβ. It liberates a large N-terminal fragment of the protein (sβAPP), which is released in the extracellular *milieu*, and a small (~12 kDa) membrane-anchored fragment called C99, which becomes the immediate substrate for γ-secretase. The γ cleavage leads to the generation of Aβ and a small cytosolic fragment called APP intracellular domain (AICD). Aβ is mostly released in the extracellular *milieu* where it tends to oligomerize and aggregate in the form of oligomers, fibrils and, ultimately, plaques

(Selkoe, 1999). AICD, instead, is released in the cytosol and rapidly translocates to the nucleus where it exerts transcriptional functions (Muller et al., 2008). Both animal and cellular studies indicate that the processing of APP is crucial for the pathogenesis of AD dementia (Carlsson et al., 2009).

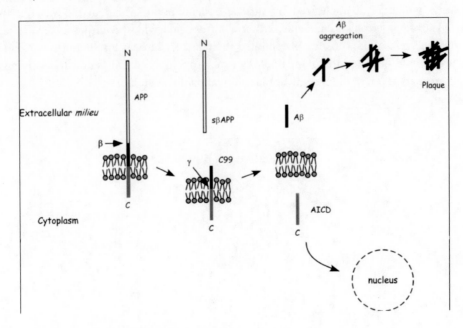

Figure 2. Generation of Aβ and AICD from APP.

APP is a type 1 membrane protein with a large extracellular domain, a single membrane-spanning domain, and a short cytoplasmic tail. The Aβ region of APP (in black) includes the first 12-14 amino acids of the membrane domain. APP is initially cleaved at the β site (amino acid 1 of the Aβ region) by BACE1. This event, which represents the rate-limiting step, liberates a large N-terminal fragment (sβAPP) that is secreted into the extracellular *milieu* and a small membrane-bound C-terminal fragment of 99 amino acids (C99; also referred to as β-APP-CTF). C99 is then cleaved by γ-secretase to liberate the Aβ as well as the AICD peptides. AICD is a transcription factor and rapidly translocates to the nucleus. The transcriptional activity of AICD seems to require co-partners such as Fe65 and Tip60 (Chang and Suh, 2010). In contrast to AICD, the Aβ peptides are released in the extracellular *milieu* where they aggregate to form oligomers, fibrils and, ultimately, amyloid plaques (also referred to as senile plaques). Intracellular aggregates containing the Aβ peptide have also been described. However the origin of such aggregates is unknown. C99 can undergo further cleavage at the ε site generating shorter versions of the AICD fragment. In addition to the above β/γ pathway, APP can also be cleaved at the α site (between amino acids 16 and 17 of the Aβ region) precluding the generation of Aβ.

Studies from transgenic mice expressing human APP indicate that the abnormal aggregation/accumulation of Aβ in the brain represents an important step for the development of the Alzheimer form of neurodegeneration (Box 2). However, AICD, which is stoichiometrically released with Aβ, can also cause some of the features that characterize AD (Ghosal et al., 2009). Importantly, BACE1 "knock-out" animals, which are not able to cleave APP at the β site and generate Aβ or AICD, do not develop AD neuropathology ((Cai et al., 2001; Luo et al., 2001); see also Box 2). The molecular mechanisms involved in the toxicity of Aβ are still in part unknown. However, recent research indicates that small Aβ aggregates (oligomers) might act as the proximate cause of neuronal injury and synaptic loss associated with AD (Cleary et al., 2005; Haass and Steiner, 2001; Klein et al., 2001; Lambert et al., 1998; Lansbury, 1999; Puzzo et al., 2008). In addition, studies in both AD patients (Delacourte et al., 2002) and mouse models (Ghosal et al., 2009; Gotz et al., 2004) seem to suggest that Aβ and AICD can both foster the aggregation of tau into NFT. The synergistic interaction of Aβ/AICD and tau, together with the mechanisms that link them to aging, is still largely unknown (discussed below).

Box 2. Lessons From Animal Models of AD

The identification of APP as the Aβ precursor, together with the fact that specific mutations in the *APP* gene are associated with familial forms of AD (FAD), prompted the generation of transgenic mice expressing human APP harboring one or more FAD-associated mutations. The resulting animals developed amyloid plaques, together with a certain degree of synaptic loss and cognitive deficits. However, the distribution of plaques and synaptic loss did not seem to reproduce completely the classical AD phenotype. In addition, the tau pathology and the neuronal loss appeared almost completely absent, suggesting that additional biochemical/molecular events are required to develop the full spectrum of AD. To circumvent this issue several new animal models have been generated where human *APP* is accompanied by additional mutated genes. These include *PSEN1/PSEN2*, encoding presenilin1 and presenilin 2, and *MAPT*, encoding tau proteins. Mice harboring three or five FAD-associated mutations (respectively called 3X and 5X mice) in two or more genes have also been generated. They all indicate that Aβ is an essential element for the development of AD-like neuropathology and show a close relationship between Aβ and the phosphorylation/ aggregation state of tau. It has been argued that none of the above mouse models fully reproduces the classical AD-phenotype. However, it must be noted that -as far as we know- only humans develop AD, thus suggesting that there is something specific to the human brain that cannot be entirely recapitulated by simply over-expressing one or more human genes into the mouse. Of course, this also implies that we still have an incomplete knowledge of the molecular aspects of AD pathology.

Transgenic mice over-expressing human Aβ40 or Aβ42 alone -in the absence of human APP- have also been generated. These mice express the Aβ region (ending at amino acid 40 or 42) fused to the BRI protein. The cleavage of the fused protein resulted in successful production/secretion of the Aβ peptide without additional APP fragments that would result from APP processing (such as the AICD; see Figure 2). Interestingly, mice engineered to generate the 42 amino acid long peptide developed amyloid pathology (in the form of dense and diffuse plaques, and amyloid angiopathy) while those producing the shorter version of the same peptide did not, indicating that the extra two amino acids in the C-terminus of Aβ are necessary for the aggregation of Aβ.

Mice that over-express only the AICD fragment of human APP (without Aβ; see Figure 2) also develop some of the features normally associated with AD (Ghosal et al., 2009; Giliberto et al., 2008). These include abnormal activation of tau kinases and abnormal phosphorylation of mouse tau as well as synaptic deficits and increased neuronal susceptibility to exogenous stress. Importantly, these features can only be observed when AICD is over-expressed together with one of its partners, human Fe65. Therefore, AICD can participate in AD neuropathology in a way that is completely Aβ-independent.

Mice expressing the human microtubule-associated protein tau develop the typical tau pathology (tangles and neuropil threads) that is found in individuals affected by frontotemporal dementia with parkinsonism; however, they do not develop high levels of Aβ indicating that tau is not required for the formation of the plaques (Duyckaerts et al., 2008). Crossing these mice with

APP transgenic mice potentiated the tau pathology and neuronal loss but did not aggravate the plaque pathology, suggesting that Aβ acts upstream of tau in the classical AD-phenotype. However, studies from AD patients, mouse models, and cellular systems (*ex vivo*) seem also to suggest that Aβ and tau can synergistically interact, fostering their respective aggregation and the neuronal loss. Finally, the down-regulation of the *Mapt* gene in mouse models of AD can recover - at least partially- the neuronal and memory loss associated with the disease. Therefore, the above studies indicate that the relationship between Aβ and tau is more complex than previously thought and appear to suggest that additional biochemical/molecular events might act upstream of both Aβ and tau in the AD brain.

A problem that always arises with AD mice that are generated by over-expressing APP and/or presenilins is that classical NFT cannot be observed. However, it must be noted that mouse tau differs from human both in the primary sequence and in the alternative splicing of the different isoforms. As a consequence, mouse tau can be phosphorylated almost similarly to the human protein but does not generate the classical paired helical filaments or neurofibrillary tangles that are observed in patients affected by late-onset AD. NFT can only be observed when human wild-type tau is over-expressed in the absence of endogenous murine tau (on a *Mapt$^{-/-}$* background) or when a mutant form of human tau is over-expressed (independently of the murine tau background). Therefore, there is something specific to human tau that causes NFT. Similarly, over-expression of mouse APP does not lead to plaque pathology. This is likely due to the fact that mouse Aβ differs from human in three amino acids and, as a result, does not aggregate.

The identification of the gene that encodes BACE1, the rate-limiting enzyme for the biosynthesis of Aβ, prompted the generation of "knock-out" mice that lack BACE1. Further crossing with the APP mutant mice models showed that the elimination of the β cleavage of APP is sufficient to resolve the pathology, the synaptic impairment, and the cognitive phenotype of the animals. Finally, these approaches indicated that BACE1 represents a viable target for the prevention/treatment of AD since the disruption of BACE1 does not result in a strong phenotype in the mouse (Cai et al., 2001; Luo et al., 2001).

In conclusion, an extensive body of work in mice indicates that APP and tau play major roles in the pathogenesis of AD; however, they also indicate that "something else" contributes to the development of other aspects of the classical AD-neuropathology, including the gliosis and the atrophy.

Excellent reviews on the different animals models of AD can be found in Duyckaerts et al., 2008; Gotz et al., 2004; McGowan et al., 2006.

In contrast to amyloid plaques, which are extracellular protein deposits, NFT are almost exclusively observed in the cytoplasm of neurons (Figure 1). NFT appear as paired, helically twisted filaments and are made of highly stable polymers of the microtubule-binding protein tau. Tau constitutes a group of alternatively spliced proteins that can be phosphorylated at multiple sites. The degree of phosphorylation inversely correlates with binding to microtubules (Figure 3). As a result, highly phosphorylated tau proteins dissociate from microtubules and tend to polymerize into filaments to form NFT (Lee et al., 2001). In addition to AD, the abnormal accumulation of NFT is observed in non-AD frontotemporal forms of dementia (Kurz and Perneczky, 2009; Lee et al., 2001). Work in multiple AD and non-AD mice models has shown that abnormal tau metabolism can cause memory deficits (Box 2). Additionally, mutations in the gene encoding tau proteins (*MAPT*) have been associated with hereditary forms of frontotemporal dementias whereas polymorphisms in the same gene appear to act as genetic risk factors for sporadic progressive supranuclear palsy and corticobasal degeneration (Kurz and Perneczky, 2009; Lee et al., 2001). To date, no mutation associated with AD has been found in the *MAPT* gene.

That the phosphorylation status of tau is closely linked to AD neuropathology is also proven by the fact that several tau kinases have been linked to the pathogenesis of AD. Among these are the dual-specificity tyrosine-regulated kinase 1A (Dyrk1A), cyclin-dependent kinase 5 (Cdk5), and glycogen synthase kinase-3β (GSK3β). Dyrk1A is highly expressed in memory-forming and -retrieving areas of the brain (Becker et al., 1998) and the *DYRK1A* gene maps to human chromosome 21 within the Down syndrome (DS) critical region. This is particularly important because DS patients are known to have increased risk of AD, and Dyrk1A is abnormally expressed in the brain of both AD and DS patients (Ferrer et al., 2005). Cdk5 is highly expressed in post-mitotic neurons and has been implicated in AD neuropathology (Wesierska-Gadek et al., 2006). Importantly, the levels of the Cdk5 regulatory cyclins p39 and p35 are increased in late-onset AD brains (Tseng et al., 2002). GSK3β is a serine/threonine kinase that appears to play important roles in both early- and late-onset AD, and is directly modulated by the other two tau kinases, Dyrk1A and Cdk5 (Ballatore et al., 2007; Hooper et al., 2008; Lee et al., 2001; Wesierska-Gadek et al., 2006).

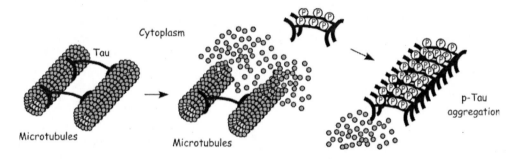

Figure 3. Hyperphosphorylated tau aggregates into neurofibrillary tangles.

Tau proteins ensure microtubule-dependent cellular functions by binding and stabilizing the microtubules. In general, the degree of phosphorylation inversely correlates with binding. As a result, phosphorylated tau proteins dissociate from microtubules and aggregate to form filaments, which ultimately result in NFT. This process destabilizes microtubules and interferes with their functions. The microtubule-binding as well as pro-aggregating properties of tau also depend on the differential splicing of the *MAPT* gene, which results in different isoforms. The pattern of phosphorylation and splicing of tau differs in the fetal and adult brain and in the central and peripheral nervous system. Finally, it also differs among mammals, probably explaining why only humans develop classical NFT.

In addition to AD brains, amyloid/senile plaques and NFT are also present in the hippocampus and neocortex of cognitively normal aged individuals. Although still under debate, it is now becoming apparent that the neuronal loss and the number of NFT correlate more closely with the cognitive profile than the amyloid plaques (Terry et al., 1991). Importantly, the neuronal/synaptic loss of the hippocampal formation, entorhinal cortex, and entorhinal-hippocampal association system appears to be the only feature able to distinguish normal aging from mild cognitive impairment (MCI) and the early stages of AD (Yankner et al., 2008). Finally, with increasing age, normal individuals show a steady and progressive increase in the frequency of high NFT stages (Thal et al., 2004) and in the degree of synaptic loss (Terry and Katzman, 2001) underscoring a strong relationship between aging and AD pathology, and suggesting that a continuum exists between cognitively normal aging and AD (Yankner et al., 2008).

IGF-1R SIGNALING AND AGING

Studies performed in model organisms such as the yeast *Saccharomyces cerevisiae*, the nematode *Caenorhabditis elegans*, the fruit fly *Drosophila melanogaster*, and the laboratory mouse have revealed that the signaling pathway acting downstream of IGF-1R plays a major role in controlling the lifespan of the organism (Bartke, 2008; Brown-Borg, 2003; Kenyon, 2005; Narasimhan et al., 2009; Puglielli, 2008; Tissenbaum and Guarente, 2002). Specifically, increased IGF-1R signaling accelerates aging and shortens lifespan whereas a partial block of IGF-1R signaling achieves the opposite effects. In addition to the signaling activity of the receptor, mechanisms controlling circulating levels and functional availability of the ligand, IGF1, also appear to predict longevity (Leduc et al., 2010). A partial block of IGF-1R signaling is also achieved by caloric restriction, which extends the maximum lifespan and delays many biological changes that are associated with aging (Anderson and Weindruch, 2010; Narasimhan et al., 2009). In humans, genetic variations that cause reduced IGF-1R signaling appear to be beneficial for old age survival and preservation of cognitive functions (Bonafe et al., 2003; Suh et al., 2008; van Heemst et al., 2005), suggesting that the mechanisms regulating lifespan and aging *via* this pathway are evolutionary conserved.

The first evidence linking IGF-1R to lifespan of the organism came from *C. elegans* where a mutation causing genetic insufficiency of *daf-2*, the worm version of mammalian *IGF1R*, was shown to double the maximum lifespan of the organism without any evident pathological manifestation (Kenyon et al., 1993; Kimura et al., 1997). Later on, a similar effect was also observed with *D. melanogaster* (Tatar et al., 2001) and the laboratory mouse (Holzenberger et al., 2003). The functional mammalian IGF-1R has two α- and two β-subunits (Figure 4). The α-subunits are entirely extracellular and contain the ligand-binding domain; in contrast, the β-subunits are largely intracellular and have one single transmembrane-domain, a tyrosine kinase domain and several regulatory phosphorylation sites (Jones and Clemmons, 1995; Ullrich et al., 1986; Ward et al., 2001). The transmembrane-domain anchors the functional receptor to the cell surface whereas the tyrosine kinase domain transduces the signal to cytosolic interactors. Upon binding to its ligand, IGF-1R undergoes auto-phosphorylation, recruits intracellular adaptor proteins called insulin/IGF1 receptor substrates (IRS) and activates the phosphatidylinositol 3-kinase (PI3K). PI3K catalyzes the conversion of phosphatidylinositol (3,4)biphosphate (PIP2) into phosphatidyilinositol (3,4,5)triphosphate (PIP3), which serves to recruit and activate protein kinase B (PKB)/Akt (Bondy and Cheng, 2004; Puglielli, 2008). An additional signaling pathway that apparently stems from IGF-1R itself utilizes Ras/MEK/ERK signaling. IGF-1R signaling is negatively regulated by the phosphate and tensin homologue deleted on chromosome ten (PTEN) anti-oncogene, which converts PIP3 back to PIP2 and down-regulates the expression levels of IGF1-R (Puglielli, 2008).

That the signaling pathway downstream of IGF-1R was directly linked to the regulation of lifespan of the organism was initially proven in *D. melanogaster* and *C. elegans* when mutations causing genetic insufficiency of *chico*, encoding the common fly version of IRS, and *age-1*, encoding the worm version of PI3K, were shown to extend lifespan. Importantly, the use of molecular biology in *C. elegans* served to prove that *daf-2* was acting upstream of *age-1*. Finally, changes in lifespan of the organism were also found in worms bearing a mutation in *daf-18*, which encodes the worm version of PTEN. The downstream output of the above genetic pathway appears to be daf-16, which encodes the worm version of mammalian FOXO. Therefore, powerful genetics and molecular biological approaches revealed a common signaling pathway that acts downstream of IGF-1R to regulate lifespan (Figure 4) (Kenyon, 2005; Puglielli, 2008; Tissenbaum and Guarente, 2002). In addition to serving as the apparent last output of IGF-1R signaling, daf-16/FOXO also appears to integrate metabolic as well as reproductive- and lipophilic-hormone signaling (Berman and Kenyon, 2006; Gerisch et al., 2001; Hsin and Kenyon, 1999; Motola et al., 2006; Rottiers et al., 2006). Therefore, cross-talk between different signaling pathways allows a fine tuning of many molecular events that are involved in the regulation of lifespan and age-associated events.

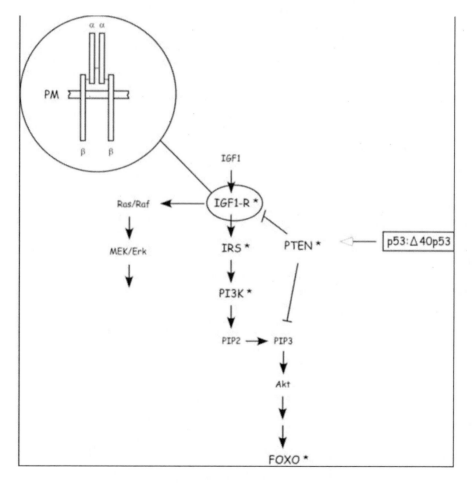

Figure 4. Schematic view of the IGF-1R signaling pathway.

The pathway is described in the appropriate section of this chapter. IGF-1R is structurally organized as two α and β subunits stabilized by disulfide bonds. Asterisks indicate elements of the pathway that have been implicated in lifespan modulation (described in the appropriate section of this chapter). In mammals, IGF-1R signaling is tightly regulated by the expression levels of p53 and Δ40p53 (p44).

In mammals, the initial evidence supporting the direct involvement of IGF-1R signaling in the regulation of lifespan came from the Ames and Snell dwarf mice, which have a defect in the pituitary gland that results in deficiency of the growth hormone (GH)/IGF1 axis (Barbieri et al., 2003; Bartke, 2008). Circulating IGF1 is secreted by the liver in response to GH. Therefore, animals that are deficient in GH secretion are also deficient in IGF1 secretion. This is not limited to the above Ames and Snell mice; it includes both Little, which are deficient in GH secretion, and $Ghr^{-/-}$ mice. Importantly, $Ghr^{-/-}$ mice lack the GH receptor and are unable to respond to GH signaling. As a result, they have high levels of circulating GH but undetectable levels of IGF1 and live ~25% longer than wild-type controls. Reduced IGF1 availability is also observed in the Midi and PAPP-A$^{-/-}$ mice, which display a 20 to 30% increase in lifespan. Finally, as mentioned before, mice with a heterozygous disruption of IGF-1R are resistant to IGF1 signaling, live longer and display resistance to age-associated stressors (Holzenberger et al., 2003). This "longevity assurance activity" is not limited to the receptor; in fact, reduced IGF-1R signaling in the $Irs1^{-/-}$ (Selman et al., 2008) and $Irs2^{+/-}$ (Taguchi et al., 2007) mice also increases lifespan. Therefore, the signaling cascade immediately downstream of IGF-1R appears to affect lifespan in mammals. Importantly, heterozygous disruption of $Irs2$ in the brain is sufficient to increase the lifespan of the mouse implicating the brain itself in the regulation of aging and/or age-associated manifestations. The important role of the brain in the control of aging is not completely novel as it was already known that genetic manipulations limited to neuronal cells could influence the lifespan of other model organisms (Bartke, 2008; Kenyon, 2005).

Recent studies in humans have identified polymorphisms in both $IGF1R$, encoding human IGF-1R, and $PI3KCB$, encoding the catalytic subunit of human PI3K, that are associated with longevity (Bonafe et al., 2003). Additionally, genetic variations that result in overall reduced IGF-1R signaling appear to be beneficial for old age survival and preservation of cognitive functions in different ethnic groups (Bonafe et al., 2003; Paolisso et al., 1997; Suh et al., 2008; van Heemst et al., 2005), indicating that the molecular events acting through this pathway are relevant for human physiology and pathology. Whether these effects are due to endocrine, paracrine or autocrine functions of IGF1 is still a matter of debate (Puglielli, 2008). It is worth noting that the anabolic functions of IGF1 during adulthood have been associated with increased risk of cancer development (Barbieri et al., 2003). Perhaps, reduced levels of IGF-1 might result into a reduced mitogenic stimulus for cells and tissues and contribute to decreased age-associated pathologies (see below).

THE LONGEVITY-ASSURANCE ACTIVITY OF THE TUMOR-SUPPRESSOR GENE *TP53*

The p53 protein is best known for its tumor suppressor activity. In fact, a large number of mutations associated with human cancers are found in the p53 gene (*TP53*). However, the last decade has surprisingly revealed a more complicated network of biological features of this protein that intermingle with aging and age-associated events such as cognitive decline and AD. Through a combination of alternative promoter usage, alternative splicing and alternative initiation of translation, the *TP53* gene generates at least nine different p53 isoforms, p53α, p53β, p53γ, Δ40p53α, Δ40p53β, Δ40p53γ, Δ133p53α, Δ133p53β, and Δ133p53γ (Figure 5).

Δ40p53 (also known as DNp53 or p44) and Δ133p53 are N-truncated versions of p53. Δ40p53 is missing the first 39 amino acids, whereas Δ133p53 is missing the first 132 amino acids. As a result, Δ40p53 lacks the first transactivation domain of full-length p53 but retains the second plus the DNA-binding domain, whereas Δ133p53 lacks both transactivation domains but retains most of the DNA-binding region. Additional splicing at the C-terminus generates three different C-terminal truncations (α, β, and γ), therefore adding more complexity to the biology of *TP53* (Bourdon, 2007; Bourdon et al., 2005; Courtois et al., 2002; Ghosh et al., 2004; Yin et al., 2002).

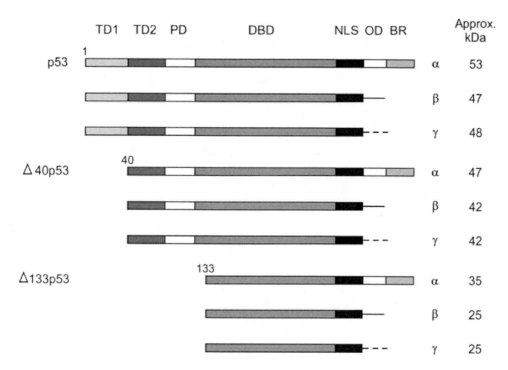

Figure 5. Domain organization of p53 isoforms.

TD, transactivation domain; PD, proline-rich domain; DBD, DNA-binding domain; NLS, nuclear localization signal; OD, oligomerization domain; BR, basic repeat domain.

In 2002, an unexpected targeting event resulted in mice over-expressing a truncated version of p53 (referred to as p53[m]) that was missing both transactivation domains and a large part of the DNA-binding domain (Tyner et al., 2002). The animals (p53[+/m]) displayed a shorter lifespan and features that resembled an accelerated form of aging. This was interpreted as caused by a "gain-of-function" of the m-allele requiring the presence of the full-length protein. In fact, genetic disruption of endogenous wild-type p53 prevented the progeroid phenotype of p53[+/m] animals. The "gain-of-function" conclusion was supported by previous work in the fruit fly where over-expression of a dominant-negative version of the fly p53 (Dmp53) in neurons increased the lifespan (Bauer et al., 2005). Only two years later, in 2004, Maier *et al.* generated a mouse that over-expressed Δ40p53 (p44[+/+]) and reported a similar progeroid phenotype (Maier et al., 2004). This time it was not a non-existing truncated version of p53 but a naturally occurring short isoform of p53, thus conferring new interest in the p53-aging connection. As with p53[+/m] mice, the progeroid phenotype required the presence of endogenous p53 (Maier et al., 2004). Therefore, the tumor suppressor gene p53 appears to have a previously unknown longevity-assurance activity that impacts aging and age-associated manifestations. Interestingly, both p53$^{+/m}$ and p44$^{+/+}$ animals are resistant to tumor formation (Maier et al., 2004; Tyner et al., 2002). This is very similar to the "Super-p53" and the Mdm2$^{puro/\Delta7-12}$ mice (Garcia-Cao et al., 2002; Mendrysa et al., 2006), both displaying increased p53 activity. However, neither the "Super-p53" nor the Mdm2$^{puro/\Delta7-12}$ mice display progeroid features or short lifespan (Garcia-Cao et al., 2002; Mendrysa et al., 2006). Therefore, the tumor-suppressor and the longevity-assurance activities of p53 are functionally separated. Both Δ40p53 and p53^{m} are missing a segment of the N-terminus of the full length protein. Perhaps the regulatory elements on the N-terminal tail are responsible for a natural equilibrium between the tumor-suppressor and longevity-assurance activities of the p53 protein. Studies with the naturally occurring Δ40p53 indicate that this isoform has a longer half-life (9.5 hours instead of the 30 minutes of the full-length protein) and can oligomerize with p53 (Courtois et al., 2002; Ghosh et al., 2004; Powell et al., 2008) forming hetero-complexes that might be more active or even differently active (Campisi, 2004; Scrable et al., 2005). Perhaps the p53:Δ40p53 hetero-complexes differ from the p53:p53 homo-complexes in promoter affinity, thus explaining the different phenotypes of the p44$^{+/+}$ and Super-p53 mice. At the molecular level this might involve recruitment of different co-activators or changes in the post-translational modifications of the Δ40p53/p53 protein (Grover et al., 2009; Murray-Zmijewski et al., 2008).

In addition to p53$^{+/m}$ and p44$^{+/+}$, other mouse models display an accelerated aging phenotype that appears to involve altered activity of p53: *Ku80$^{-/-}$* (Lim et al., 2000; Vogel et al., 1999), *Zmpste24$^{-/-}$* (Varela et al., 2005), *mTR$^{-/-}$* (Chin et al., 1999; Rudolph et al., 1999) and *Brca1$^{\Delta11/\Delta11}$* (Cao et al., 2006; Cao et al., 2003) mice. Ku80 is a DNA repair protein; Zmpste24 is a metalloproteinase involved in the maturation of laminin A, which is an integral component of the nuclear envelope; mTR is a telomerase; BRCA1 is a tumor-suppressor protein. Although they differ in functions, somehow they all implicate p53 as the common culprit of the common aging phenotype (Rodier et al., 2007). Together with p53$^{+/m}$ and p44$^{+/+}$, the above mouse models also seem to suggest that a premature cellular senescence is part (if not the cause itself) of the progeroid phenotype: premature senescence means premature arrest of the replicative potential of somatic cells, which in turn results in premature aging of the organism. One of the leading theories of aging suggests that during evolution complex organisms have been able to "insert" mitotically competent cells among post-mitotic cells in the different somatic tissues. Post-mitotic cells perform the normal functions of the tissue but are unable to divide and grow. As such, they are resistant to tumor transformation but, when severely damaged, they simply die. In contrast, mitotically-competent cells are able to divide, grow and differentiate, thus ensuring renewal, repair or even regeneration of the tissue. However, since they can divide they can also transfer possible DNA damage to the progeny. As such, they are susceptible to hyperproliferation and tumor transformation. In conclusion, the ability of complex organisms to repair their tissues allows them to live longer but also confers them increased susceptibility to cancer (Campisi, 2005a; Campisi, 2005b). This theory would provide great power to "gate-keeper" genes, which regulate the cell-cycle progression and mitotic potential of a cell. In fact, by controlling the mitotic potential of the cell they would be able to reduce the cancer susceptibility as well as the longevity of the organism. Interestingly, p53 is a tumor-suppressor gene that acts as a gate-keeper (Campisi, 2005a; Campisi, 2005b) and altered p53:Δ40p53 ratio affects proliferation, pluripotency and differentiation potential of undifferentiated embryonic stem cells (Ungewitter and Scrable, 2010). Whether the stem cell potential is indeed the key to the progeroid phenotype of the above p53-dependent mouse models remains to be proven. However, what is clear is that a previously known tumor-suppressor gene is also a longevity-assurance gene.

A MOLECULAR PATHWAY TO COGNITIVE LOSS AND NEURODEGENERATION

Mechanistically, the progeroid phenotype of the p44[+/+] transgenic mice has been associated with the IGF-1R signaling pathway. In fact, p53 and Δ40p53 can antagonistically affect IGF-1R signaling at different levels. They regulate the expression levels of IGF-1R (Maier et al., 2004; Werner et al., 1996) but also the expression levels as well as the lipid-phosphatase activity of PTEN (Maier et al., 2004; Schuster et al., 2001; Stambolic et al., 2001; see also Figure 4). As a result of the above, p44[+/+] transgenic mice have increased activation of IGF-1R signaling (Maier et al., 2004). The role of IGF-1R in the regulation of aging and age-associated events was discussed in the previous section. Whether additional and yet unknown points of contact between IGF-1R signaling and p53:Δ40p53 exist remains to be determined.

A careful analysis of p44[+/+] transgenic mice revealed that in addition to reduced lifespan, short reproductive health span, osteoporosis and premature lordokyphosis, they develop a premature and severe loss of associative as well as spatial navigational memory (Pehar et al., 2010). These cognitive functions are typically linked to amygdala and hippocampal activities, and tend to be affected in a large segment of the aging population (Grady, 2008; Hedden and Gabrieli, 2004; Hedden and Gabrieli, 2005; Yankner et al., 2008). The cognitive impairment in p44[+/+] animals is also accompanied by synaptic deficits of the same brain areas (Pehar et al., 2010). Importantly, the synaptic deficits appear to be limited to the late-component of post-synaptic transmission (see Box 1), which is typically viewed as the electrophysiological correlate of the learning decline that characterizes aging in mammals (Hedden and Gabrieli, 2005; Yankner et al., 2008). Both the cognitive decline and the synaptic deficits of p44[+/+] transgenic mice are already evident at young age (Pehar et al., 2010). Importantly, the synaptic deficits can be rescued by the haploinsufficiency of either *Igf1r* or *Mapt* (Pehar et al., 2010). The fact that reduced IGF-1R signaling can restore the synaptic deficits supports the conclusion that IGF-1R is responsible, at least in part, for the progeroid phenotype of the p44[+/+] transgenic mice. The fact that haploinsufficiency of *Mapt* could rescue the phenotype is also very interesting. In fact *Mapt* encodes the microtubule-binding protein tau, which has been linked to memory deficits in both AD and non-AD mouse models (Kurz and Perneczky, 2009; Lee et al., 2001; see also above). In addition, tau metabolism is often found altered in the brain of non-demented old individuals (Lee et al., 2001). The role of tau in neurodegeneration has been discussed previously in this chapter.

In addition to the above features, p44[+/+] transgenic animals display increased processing of APP and over-production of Aβ (Costantini et al., 2006), which is directly linked to the altered levels and activity of the β secretase BACE1 (see Figure 2; also discussed later). Although p44[+/+] single-transgenics have increased levels of endogenous Aβ (Costantini et al.,

2006), they do not develop the features that normally characterize AD neuropathology. However, it must be noted that murine Aβ differs from human in three amino acids and does not display significant pro-aggregating properties or neurotoxic functions (Duyckaerts et al., 2008). To circumvent this limitation, p44[+/+] animals were engineered to over-express a "humanized" form of mouse APP, which is a modified version of mouse APP where the Aβ region of the protein has been changed with the human sequence. As a result, the animals over-produce human Aβ. Under these conditions, they develop several of the features that characterize AD, including loss of the afferent path that connects the entorhinal cortex to the hippocampal formation, loss of synaptic terminals and widespread degeneration of the hippocampal formation, astrogliosis, hyperphosphorylation of tau and degeneration of the corpus callosum (Pehar et al., 2010). Importantly, the loss of the association fibers of the afferent path and corpus callosum is among the first alterations observed in AD patients (Hyman et al., 1984). The afferent path is a complex and (mostly) uni-directional system that connects the entorhinal cortex to the hippocampal formation and is critical for both the formation and the consolidation of new memories. The association fibers of the corpus callosum, instead, are involved in memory retrieval and in high-order activities that require information transfer between the two hemispheres. In conclusion, p44[+/+] single-transgenic mice display increased production of endogenous Aβ, hyperphosphorylation of endogenous tau and cognitive as well as synaptic deficits, thus mimicking a phenotype that is often observed in a large segment of the aging population. When engineered to over-produce human Aβ, they also develop a dramatic form of neurodegeneration that mimics AD neuropathology. These findings clearly suggest that appropriate dissection of the events driving the progeroid phenotype in p44[+/+] mice can help us decipher age-associated manifestations as well as age-associated diseases that afflict the general population.

The abnormal production of Aβ in the p44[+/+] animals has been linked to a naturally occurring and age-associated switch in neurotrophin signaling, down-stream of IGF-1R (Costantini et al., 2006; Costantini et al., 2005) (Figure 6). Specifically, the expression profile of neurotrophin receptors TrkA and p75[NTR] in the brain changes as a function of age: while the levels of TrkA decrease, those of p75[NTR] increase ((Costantini et al., 2005); reviewed in (Puglielli, 2008)). Importantly, TrkA prevents, whereas p75[NTR] stimulates the β cleavage of APP (Costantini et al., 2005). The p75[NTR]–dependent activation of Aβ generation involves neutral sphingomyelinase, a required signaling molecule down-stream of p75[NTR], and the consequent activation of the lipid second messenger ceramide (Costantini et al., 2005). Therefore, as a result of the above molecular switch, during the process of aging we observe a progressive increase in the generation of Aβ ((Costantini et al., 2005); reviewed in (Puglielli, 2008)). These events, which are age-associated in non transgenic animals (Costantini et al., 2005), occur prematurely in p44[+/+] transgenic mice as part of their progeroid phenotype (Costantini et al., 2006). However, in both cases, the increased generation of Aβ is prevented by inhibiting neutral sphingomyelinase, therefore proving that ceramide is the last output of the common signaling pathway (Costantini et al., 2006; Costantini et al., 2005). It is also worth stressing that in addition to the above work immediately focused on the molecular characterization of p44[+/+] animals, the involvement of ceramide and its up-stream activator, p75[NTR], in AD neuropathology has been described and characterized by many other groups (Capsoni and Cattaneo, 2006; Coulson, 2006; Cutler et al., 2004; Han et al., 2002; Hu et al., 2002; Rabizadeh and Bredesen, 2003; Yang et al., 2008). Finally, AD patients have a ~3-fold increase in the levels of ceramide in AD-relevant brain areas (Cutler et al., 2004; Han et al.,

2002). Further dissection of the above events has revealed that the TrkA-to-p75^{NTR} switch occurs down-stream of IGF-1R (Costantini et al., 2006). Specifically, an age-associated up-regulation of IGF-1R levels in neurons leads to activation of IGF-1R signaling, recruitment of IRS2, PI3K and activation of the second messenger PIP3 (Costantini et al., 2006). The transcriptional machinery that down-regulates TrkA while up-regulating p75^{NTR} includes Egr-1 and Hipk2 (Li et al., 2008).

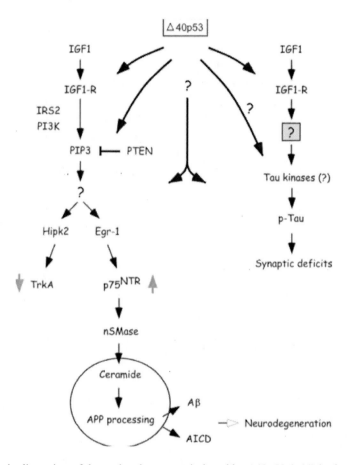

Figure 6. Schematic dissection of the molecular events induced by Δ40p53 (p44) in the mouse.

A combination of biochemistry, cell and molecular biology in several *in vitro*, *ex vivo* and *in vivo* settings has revealed that Δ40p53 acts upstream of IGF-1R signaling in regulating neurotrophin signaling and APP processing (see *left pathway*; appropriate references are found in the appropriate section of this chapter). Δ40p53 up-regulates IGF-1R while down-regulating PTEN. The lipid phosphatase activity of PTEN controls IGF-1R signaling by converting PIP3 into PIP2. Reduced levels or activity of PTEN result in increased levels of PIP3 and increased signal transduction. Although Δ40p53 has been shown to act through IGF-1R and PTEN, additional points of regulation may exist (see arrow with question mark). The lipid second messenger ceramide is the last output of the pathway that regulates APP processing. This crucial step (circled in this figure) is dissected in Figure 7. The increased processing of APP results in increased generation of Aβ and AICD, which have been linked to neurodegeneration. Whether the same pathway leads to the altered metabolism of tau in the p44$^{+/+}$ transgenic mice remains to be determined (see *right pathway*). It is possible that Δ40p53 affects tau kinases either directly or through IGF-1R signaling. The altered metabolism of tau in the p44$^{+/+}$ mice is responsible for the synaptic deficits. Question marks (?) indicate signaling steps that need further clarification.

In addition to controlling the rate of Aβ generation through the molecular pathway delineated above, IGF-1R could also affect the toxicity of the Aβ peptide. In fact, down-regulation of the IGF1-R orthologue pathway (DAF-2) in *C. elegans*, appears to reduce the toxicity induced by the Aβ peptide (Cohen et al., 2006). In addition to the work performed in $p44^{+/+}$ animals, the role of IGF-1R in AD neuropathology has been confirmed in three commonly used mouse models of AD, $APP_{695/swe}$, $APP_{695/swe};PS1^{\Delta E9}$ and Tg2576 mice (Cohen et al., 2009; Freude et al., 2009; Pehar et al., 2010). Both $APP_{695/swe}$ and $APP_{695/swe};PS1^{\Delta E9}$ animals express a chimeric mouse APP cDNA encoding the 695-amino acid isoform with a "humanized" Aβ domain that includes the familial AD-associated Swedish double mutation (K595N and M596L). $APP_{695/swe};PS1^{\Delta E9}$ animals co-express a cDNA encoding human presenilin 1 with the familial AD-associated Δ9 mutation. Finally, Tg2576 express the entire human APP 695-amino acid isoform with the Swedish double-mutation. Although the molecular link between IGF-1R signaling and Aβ has been carefully dissected, the mechanisms leading to abnormal metabolism of the microtubule binding protein tau are still unclear (Figure 6). Since tau appears to play such a fundamental role in synaptic deficits of the $p44^{+/+}$ animals (Pehar et al., 2010) as well as other AD and non-AD mouse models (Kurz and Perneczky, 2009; Lee et al., 2001), we can certainly predict that dissection of these events will help us understand general mechanisms of synaptic/cognitive deficits affecting the aging population as well as AD and non-AD forms of dementia.

The main objective of this chapter is to dissect the molecular pathway(s) that connect aging to cognition and AD. As such, the mechanistic role of p53 was evaluated under the light of its longevity-assurance activity. However, the reader should also be aware that p53 is an extremely versatile protein that can impact on different aspects of brain physiology and pathology through additional and, perhaps, equally important activities (see Box 3).

Box 3. p53 and Neurodegeneration: More than Just Aging

Two additional and, perhaps, equally important functions of p53 that can influence different aspects of brain physiology and pathology are: (*i*) the ability to regulate renewal and differentiation of neuronal precursors in key brain regions and (*ii*) the ability to induce and promote apoptotic death of neurons.

(i) Stem Cell Renewal/Differentiation

Several studies have established the presence of progenitor cells in two important regions of the adult brain, the subgranular zone (SGZ) of the dentate gyrus in the hippocampal formation and the subventricular zone (SVZ) of the lateral ventricle wall. These progenitor cells are capable of differentiating into functionally integrated neurons (reviewed in Cayre et al., 2009; Landgren and Curtis, 2010). The progenitor neurons of the SVZ can migrate to the olfactory bulb where they help maintaining plasticity of the system (Imayoshi et al., 2009; Imayoshi et al., 2008; Landgren and Curtis, 2010). In the mouse, the constant (albeit small) renewal of neurons in the olfactory bulb has been involved with the maintenance of olfactory processing as well as behavior (Imayoshi et al., 2009; Scrable et al., 2009). On the other hand, the progenitor cells of the SGZ can migrate from the subgranular to the granular layer of the dentate gyrus and contribute to the increase in the number of granule cells that occurs during adulthood, which is important for the plasticity of pre-existing neuronal circuits and for hippocampal-dependent learning and memory (Imayoshi et al., 2009). Reduced neurogenesis in the dentate gyrus has been associated with cognitive deficits; conversely, increased neurogenesis appears to improve both acquisition of new memories and retrieval of consolidated memories (Clelland et al., 2009; Coras et al., 2010; Deng et al., 2009; Dobrossy et al., 2003; Dupret et al., 2007; Knoth et al., 2010). A possible implication with temporal association functions has also been proposed (Aimone et al., 2006). Importantly, proliferative as well as migration capacities of precursor cells in the SVZ and SGZ are altered during aging and appear compromised in several pathological conditions, which include stroke, seizures, psychiatric disorders and neurodegenerative diseases, such as AD, Parkinson's disease (PD) and Huntington's disease (HD) (Cayre et al., 2009; Scrable et al., 2009; Vandenbosch et al., 2009).

p53 is abundantly expressed in the SVZ and SGZ during both development and adulthood (van Lookeren Campagne and Gill, 1998). A combination of experimental approaches has shown that p53 regulates self-renewal and differentiation of neural stem cells (NSCs) (Armesilla-Diaz et al., 2009; Meletis et al., 2006). Adult NSCs are relatively quiescent; this compensates for their limited proliferation capacities, thus ensuring long-term self-renewal (Kippin et al., 2005). The proliferation rate of NSCs seems to be regulated by p21/CIP1/WAF1, a well-known mediator of cell cycle arrest. NSCs deficient in p21/CIP1/WAF1 fail to maintain quiescence and have a higher proliferation rate. As a result, they display reduced long-term self-renewal potential and early exhaustion of proliferative capacities. Both features have been linked to decreased generation of new neurons in the aging mouse (Kippin et al., 2005). p53 is a powerful positive regulator of p21/CIP1/WAF1. Accordingly, deficiency of p53 down-regulates p21/CIP1/WAF1 (including other cell cycle regulators) and increases the proliferation rate of NSCs (Gil-Perotin et al., 2006; Piltti et al., 2006; Meletis et al., 2006). These results suggest that p53 negatively regulates self-renewal, at least in part, by regulating p21/CIP1/WAF1. In addition to the proliferation rate, p53 deficiency also alters the differentiation pattern of embryonic and adult NSCs. In fact, it increases the production of neurons while reducing the production of astrocytes (Gil-Perotin et al., 2006; Armesilla-Diaz et al., 2009). As already mentioned elsewhere in this chapter, altered p53:Δ40p53 levels affect proliferation, pluripotency and differentiation potential of undifferentiated embryonic stem cells (Ungewitter and Scrable, 2010) and result in impaired adult neurogenesis (Medrano et al., 2009). Therefore, it is likely that by affecting the self-renewal and differentiation of NSCs, p53 plays an essential role in determining the final regenerative capacity of the brain.

(ii) Apoptotic Cell Death

The induction of cell death by p53 involves transcriptional regulation of target genes and microRNAs as well as transcription-independent actions (Bennett, 1999; Chipuk and Green, 2006; He et al., 2007). Although multiple types of cell death have been involved in the progression of neurodegenerative diseases, both the deregulation of p53 and the induction of apoptosis have been proposed as leading mechanisms. Apoptosis can proceed through three major signaling cascades: the extrinsic (or death receptor-mediated) pathway, the intrinsic (or mitochondrial) pathway, and the reticular (or endoplasmic reticulum) pathway. Depending on the cell type, cellular context and apoptotic stimuli, p53 can modulate the intrinsic and extrinsic apoptotic pathways through transcriptional-dependent and -independent mechanisms. Characterized mechanisms that have been implicated with the extrinsic pathway include transcriptional regulation (Muller et al., 1998; Sheikh et al., 1998) as well as trafficking (Bennett et al., 1998) of death receptors, such as Fas, DR5 and TNF-R1. Characterized mechanisms that have been implicated with the intrinsic pathway include increased expression of several pro-apoptotic proteins from the Bcl-2 family (such as Bax, Bid, Noxa and Puma), decreased expression of Bcl-2, increased expression of mitochondrial proteins that promote the permeabilization of the mitochondrial outer membrane (such as ferredoxin reductase, proline oxidase, p53AIP1 and mtCLIC), and increased mitochondrial oxidative stress (due to the down-regulation of MnSOD or the suppression of Nrf2-mediated transcription of antioxidant response genes) (Bennett, 1999; Chipuk and Green, 2006; Faraonio et al., 2006; Galluzzi et al., 2008; Holley et al., 2010). In addition, p53 can also act directly in the cytoplasm, by affecting the oligomerization of proapoptotic multidomain Bcl-2 proteins such as Bax and Bak, and at the mitochondria, by affecting the permeabilization of the outer membrane through the interaction with Bcl-2 and Bcl-xL (Chipuk and Green, 2006; Vaseva and Moll, 2009). The involvement of p53 in the regulation of the reticular pathway is not clearly established. However, the identification of the ER protein scotin, as a p53-inducible proapoptotic gene (Bourdon et al., 2002), appears to implicate the ER with p53-dependent apoptosis.

The induction of apoptosis by p53 has been implicated in several neurodegenerative diseases. Increased levels of p53 have been detected in vulnerable central nervous system regions of AD, PD, HD and Amyotrophic Lateral Sclerosis (ALS) patients (Bae et al., 2005; de la Monte et al., 1997; Eve et al., 2007; Martin, 2000; Mogi et al., 2007). Obviously, increased levels of p53 do not necessarily lead to p53-mediated apoptotic signaling. Indeed, a delicate balance between pro-death and pro-survival signals will determine the final fate of the cell. The identification of different cofactors that regulate the selection of p53-target genes indicates that there are several mechanisms by which the apoptotic response can be regulated (Vousden and Lu, 2002). Nevertheless, studies performed in animal models support the participation of p53 in the neuronal loss observed in AD (LaFerla et al., 1996), PD (Duan et al., 2002; Mandir et al., 2002; Trimmer et al., 1996) and HD (Bae et al., 2005; Ryan et al., 2006). Interestingly, a study performed in a mouse model of spinocerebellar ataxia 1 (SCA1) suggested that p53 might be involved in the progression of the disease through a mechanism that is unrelated to its pro-apoptotic functions (Shahbazian et al., 2001). On the other hand, two different studies showed that p53 deficiency does not affect disease progression in the G93A mutant-SOD1 ALS mouse model (Kuntz et al., 2000; Prudlo et al., 2000). Therefore, the picture is more complex than expected and, perhaps, crosstalk between the different members of the p53 family could affect the final outcome of p53 deficiency.

It is also worth mentioning that various proteins known to have an essential pathogenic role in different neurodegenerative diseases regulate and/or are regulated by p53 activity. In the case of HD, several putative p53-response elements have been identified in mouse and human huntingtin genes. As a result, activation of p53 increases wild-type and mutant huntingtin protein expression *in vitro* and *in vivo* (Feng et al., 2006; Ryan et al., 2006). In the case of PD, several proteins associated with familial forms of the disease, including α-synuclein, parkin and DJ-1, have been shown to modulate the expression and activity of p53 (reviewed in Alves da Costa and Checler, 2011). In the case of AD, an intricate cross-talk exists between p53 and members of the γ-secretase complex. All the members of the γ-secretase complex control p53-mediated cell death through both γ-secretase activity-dependent and -independent mechanisms. At the same time, p53 plays a central role in the regulation of the expression of the members of the γ-secretase complex while AICD, the final end-point of γ-secretase activity, regulate p53 itself (reviewed in Checler et al., 2010). Finally, the survival motor neuron (SMN) protein, which is mutated in spinal muscular atrophy (SMA), interacts directly with p53 and sequesters the transcription factor in nuclear bodies (Young et al., 2002). SMN mutations that are linked to SMA reduce p53 binding, and the degree of inhibition correlates with disease severity (Young et al., 2002). This complex functional cross-talk between pathogenic proteins (or protein derivatives) and p53 clearly suggest that the p53 family plays a pivotal role in the molecular mechanisms of neurodegeneration.

BIOCHEMICAL DISSECTION OF APP PROCESSING
DOWN-STREAM OF IGF-1R

A combination of *in vitro*, *ex vivo* and *in vivo* approaches has clearly shown that the lipid second messenger ceramide plays a crucial role in the above events ((Costantini et al., 2006; Costantini et al., 2005; Puglielli et al., 2003); also reviewed in (Puglielli, 2008)). In fact, ceramide acts as the last output of the signaling pathway that connects IGF-1R to the processing of APP by BACE1 (see Figure 6). Specifically, ceramide regulates a novel form of post-translational regulation of BACE1 involving transient lysine acetylation of the nascent protein in the lumen of the endoplasmic reticulum (ER) and deacetylation in the lumen of the Golgi apparatus (Figure 7). Correctly acetylated intermediates of nascent BACE1 are able to reach the Golgi apparatus and complete maturation whereas non-acetylated intermediates are retained in the ER Golgi Intermediate Compartment (ERGIC) and directed toward degradation (Costantini et al., 2007; Jonas et al., 2008; Jonas et al., 2010; Ko and Puglielli, 2009). In p44$^{+/+}$ mice, the increased activation of ceramide, down-stream of IGF-1R, leads to a more efficient acetylation of nascent BACE1 and to increased levels of mature BACE1. As a result, the animals over-produce Aβ (Costantini et al., 2006).

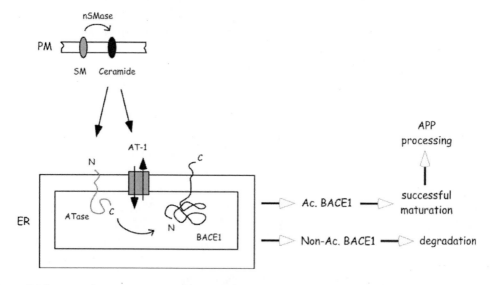

Figure 7. The second messenger ceramide controls BACE1 acetylation in the ER.

These events are described in the appropriate section of this chapter.

BACE1 is a type I membrane protein. It has one single membrane-spanning domain, a large N-terminal ectodomain that faces the lumen of the ER during biogenesis and a short C-terminal tail that faces the cytosol. During translocation along the secretory pathway, BACE1 undergoes different forms of post-translational modifications. The N^{ε}-lysine acetylation occurs in the lumen of the ER and requires an ER membrane acetylCoA transporter (called AT-1) and two acetylCoA:lysine acetyltransferases (called ATase1 and ATase2) (Jonas et al., 2010; Ko and Puglielli, 2009) (Figure 7). AcetylCoA serves as the common donor of the acetyl group for the acetylation of the ε-amino group of lysine residues. It is generated in the cytosol and is highly charged. Because of its charged nature, it is unable to cross a lipid bilayer and is completely impermeable to biological membranes. Therefore, by ensuring active translocation of acetylCoA from the cytosol into the lumen of the ER, AT-1 represents an essential biochemical component of the ER-based acetylation machinery.

AT-1 (also referred to as solute carrier family 33 member 1, SLC33A1) is a 549 amino acids-long protein with eleven to twelve predicted transmembrane domains (Hirabayashi et al., 2004; Kanamori et al., 1997; Lin et al., 2008). It is inhibited with high specificity by extra-lumenal/cytosolic Coenzyme A (CoA), which serves as the carrier of the acetyl group. This behavior is reminiscent of Golgi membrane nucleotide-sugar transporters, mitochondria and ER membrane ATP transporters, and the Golgi membrane adenosine 3'-phosphate 5'-phosphosulfate transporter, which are all inhibited by the corresponding carrier/antiporter molecules (Berninsone and Hirschberg, 2000; Hirschberg et al., 1998). Because of the similarities with the above membrane transporters, it is likely that AT-1 acts as an antiporter. The antiporter mechanism would provide the energy for the carrier-mediated translocation by coupling the entry of extra-lumenal/cytosolic acetyl-CoA with the exit of endolumenal/ER CoA, which results from the reaction of lysine acetylation. Over-expression of AT-1 in cellular systems leads to a more efficient acetylation of nascent BACE1 and increased levels of mature BACE1, thus resulting in more efficient processing of APP and increased production of Aβ (Jonas et al., 2010). Importantly, AT-1 is up-regulated when cells are treated with exogenous ceramide and under conditions that are typically associated with increased levels of endogenous ceramide (Jonas et al., 2010). Particularly, AT-1 is found up-regulated in the brain of p44$^{+/+}$ transgenic mice, in the brain of late-onset/sporadic AD patients, and in aging neurons (Jonas et al., 2010). Surprisingly, AT-1 is also up-regulated in the brain of patients affected by ALS (Jiang et al., 2007) and is mutated in patients affected by autosomal dominant spastic paraplegia-42 (SPG42) (Lin et al., 2008) suggesting a broader role in neurodegenerative diseases.

The two acetylCoA:lysine acetyltransferases, ATase1 and ATase2, are 86% identical at the protein level. Both have one short cytosolic tail, a single transmembrane segment and a larger endolumenal domain with the catalytic activity (Ko and Puglielli, 2009). Both enzymes are able to acetylate BACE1 in vitro and in vivo and are up-regulated by ceramide treatment (Ko and Puglielli, 2009). Importantly, down-regulation of either ATase1 or ATase2 results in reduced levels of BACE1 (Ko and Puglielli, 2009) suggesting that they constitute valid targets for therapeutic purposes. This is in sharp contrast with AT-1, which appears to be essential for cell viability (Jonas et al., 2010). At the mechanistic level, both ATases appear to act as modifying enzymes as well as chaperones (Ko and Puglielli, 2009). The chaperone-like activity might serve to protect the acetylated intermediates of nascent BACE1 from degradation in the ERGIC (Jonas et al., 2008; Ko and Puglielli, 2009). This would explain why only the loss-of-acetylation mutant forms of BACE1 are retained and disposed of in the ERGIC (Jonas et al., 2008).

CONCLUSION

The surprising discovery that the lifespan of an organism is regulated by specific signaling pathways has spurred active interest toward the understanding of the molecular and biochemical events associated with aging. In fact, careful dissection of these events could offer multiple points of intervention for the prevention of the many age-associated diseases and disabilities that affect the growing aging population. Recent work aimed at bridging the gap between aging research and neuroscience has revealed that one of the signaling pathways involved in the regulation of lifespan in the organism is also implicated in the cognitive decline that accompanies aging and in the pathogenesis of AD dementia. This pathway, which acts down-stream of IGF-1R, has been dissected with a combination of *in vitro*, *ex vivo* and *in vivo* approaches. Appropriate animal models have also been generated. Collectively, these studies suggest that pharmacologic strategies targeting IGF-1R signaling should be explored for the treatment and/or prevention of both the cognitive decline that accompanies aging and the AD form of neuropathology.

Dissection of the events connecting IGF-1R signaling to AD has also revealed the existence of a biochemical machinery that controls a transient form of lysine acetylation in the lumen of the ER. Surprisingly, this machinery appears to be deregulated in different neurodegenerative diseases, suggesting a more global implication in the homeostasis of the central nervous system. Therefore, research that was initially aimed at the understanding of aging itself is now helping us understand molecular events that underlie complex neurodegenerative diseases. Further biochemical and molecular assessment of the events described in this chapter might result in additional surprises.

ACKNOWLEDGMENTS

The Authors wish to thank Dr. Corinna Burger, Dr. Rozalyn M. Anderson and Dr. James S. Malter for helpful discussion and critical reading of an early version of this chapter. L.P. is supported by the National Institute of Health and the Department of Veterans Affairs. While preparing the reference section, the Authors tried to maintain a balance between original papers and reviews on selected topics. However, due to the large number of references, they apologize to those whose work could not be included in the final version of this chapter.

DISCLOSURE STATEMENT

The authors have no conflict of interest to disclose.

REVIEWED BY

Dr. Corinna Burger, Dr. Rozalyn M. Anderson and Dr. James S. Malter.

REFERENCES

Aimone, J.B., Wiles, J. and Gage, F.H. (2006) Potential role for adult neurogenesis in the encoding of time in new memories. *Nat Neurosci* 9, 723-7.

Alexander, G. E., Chen, K., Aschenbrenner, M., Merkley, T. L., Santerre-Lemmon, L. E., Shamy, J. L., Skaggs, W. E., Buonocore, M. H., Rapp, P. R. and Barnes, C. A. (2008) Age-related regional network of magnetic resonance imaging gray matter in the rhesus macaque. *J Neurosci* 28, 2710-8.

Alves da Costa, C. and Checler, F. (2011) Apoptosis in Parkinson's disease: Is p53 the missing link between genetic and sporadic Parkinsonism? *Cell Signal* 23, 963-8.

Anderson, R. M. and Weindruch, R. (2010) Metabolic reprogramming, caloric restriction and aging. *Trends Endocrinol Metab* 21, 134-41.

Armesilla-Diaz, A., Bragado, P., Del Valle, I., Cuevas, E., Lazaro, I., Martin, C., Cigudosa, J.C. and Silva, A. (2009) p53 regulates the self-renewal and differentiation of neural precursors. *Neuroscience* 158, 1378-89.

Bae, B.I., Xu, H., Igarashi, S., Fujimuro, M., Agrawal, N., Taya, Y., Hayward, S.D., Moran, T.H., Montell, C., Ross, C.A., Snyder, S.H. and Sawa, A. (2005) p53 mediates cellular dysfunction and behavioral abnormalities in Huntington's disease. *Neuron* 47, 29-41.

Ballatore, C., Lee, V. M. and Trojanowski, J. Q. (2007) Tau-mediated neurodegeneration in Alzheimer's disease and related disorders. *Nat Rev Neurosci* 8 663-72.

Barbieri, M., Bonafe, M., Franceschi, C. and Paolisso, G. (2003) Insulin/IGF-I-signaling pathway: an evolutionarily conserved mechanism of longevity from yeast to humans. *Am J Physiol Endocrinol Metab* 285, E1064-71.

Barnes, C. A., Rao, G. and Houston, F. P. (2000a) LTP induction threshold change in old rats at the perforant path--granule cell synapse. *Neurobiol Aging* 21, 613-20.

Barnes, C. A., Rao, G. and Orr, G. (2000b) Age-related decrease in the Schaffer collateral-evoked EPSP in awake, freely behaving rats. *Neural Plast* 7, 167-78.

Bartke, A. (2008) Impact of reduced insulin-like growth factor-1/insulin signaling on aging in mammals: novel findings. *Aging Cell* 7, 285-90.

Bauer, J. H., Poon, P. C., Glatt-Deeley, H., Abrams, J. M. and Helfand, S. L. (2005) Neuronal expression of p53 dominant-negative proteins in adult Drosophila melanogaster extends life span. *Curr Biol* 15, 2063-8.

Becker, W., Weber, Y., Wetzel, K., Eirmbter, K., Tejedor, F. J. and Joost, H. G. (1998) Sequence characteristics, subcellular localization, and substrate specificity of DYRK-related kinases, a novel family of dual specificity protein kinases. *J Biol Chem* 273, 25893-902.

Bennett, M., Macdonald, K., Chan, S.W., Luzio, J.P., Simari, R. and Weissberg, P. (1998) Cell surface trafficking of Fas: a rapid mechanism of p53-mediated apoptosis. *Science* 282, 290-3.

Bennett, M.R. (1999) Mechanisms of p53-induced apoptosis. *Biochem Pharmacol* 58, 1089-95.

Berman, J. R. and Kenyon, C. (2006) Germ-cell loss extends C. elegans life span through regulation of DAF-16 by kri-1 and lipophilic-hormone signaling. *Cell* 124, 1055-68.

Berninsone, P. M. and Hirschberg, C. B. (2000) Nucleotide sugar transporters of the Golgi apparatus. *Curr Opin Struct Biol* 10, 542-7.

Blalock, E. M., Chen, K. C., Sharrow, K., Herman, J. P., Porter, N. M., Foster, T. C. and Landfield, P. W. (2003) Gene microarrays in hippocampal aging: statistical profiling identifies novel processes correlated with cognitive impairment. *J Neurosci* 23, 3807-19.

Blalock, E. M., Geddes, J. W., Chen, K. C., Porter, N. M., Markesbery, W. R. and Landfield, P. W. (2004) Incipient Alzheimer's disease: microarray correlation analyses reveal major transcriptional and tumor suppressor responses. *Proc Natl Acad Sci U S A* 101, 2173-8.

Bonafe, M., Barbieri, M., Marchegiani, F., Olivieri, F., Ragno, E., Giampieri, C., Mugianesi, E., Centurelli, M., Franceschi, C. and Paolisso, G. (2003) Polymorphic variants of insulin-like growth factor I (IGF-I) receptor and phosphoinositide 3-kinase genes affect IGF-I plasma levels and human longevity: cues for an evolutionarily conserved mechanism of life span control. *J Clin Endocrinol Metab* 88, 3299-304.

Bondy, C. A. and Cheng, C. M. (2004) Signaling by insulin-like growth factor 1 in brain. *Eur J Pharmacol* 490, 25-31.

Bourdon, J. C. (2007) p53 and its isoforms in cancer. *Br J Cancer* 97, 277-82.

Bourdon, J. C., Fernandes, K., Murray-Zmijewski, F., Liu, G., Diot, A., Xirodimas, D. P., Saville, M. K. and Lane, D. P. (2005) p53 isoforms can regulate p53 transcriptional activity. *Genes Dev* 19, 2122-37.

Bourdon, J.C., Renzing, J., Robertson, P.L., Fernandes, K.N. and Lane, D.P. (2002) Scotin, a novel p53-inducible proapoptotic protein located in the ER and the nuclear membrane. *J Cell Biol* 158, 235-46.

Brown-Borg, H. M. (2003) Hormonal regulation of aging and life span. *Trends Endocrinol Metab* 14, 151-3.

Butler, R. N., Sprott, R., Warner, H., Bland, J., Feuers, R., Forster, M., Fillit, H., Harman, S. M., Hewitt, M., Hyman, M. et al. (2004a) Biomarkers of aging: from primitive organisms to humans. *J Gerontol A Biol Sci Med Sci* 59, B560-7.

Butler, R. N., Warner, H. R., Williams, T. F., Austad, S. N., Brody, J. A., Campisi, J., Cerami, A., Cohen, G., Cristofalo, V. J., Drachman, D. A. et al. (2004b) The aging factor in health and disease: the promise of basic research on aging. *Aging Clin Exp Res* 16, 104-11; discussion 111-2.

Cai, H., Wang, Y., McCarthy, D., Wen, H., Borchelt, D. R., Price, D. L. and Wong, P. C. (2001) BACE1 is the major beta-secretase for generation of Abeta peptides by neurons. *Nat Neurosci* 4, 233-4.

Campisi, J. (2004) Fragile fugue: p53 in aging, cancer and IGF signaling. *Nat Med* 10, 231-2.

Campisi, J. (2005a) Senescent cells, tumor suppression, and organismal aging: good citizens, bad neighbors. *Cell* 120, 513-22.

Campisi, J. (2005b) Suppressing cancer: the importance of being senescent. *Science* 309, 886-7.

Cao, L., Kim, S., Xiao, C., Wang, R. H., Coumoul, X., Wang, X., Li, W. M., Xu, X. L., De Soto, J. A., Takai, H. et al. (2006) ATM-Chk2-p53 activation prevents tumorigenesis at an expense of organ homeostasis upon Brca1 deficiency. *Embo J* 25, 2167-77.

Cao, L., Li, W., Kim, S., Brodie, S. G. and Deng, C. X. (2003) Senescence, aging, and malignant transformation mediated by p53 in mice lacking the Brca1 full-length isoform. *Genes Dev* 17, 201-13.

Capsoni, S. and Cattaneo, A. (2006) On the Molecular Basis Linking Nerve Growth Factor (NGF) to Alzheimer's Disease. *Cell Mol Neurobiol* 26, 619-33.

Carlsson, C. M., Gleason, C. E., Puglielli, L., and Asthana, S. Dementia Including Alzheimer's Disease. In: Halter JB. *et al* (Editors) Hazzard's Geriatric Medicine and Gerontology - 6th Ed. New York: McGraw-Hill; 2009: 797-811.

Cayre, M., Canoll, P. and Goldman, J.E. (2009) Cell migration in the normal and pathological postnatal mammalian brain. *Prog Neurobiol* 88, 41-63.

Chang, K. A. and Suh, Y. H. (2010) Possible roles of amyloid intracellular domain of amyloid precursor protein. *BMB Rep* 43, 656-63.

Checler, F., Dunys, J., Pardossi-Piquard, R. and Alves da Costa, C. (2010) p53 is regulated by and regulates members of the gamma-secretase complex. *Neurodegener Dis* 7, 50-5.

Chin, L., Artandi, S. E., Shen, Q., Tam, A., Lee, S. L., Gottlieb, G. J., Greider, C. W. and DePinho, R. A. (1999) p53 deficiency rescues the adverse effects of telomere loss and cooperates with telomere dysfunction to accelerate carcinogenesis. *Cell* 97, 527-38.

Chipuk, J.E. and Green, D.R. (2006) Dissecting p53-dependent apoptosis. *Cell Death Differ* 13, 994-1002.

Cleary, J. P., Walsh, D. M., Hofmeister, J. J., Shankar, G. M., Kuskowski, M. A., Selkoe, D. J. and Ashe, K. H. (2005) Natural oligomers of the amyloid-beta protein specifically disrupt cognitive function. *Nat Neurosci* 8, 79-84.

Clelland, C.D., Choi, M., Romberg, C., Clemenson, G.D., Jr., Fragniere, A., Tyers, P., Jessberger, S., Saksida, L.M., Barker, R.A., Gage, F.H. and Bussey, T.J. (2009) A functional role for adult hippocampal neurogenesis in spatial pattern separation. *Science* 325, 210-3.

Cohen, E., Bieschke, J., Perciavalle, R. M., Kelly, J. W. and Dillin, A. (2006) Opposing activities protect against age-onset proteotoxicity. *Science* 313, 1604-10.

Cohen, E., Paulsson, J. F., Blinder, P., Burstyn-Cohen, T., Du, D., Estepa, G., Adame, A., Pham, H. M., Holzenberger, M., Kelly, J. W. et al. (2009) Reduced IGF-1 signaling delays age-associated proteotoxicity in mice. *Cell* 139, 1157-69.

Coras, R., Siebzehnrubl, F.A., Pauli, E., Huttner, H.B., Njunting, M., Kobow, K., Villmann, C., Hahnen, E., Neuhuber, W., Weigel, D., Buchfelder, M., Stefan, H., Beck, H., Steindler, D.A. and Blumcke, I. (2010) Low proliferation and differentiation capacities of adult hippocampal stem cells correlate with memory dysfunction in humans. *Brain* 133, 3359-72.

Costantini, C., Ko, M. H., Jonas, M. C. and Puglielli, L. (2007) A reversible form of lysine acetylation in the ER and Golgi lumen controls the molecular stabilization of BACE1. *Biochem J* 407, 383-95.

Costantini, C., Scrable, H. and Puglielli, L. (2006) An aging pathway controls the TrkA to p75(NTR) receptor switch and amyloid beta-peptide generation. *Embo J* 25, 1997-2006.

Costantini, C., Weindruch, R., Della Valle, G. and Puglielli, L. (2005) A TrkA-to-p75NTR molecular switch activates amyloid beta-peptide generation during aging. *Biochem J* 391, 59-67.

Coulson, E. J. (2006) Does the p75 neurotrophin receptor mediate Abeta-induced toxicity in Alzheimer's disease? *J Neurochem* 98, 654-60.

Courtois, S., Verhaegh, G., North, S., Luciani, M. G., Lassus, P., Hibner, U., Oren, M. and Hainaut, P. (2002) DeltaN-p53, a natural isoform of p53 lacking the first transactivation domain, counteracts growth suppression by wild-type p53. *Oncogene* 21, 6722-8.

Crozier, R. A., Philpot, B. D., Sawtell, N. B., and Bear, M. F. Long-Term Plasticity of Glutamatergic Synaptic Transmission in the Cerebral Cortex. In: Gazzaniga, M.S. (Editor) The Cognitive Neurosciences III. Cambridge, MA: The MIT Press; 2004: 109-126.

Cutler, R. G., Kelly, J., Storie, K., Pedersen, W. A., Tammara, A., Hatanpaa, K., Troncoso, J. C. and Mattson, M. P. (2004) Involvement of oxidative stress-induced abnormalities in ceramide and cholesterol metabolism in brain aging and Alzheimer's disease. *Proc Natl Acad Sci U S A* 101, 2070-5.

De Strooper, B. (2003) Aph-1, Pen-2, and Nicastrin with Presenilin generate an active gamma-Secretase complex. *Neuron* 38, 9-12.

Delacourte, A., Sergeant, N., Champain, D., Wattez, A., Maurage, C. A., Lebert, F., Pasquier, F. and David, J. P. (2002) Nonoverlapping but synergetic tau and APP pathologies in sporadic Alzheimer's disease. *Neurology* 59, 398-407.

de la Monte, S.M., Sohn, Y.K. and Wands, J.R. (1997) Correlates of p53- and Fas (CD95)- mediated apoptosis in Alzheimer's disease. *J Neurol Sci* 152, 73-83.

Deng, W., Saxe, M.D., Gallina, I.S. and Gage, F.H. (2009) Adult-born hippocampal dentate granule cells undergoing maturation modulate learning and memory in the brain. *J Neurosci* 29, 13532-42.

Dobrossy, M.D., Drapeau, E., Aurousseau, C., Le Moal, M., Piazza, P.V. and Abrous, D.N. (2003) Differential effects of learning on neurogenesis: learning increases or decreases the number of newly born cells depending on their birth date. *Mol Psychiatry* 8, 974-82.

Duan, W., Zhu, X., Ladenheim, B., Yu, Q.S., Guo, Z., Oyler, J., Cutler, R.G., Cadet, J.L., Greig, N.H. and Mattson, M.P. (2002) p53 inhibitors preserve dopamine neurons and motor function in experimental parkinsonism. *Ann Neurol* 52, 597-606.

Dumitriu, D., Hao, J., Hara, Y., Kaufmann, J., Janssen, W. G., Lou, W., Rapp, P. R. and Morrison, J. H. (2010) Selective changes in thin spine density and morphology in monkey prefrontal cortex correlate with aging-related cognitive impairment. *J Neurosci* 30, 7507-15.

Dupret, D., Fabre, A., Dobrossy, M.D., Panatier, A., Rodriguez, J.J., Lamarque, S., Lemaire, V., Oliet, S.H., Piazza, P.V. and Abrous, D.N. (2007) Spatial learning depends on both the addition and removal of new hippocampal neurons. *PLoS Biol* 5, e214.

Duyckaerts, C., Potier, M. C. and Delatour, B. (2008) Alzheimer disease models and human neuropathology: similarities and differences. *Acta Neuropathol* 115, 5-38.

Erraji-Benchekroun, L., Underwood, M. D., Arango, V., Galfalvy, H., Pavlidis, P., Smyrniotopoulos, P., Mann, J. J. and Sibille, E. (2005) Molecular aging in human prefrontal cortex is selective and continuous throughout adult life. *Biol Psychiatry* 57, 549-58.

Eve, D.J., Dennis, J.S. and Citron, B.A. (2007) Transcription factor p53 in degenerating spinal cords. *Brain Res* 1150, 174-181.

Faraonio, R., Vergara, P., Di Marzo, D., Pierantoni, M.G., Napolitano, M., Russo, T. and Cimino, F. (2006) p53 suppresses the Nrf2-dependent transcription of antioxidant response genes. *J Biol Chem* 281, 39776-84.

Feng, Z., Jin, S., Zupnick, A., Hoh, J., de Stanchina, E., Lowe, S., Prives, C. and Levine, A.J. (2006) p53 tumor suppressor protein regulates the levels of huntingtin gene expression. *Oncogene* 25, 1-7.

Ferrer, I., Barrachina, M., Puig, B., Martinez de Lagran, M., Marti, E., Avila, J. and Dierssen, M. (2005) Constitutive Dyrk1A is abnormally expressed in Alzheimer disease, Down syndrome, Pick disease, and related transgenic models. *Neurobiol Dis* 20, 392-400.

Ferri, C. P., Prince, M., Brayne, C., Brodaty, H., Fratiglioni, L., Ganguli, M., Hall, K., Hasegawa, K., Hendrie, H., Huang, Y. et al. (2005) Global prevalence of dementia: a Delphi consensus study. *Lancet* 366, 2112-7.

Fraser, H. B., Khaitovich, P., Plotkin, J. B., Paabo, S. and Eisen, M. B. (2005) Aging and gene expression in the primate brain. *PLoS Biol* 3, e274.

Freude, S., Hettich, M. M., Schumann, C., Stohr, O., Koch, L., Kohler, C., Udelhoven, M., Leeser, U., Muller, M., Kubota, N. et al. (2009) Neuronal IGF-1 resistance reduces Abeta accumulation and protects against premature death in a model of Alzheimer's disease. *Faseb J* 23, 3315-24.

Galluzzi, L., Morselli, E., Kepp, O., Tajeddine, N. and Kroemer, G. (2008) Targeting p53 to mitochondria for cancer therapy. *Cell Cycle* 7, 1949-55.

Garcia-Cao, I., Garcia-Cao, M., Martin-Caballero, J., Criado, L. M., Klatt, P., Flores, J. M., Weill, J. C., Blasco, M. A. and Serrano, M. (2002) "Super p53" mice exhibit enhanced DNA damage response, are tumor resistant and age normally. *Embo J* 21, 6225-35.

Gerisch, B., Weitzel, C., Kober-Eisermann, C., Rottiers, V. and Antebi, A. (2001) A hormonal signaling pathway influencing C. elegans metabolism, reproductive development, and life span. *Dev Cell* 1, 841-51.

Ghosal, K., Vogt, D. L., Liang, M., Shen, Y., Lamb, B. T. and Pimplikar, S. W. (2009) Alzheimer's disease-like pathological features in transgenic mice expressing the APP intracellular domain. *Proc Natl Acad Sci U S A* 106, 18367-72.

Ghosh, A., Stewart, D. and Matlashewski, G. (2004) Regulation of human p53 activity and cell localization by alternative splicing. *Mol Cell Biol* 24, 7987-97.

Gil-Perotin, S., Marin-Husstege, M., Li, J., Soriano-Navarro, M., Zindy, F., Roussel, M.F., Garcia-Verdugo, J.M. and Casaccia-Bonnefil, P. (2006) Loss of p53 induces changes in the behavior of subventricular zone cells: implication for the genesis of glial tumors. *J Neurosci* 26, 1107-16.

Giliberto, L., Zhou, D., Weldon, R., Tamagno, E., De Luca, P., Tabaton, M. and D'Adamio, L. (2008) Evidence that the Amyloid beta Precursor Protein-intracellular domain lowers the stress threshold of neurons and has a "regulated" transcriptional role. *Mol Neurodegener* 3, 12.

Gotz, J., Streffer, J. R., David, D., Schild, A., Hoerndli, F., Pennanen, L., Kurosinski, P. and Chen, F. (2004) Transgenic animal models of Alzheimer's disease and related disorders: histopathology, behavior and therapy. *Mol Psychiatry* 9, 664-83.

Grady, C. L. (2008) Cognitive neuroscience of aging. *Ann N Y Acad Sci* 1124, 127-44.

Grover, R., Candeias, M. M., Fahraeus, R. and Das, S. (2009) p53 and little brother p53/47: linking IRES activities with protein functions. *Oncogene* 28, 2766-72.

Haass, C. and Steiner, H. (2001) Protofibrils, the unifying toxic molecule of neurodegenerative disorders? *Nat Neurosci* 4, 859-60.

Han, X., D, M. H., McKeel, D. W., Jr., Kelley, J. and Morris, J. C. (2002) Substantial sulfatide deficiency and ceramide elevation in very early Alzheimer's disease: potential role in disease pathogenesis. *J Neurochem* 82, 809-18.

He, L., He, X., Lowe, S.W. and Hannon, G.J. (2007) microRNAs join the p53 network-- another piece in the tumour-suppression puzzle. *Nat Rev Cancer* 7, 819-22.

Hebert, L. E., Scherr, P. A., Bienias, J. L., Bennett, D. A. and Evans, D. A. (2003) Alzheimer disease in the US population: prevalence estimates using the 2000 census. *Arch Neurol* 60, 1119-22.

Hedden, T. and Gabrieli, J. D. (2004) Insights into the ageing mind: a view from cognitive neuroscience. *Nat Rev Neurosci* 5, 87-96.

Hedden, T. and Gabrieli, J. D. (2005) Healthy and pathological processes in adult development: new evidence from neuroimaging of the aging brain. *Curr Opin Neurol* 18, 740-7.

Hirabayashi, Y., Kanamori, A., Nomura, K. H. and Nomura, K. (2004) The acetyl-CoA transporter family SLC33. *Pflugers Arch* 447, 760-2.

Hirschberg, C. B., Robbins, P. W. and Abeijon, C. (1998) Transporters of nucleotide sugars, ATP, and nucleotide sulfate in the endoplasmic reticulum and Golgi apparatus. *Annu Rev Biochem* 67, 49-69.

Holley, A.K., Dhar, S.K. and St Clair, D.K. (2010) Manganese superoxide dismutase versus p53: the mitochondrial center. *Ann N Y Acad Sci* 1201, 72-8.

Holzenberger, M., Dupont, J., Ducos, B., Leneuve, P., Geloen, A., Even, P. C., Cervera, P. and Le Bouc, Y. (2003) IGF-1 receptor regulates lifespan and resistance to oxidative stress in mice. *Nature* 421, 182-7.

Hooper, C., Killick, R. and Lovestone, S. (2008) The GSK3 hypothesis of Alzheimer's disease. *J Neurochem* 104, 1433-9.

Hsin, H. and Kenyon, C. (1999) Signals from the reproductive system regulate the lifespan of C. elegans. *Nature* 399, 362-6.

Hu, X. Y., Zhang, H. Y., Qin, S., Xu, H., Swaab, D. F. and Zhou, J. N. (2002) Increased p75(NTR) expression in hippocampal neurons containing hyperphosphorylated tau in Alzheimer patients. *Exp Neurol* 178, 104-11.

Hyman, B. T., Van Hoesen, G. W., Damasio, A. R. and Barnes, C. L. (1984) Alzheimer's disease: cell-specific pathology isolates the hippocampal formation. *Science* 225, 1168-70.

Imayoshi, I., Sakamoto, M., Ohtsuka, T. and Kageyama, R. (2009) Continuous neurogenesis in the adult brain. *Dev Growth Differ* 51, 379-86.

Imayoshi, I., Sakamoto, M., Ohtsuka, T., Takao, K., Miyakawa, T., Yamaguchi, M., Mori, K., Ikeda, T., Itohara, S. and Kageyama, R. (2008) Roles of continuous neurogenesis in the structural and functional integrity of the adult forebrain. *Nat Neurosci* 11, 1153-61.

Jiang, C. H., Tsien, J. Z., Schultz, P. G. and Hu, Y. (2001) The effects of aging on gene expression in the hypothalamus and cortex of mice. *Proc Natl Acad Sci U S A* 98, 1930-4.

Jiang, Y. M., Yamamoto, M., Tanaka, F., Ishigaki, S., Katsuno, M., Adachi, H., Niwa, J., Doyu, M., Yoshida, M., Hashizume, Y. et al. (2007) Gene expressions specifically detected in motor neurons (dynactin 1, early growth response 3, acetyl-CoA transporter, death receptor 5, and cyclin C) differentially correlate to pathologic markers in sporadic amyotrophic lateral sclerosis. *J Neuropathol Exp Neurol* 66, 617-27.

Jonas, M. C., Costantini, C. and Puglielli, L. (2008) PCSK9 is required for the disposal of non-acetylated intermediates of the nascent membrane protein BACE1. *EMBO Rep* 9, 916-22.

Jonas, M. C., Pehar, M. and Puglielli, L. (2010) AT-1 is the ER membrane acetyl-CoA transporter and is essential for cell viability. *J Cell Sci* 123, 3378-88.

Jones, J. I. and Clemmons, D. R. (1995) Insulin-like growth factors and their binding proteins: biological actions. *Endocr Rev* 16, 3-34.

Kanamori, A., Nakayama, J., Fukuda, M. N., Stallcup, W. B., Sasaki, K., Fukuda, M. and Hirabayashi, Y. (1997) Expression cloning and characterization of a cDNA encoding a novel membrane protein required for the formation of O-acetylated ganglioside: a putative acetyl-CoA transporter. *Proc Natl Acad Sci U S A* 94, 2897-902.

Kenyon, C. (2005) The plasticity of aging: insights from long-lived mutants. *Cell* 120, 449-60.

Kenyon, C., Chang, J., Gensch, E., Rudner, A. and Tabtiang, R. (1993) A C. elegans mutant that lives twice as long as wild type. *Nature* 366, 461-4.

Kimura, K. D., Tissenbaum, H. A., Liu, Y. and Ruvkun, G. (1997) daf-2, an insulin receptor-like gene that regulates longevity and diapause in Caenorhabditis elegans. *Science* 277, 942-6.

Kippin, T.E., Martens, D.J. and van der Kooy, D. (2005) p21 loss compromises the relative quiescence of forebrain stem cell proliferation leading to exhaustion of their proliferation capacity. *Genes Dev* 19, 756-67.

Klein, W. L., Krafft, G. A. and Finch, C. E. (2001) Targeting small Abeta oligomers: the solution to an Alzheimer's disease conundrum? *Trends Neurosci* 24, 219-24.

Knoth, R., Singec, I., Ditter, M., Pantazis, G., Capetian, P., Meyer, R.P., Horvat, V., Volk, B. and Kempermann, G. (2010) Murine features of neurogenesis in the human hippocampus across the lifespan from 0 to 100 years. *PLoS One*, 5, e8809.

Ko, M. H. and Puglielli, L. (2009) Two endoplasmic eeticulum (ER)/ER Golgi intermediate compartment-based lysine acetyltransferases post-translationally regulate BACE1 levels. *J Biol Chem* 284, 2482-92.

Kuntz, C.t., Kinoshita, Y., Beal, M.F., Donehower, L.A. and Morrison, R.S. (2000) Absence of p53: no effect in a transgenic mouse model of familial amyotrophic lateral sclerosis. *Exp Neurol* 165, 184-90.

Kurz, A. and Perneczky, R. (2009) Neurobiology of cognitive disorders. *Curr Opin Psychiatry* 22, 546-51.

LaFerla, F.M., Hall, C.K., Ngo, L. and Jay, G. (1996) Extracellular deposition of beta-amyloid upon p53-dependent neuronal cell death in transgenic mice. *J Clin Invest* 98, 1626-32.

Lambert, M. P., Barlow, A. K., Chromy, B. A., Edwards, C., Freed, R., Liosatos, M., Morgan, T. E., Rozovsky, I., Trommer, B., Viola, K. L. et al. (1998) Diffusible, nonfibrillar ligands derived from Abeta1-42 are potent central nervous system neurotoxins. *Proc Natl Acad Sci U S A* 95, 6448-53.

Landfield, P. W., Braun, L. D., Pitler, T. A., Lindsey, J. D. and Lynch, G. (1981) Hippocampal aging in rats: a morphometric study of multiple variables in semithin sections. *Neurobiol Aging* 2, 265-75.

Landfield, P. W. and Lynch, G. (1977) Impaired monosynaptic potentiation in in vitro hippocampal slices from aged, memory-deficient rats. *J Gerontol* 32, 523-33.

Landgren, H. and Curtis, M.A. (2010) Locating and labeling neural stem cells in the brain. *J Cell Physiol* 226, 1-7.

Lansbury, P. T., Jr. (1999) Evolution of amyloid: what normal protein folding may tell us about fibrillogenesis and disease. *Proc Natl Acad Sci U S A* 96, 3342-4.

Leduc, M. S., Hageman, R. S., Meng, Q., Verdugo, R. A., Tsaih, S. W., Churchill, G. A., Paigen, B. and Yuan, R. (2010) Identification of genetic determinants of IGF-1 levels and longevity among mouse inbred strains. *Aging Cell* 9, 823-36.

Lee, C. K., Weindruch, R. and Prolla, T. A. (2000) Gene-expression profile of the ageing brain in mice. *Nat Genet* 25, 294-7.

Lee, V. M., Goedert, M. and Trojanowski, J. Q. (2001) Neurodegenerative tauopathies. *Annu Rev Neurosci* 24, 1121-59.

Li, H., Costantini, C., Scrable, H., Weindruch, R. and Puglielli, L. (2008) Egr-1 and Hipk2 are required for the TrkA to p75(NTR) switch that occurs downstream of IGF1-R. *Neurobiol Aging* 30, 2010-20.

Lim, D. S., Vogel, H., Willerford, D. M., Sands, A. T., Platt, K. A. and Hasty, P. (2000) Analysis of ku80-mutant mice and cells with deficient levels of p53. *Mol Cell Biol* 20, 3772-80.

Lin, P., Li, J., Liu, Q., Mao, F., Qiu, R., Hu, H., Song, Y., Yang, Y., Gao, G., Yan, C. et al. (2008) A missense mutation in SLC33A1, which encodes the acetyl-CoA transporter, causes autosomal-dominant spastic paraplegia (SPG42). *Am J Hum Genet* 83, 752-9.

Logan, J. M., Sanders, A. L., Snyder, A. Z., Morris, J. C. and Buckner, R. L. (2002) Under-recruitment and nonselective recruitment: dissociable neural mechanisms associated with aging. *Neuron* 33, 827-40.

Lu, T., Pan, Y., Kao, S. Y., Li, C., Kohane, I., Chan, J. and Yankner, B. A. (2004) Gene regulation and DNA damage in the ageing human brain. *Nature* 429, 883-91.

Luebke, J., Barbas, H. and Peters, A. (2010) Effects of normal aging on prefrontal area 46 in the rhesus monkey. *Brain Res Rev* 62, 212-32.

Luo, Y., Bolon, B., Kahn, S., Bennett, B. D., Babu-Khan, S., Denis, P., Fan, W., Kha, H., Zhang, J., Gong, Y. et al. (2001) Mice deficient in BACE1, the Alzheimer's beta-secretase, have normal phenotype and abolished beta-amyloid generation. *Nat Neurosci* 4, 231-2.

Mandir, A.S., Simbulan-Rosenthal, C.M., Poitras, M.F., Lumpkin, J.R., Dawson, V.L., Smulson, M.E. and Dawson, T.M. (2002) A novel in vivo post-translational modification of p53 by PARP-1 in MPTP-induced parkinsonism. *J Neurochem* 83, 186-92.

Maier, B., Gluba, W., Bernier, B., Turner, T., Mohammad, K., Guise, T., Sutherland, A., Thorner, M. and Scrable, H. (2004) Modulation of mammalian life span by the short isoform of p53. *Genes Dev* 18, 306-19.

Martin, L.J. (2000) p53 is abnormally elevated and active in the CNS of patients with amyotrophic lateral sclerosis. *Neurobiol Dis* 7, 613-22.

McGowan, E., Eriksen, J. and Hutton, M. (2006) A decade of modeling Alzheimer's disease in transgenic mice. *Trends Genet* 22, 281-9.

Medrano, S., Burns-Cusato, M., Atienza, M.B., Rahimi, D. and Scrable, H. (2009) Regenerative capacity of neural precursors in the adult mammalian brain is under the control of p53. *Neurobiol Aging* 30, 483-97.

Meletis, K., Wirta, V., Hede, S.M., Nister, M., Lundeberg, J. and Frisen, J. (2006) p53 suppresses the self-renewal of adult neural stem cells. *Development* 133, 363-9.

Mendrysa, S. M., O'Leary, K. A., McElwee, M. K., Michalowski, J., Eisenman, R. N., Powell, D. A. and Perry, M. E. (2006) Tumor suppression and normal aging in mice with constitutively high p53 activity. *Genes Dev* 20, 16-21.

Mogi, M., Kondo, T., Mizuno, Y. and Nagatsu, T. (2007) p53 protein, interferon-gamma, and NF-kappaB levels are elevated in the parkinsonian brain. *Neurosci Lett* 414, 94-7.

Motola, D. L., Cummins, C. L., Rottiers, V., Sharma, K. K., Li, T., Li, Y., Suino-Powell, K., Xu, H. E., Auchus, R. J., Antebi, A. et al. (2006) Identification of ligands for DAF-12 that govern dauer formation and reproduction in C. elegans. *Cell* 124, 1209-23.

Muller, M., Wilder, S., Bannasch, D., Israeli, D., Lehlbach, K., Li-Weber, M., Friedman, S.L., Galle, P.R., Stremmel, W., Oren, M. and Krammer, P.H. (1998) p53 activates the CD95 (APO-1/Fas) gene in response to DNA damage by anticancer drugs. *J Exp Med* 188, 2033-45.

Muller, T., Meyer, H. E., Egensperger, R. and Marcus, K. (2008) The amyloid precursor protein intracellular domain (AICD) as modulator of gene expression, apoptosis, and cytoskeletal dynamics-relevance for Alzheimer's disease. *Prog Neurobiol* 85, 393-406.

Murray-Zmijewski, F., Slee, E. A. and Lu, X. (2008) A complex barcode underlies the heterogeneous response of p53 to stress. *Nat Rev Mol Cell Biol* 9, 702-12.

Narasimhan, S. D., Yen, K. and Tissenbaum, H. A. (2009) Converging pathways in lifespan regulation. *Curr Biol* 19, R657-66.

Paolisso, G., Ammendola, S., Del Buono, A., Gambardella, A., Riondino, M., Tagliamonte, M. R., Rizzo, M. R., Carella, C. and Varricchio, M. (1997) Serum levels of insulin-like growth factor-I (IGF-I) and IGF-binding protein-3 in healthy centenarians: relationship with plasma leptin and lipid concentrations, insulin action, and cognitive function. *J Clin Endocrinol Metab* 82, 2204-9.

Pehar, M., O'Riordan, K. J., Burns-Cusato, M., Andrzejewski, M. E., del Alcazar, C. G., Burger, C., Scrable, H. and Puglielli, L. (2010) Altered longevity-assurance activity of p53:p44 in the mouse causes memory loss, neurodegeneration and premature death. *Aging Cell* 9, 174-90.

Persson, J., Sylvester, C. Y., Nelson, J. K., Welsh, K. M., Jonides, J. and Reuter-Lorenz, P. A. (2004) Selection requirements during verb generation: differential recruitment in older and younger adults. *Neuroimage* 23, 1382-90.

Piltti, K., Kerosuo, L., Hakanen, J., Eriksson, M., Angers-Loustau, A., Leppa, S., Salminen, M., Sariola, H. and Wartiovaara, K. (2006) E6/E7 oncogenes increase and tumor suppressors decrease the proportion of self-renewing neural progenitor cells. *Oncogene* 25, 4880-9.

Powell, D. J., Hrstka, R., Candeias, M., Bourougaa, K., Vojtesek, B. and Fahraeus, R. (2008) Stress-dependent changes in the properties of p53 complexes by the alternative translation product p53/47. *Cell Cycle* 7, 950-9.

Prudlo, J., Koenig, J., Graser, J., Burckhardt, E., Mestres, P., Menger, M. and Roemer, K. (2000) Motor neuron cell death in a mouse model of FALS is not mediated by the p53 cell survival regulator. *Brain Res* 879, 183-7.

Puglielli, L. (2008) Aging of the brain, neurotrophin signaling, and Alzheimer's disease: is IGF1-R the common culprit? *Neurobiol Aging* 29, 795-811.

Puglielli, L., Ellis, B. C., Saunders, A. J. and Kovacs, D. M. (2003) Ceramide stabilizes beta-site amyloid precursor protein-cleaving enzyme 1 and promotes amyloid beta-peptide biogenesis. *J Biol Chem* 278, 19777-83.

Puzzo, D., Privitera, L., Leznik, E., Fa, M., Staniszewski, A., Palmeri, A. and Arancio, O. (2008) Picomolar amyloid-beta positively modulates synaptic plasticity and memory in hippocampus. *J Neurosci* 28, 14537-45.

Rabizadeh, S. and Bredesen, D. E. (2003) Ten years on: mediation of cell death by the common neurotrophin receptor p75(NTR). *Cytokine Growth Factor Rev* 14, 225-39.

Rodier, F., Campisi, J. and Bhaumik, D. (2007) Two faces of p53: aging and tumor suppression. *Nucleic Acids Res* 35, 7475-84.

Rottiers, V., Motola, D. L., Gerisch, B., Cummins, C. L., Nishiwaki, K., Mangelsdorf, D. J. and Antebi, A. (2006) Hormonal control of C. elegans dauer formation and life span by a Rieske-like oxygenase. *Dev Cell* 10, 473-82.

Rudolph, K. L., Chang, S., Lee, H. W., Blasco, M., Gottlieb, G. J., Greider, C. and DePinho, R. A. (1999) Longevity, stress response, and cancer in aging telomerase-deficient mice. *Cell* 96, 701-12.

Ryan, A.B., Zeitlin, S.O. and Scrable, H. (2006) Genetic interaction between expanded murine Hdh alleles and p53 reveal deleterious effects of p53 on Huntington's disease pathogenesis. *Neurobiol Dis* 24, 419-27.

Schuster, N., Gotz, C., Faust, M., Schneider, E., Prowald, A., Jungbluth, A. and Montenarh, M. (2001) Wild-type p53 inhibits protein kinase CK2 activity. *J Cell Biochem* 81, 172-83.

Scrable, H., Sasaki, T. and Maier, B. (2005) DeltaNp53 or p44: priming the p53 pump. *Int J Biochem Cell Biol* 37, 913-9.

Scrable, H., Burns-Cusato, M. and Medrano, S. (2009) Anxiety and the aging brain: stressed out over p53? *Biochim Biophys Acta* 1790, 1587-91.

Selkoe, D. J. (1999) Translating cell biology into therapeutic advances in Alzheimer's disease. *Nature* 399, A23-31.

Selkoe, D. J. (2002) Alzheimer's disease is a synaptic failure. *Science* 298, 789-91.

Selman, C., Lingard, S., Choudhury, A. I., Batterham, R. L., Claret, M., Clements, M., Ramadani, F., Okkenhaug, K., Schuster, E., Blanc, E. et al. (2008) Evidence for lifespan extension and delayed age-related biomarkers in insulin receptor substrate 1 null mice. *Faseb J* 22, 807-18.

Shahbazian, M.D., Orr, H.T. and Zoghbi, H.Y. (2001) Reduction of Purkinje cell pathology in SCA1 transgenic mice by p53 deletion. *Neurobiol Dis* 8, 974-81.

Sheikh, M.S., Burns, T.F., Huang, Y., Wu, G.S., Amundson, S., Brooks, K.S., Fornace, A.J., Jr. and el-Deiry, W.S. (1998) p53-dependent and -independent regulation of the death receptor KILLER/DR5 gene expression in response to genotoxic stress and tumor necrosis factor alpha. *Cancer Res* 58, 1593-8.

Small, S. A., Chawla, M. K., Buonocore, M., Rapp, P. R. and Barnes, C. A. (2004) Imaging correlates of brain function in monkeys and rats isolates a hippocampal subregion differentially vulnerable to aging. *Proc Natl Acad Sci U S A* 101, 7181-6.

Stambolic, V., MacPherson, D., Sas, D., Lin, Y., Snow, B., Jang, Y., Benchimol, S. and Mak, T. W. (2001) Regulation of PTEN transcription by p53. *Mol Cell* 8, 317-25.

Suh, Y., Atzmon, G., Cho, M. O., Hwang, D., Liu, B., Leahy, D. J., Barzilai, N. and Cohen, P. (2008) Functionally significant insulin-like growth factor I receptor mutations in centenarians. *Proc Natl Acad Sci U S A* 105, 3438-42.

Taguchi, A., Wartschow, L. M. and White, M. F. (2007) Brain IRS2 signaling coordinates life span and nutrient homeostasis. *Science* 317, 369-72.

Tatar, M., Kopelman, A., Epstein, D., Tu, M. P., Yin, C. M. and Garofalo, R. S. (2001) A mutant Drosophila insulin receptor homolog that extends life-span and impairs neuroendocrine function. *Science* 292, 107-10.

Terry, R. D. and Katzman, R. (2001) Life span and synapses: will there be a primary senile dementia? *Neurobiol Aging* 22, 347-8; discussion 353-4.

Terry, R. D., Masliah, E., Salmon, D. P., Butters, N., DeTeresa, R., Hill, R., Hansen, L. A. and Katzman, R. (1991) Physical basis of cognitive alterations in Alzheimer's disease: synapse loss is the major correlate of cognitive impairment. *Ann Neurol* 30, 572-80.

Thal, D. R., Del Tredici, K. and Braak, H. (2004) Neurodegeneration in normal brain aging and disease. *Sci Aging Knowledge Environ* 2004, pe26.

Tissenbaum, H. A. and Guarente, L. (2002) Model organisms as a guide to mammalian aging. *Dev Cell* 2, 9-19.

Trimmer, P.A., Smith, T.S., Jung, A.B. and Bennett, J.P., Jr. (1996) Dopamine neurons from transgenic mice with a knockout of the p53 gene resist MPTP neurotoxicity. *Neurodegeneration* 5, 233-9.

Tseng, H. C., Zhou, Y., Shen, Y. and Tsai, L. H. (2002) A survey of Cdk5 activator p35 and p25 levels in Alzheimer's disease brains. *FEBS Lett* 523, 58-62.

Tyner, S. D., Venkatachalam, S., Choi, J., Jones, S., Ghebranious, N., Igelmann, H., Lu, X., Soron, G., Cooper, B., Brayton, C. et al. (2002) p53 mutant mice that display early ageing-associated phenotypes. *Nature* 415, 45-53.

Ullrich, A., Gray, A., Tam, A. W., Yang-Feng, T., Tsubokawa, M., Collins, C., Henzel, W., Le Bon, T., Kathuria, S., Chen, E. et al. (1986) Insulin-like growth factor I receptor primary structure: comparison with insulin receptor suggests structural determinants that define functional specificity. *Embo J* 5, 2503-12.

Ungewitter, E. and Scrable, H. (2010) Delta40p53 controls the switch from pluripotency to differentiation by regulating IGF signaling in ESCs. *Genes Dev* 24, 2408-19.

van Heemst, D., Beekman, M., Mooijaart, S. P., Heijmans, B. T., Brandt, B. W., Zwaan, B. J., Slagboom, P. E. and Westendorp, R. G. (2005) Reduced insulin/IGF-1 signalling and human longevity. *Aging Cell* 4, 79-85.

van Lookeren Campagne, M. and Gill, R. (1998) Tumor-suppressor p53 is expressed in proliferating and newly formed neurons of the embryonic and postnatal rat brain: comparison with expression of the cell cycle regulators p21Waf1/Cip1, p27Kip1, p57Kip2, p16Ink4a, cyclin G1, and the proto-oncogene Bax. *J Comp Neurol* 397, 181-198.

Vandenbosch, R., Borgs, L., Beukelaers, P., Belachew, S., Moonen, G., Nguyen, L. and Malgrange, B. (2009) Adult neurogenesis and the diseased brain. *Curr Med Chem* 16, 652-66.

Varela, I., Cadinanos, J., Pendas, A. M., Gutierrez-Fernandez, A., Folgueras, A. R., Sanchez, L. M., Zhou, Z., Rodriguez, F. J., Stewart, C. L., Vega, J. A. et al. (2005) Accelerated ageing in mice deficient in Zmpste24 protease is linked to p53 signalling activation. *Nature* 437, 564-8.

Vaseva, A.V. and Moll, U.M. (2009) The mitochondrial p53 pathway. *Biochim Biophys Acta* 1787, 414-20.

Vogel, H., Lim, D. S., Karsenty, G., Finegold, M. and Hasty, P. (1999) Deletion of Ku86 causes early onset of senescence in mice. *Proc Natl Acad Sci U S A* 96, 10770-5.

Vousden, K.H. and Lu, X. (2002) Live or let die: the cell's response to p53. *Nat Rev Cancer*, 2, 594-604.

Ward, C. W., Garrett, T. P., McKern, N. M., Lou, M., Cosgrove, L. J., Sparrow, L. G., Frenkel, M. J., Hoyne, P. A., Elleman, T. C., Adams, T. E. et al. (2001) The three dimensional structure of the type I insulin-like growth factor receptor. *Mol Pathol* 54, 125-32.

Werner, H., Karnieli, E., Rauscher, F. J. and LeRoith, D. (1996) Wild-type and mutant p53 differentially regulate transcription of the insulin-like growth factor I receptor gene. *Proc Natl Acad Sci U S A* 93, 8318-23.

Wesierska-Gadek, J., Strosznajder, J. B. and Schmid, G. (2006) Interplay between the p53 tumor suppressor protein family and Cdk5: novel therapeutic approaches for the treatment of neurodegenerative diseases using selective Cdk inhibitors. *Mol Neurobiol* 34, 27-50.

Wimo, A., Winblad, B. and Jonsson, L. (2007) An estimate of the total worldwide societal costs of dementia in 2005. *Alzheimers Dement* 3, 81-91.

World Alzheimer Report. 2009 Executive Summary:
http://www.alz.org/national/documents/report_summary_2009worldalzheimerreport.pdf.

Yang, T., Knowles, J. K., Lu, Q., Zhang, H., Arancio, O., Moore, L. A., Chang, T., Wang, Q., Andreasson, K., Rajadas, J. et al. (2008) Small molecule, non-peptide p75 ligands inhibit Abeta-induced neurodegeneration and synaptic impairment. *PLoS One* 3, e3604.

Yankner, B. A., Lu, T. and Loerch, P. (2008) The aging brain. *Annu Rev Pathol* 3, 41-66.

Yin, Y., Stephen, C. W., Luciani, M. G. and Fahraeus, R. (2002) p53 stability and activity is regulated by Mdm2-mediated induction of alternative p53 translation products. *Nat Cell Biol* 4, 462-7.

Young, P.J., Day, P.M., Zhou, J., Androphy, E.J., Morris, G.E. and Lorson, C.L. (2002) A direct interaction between the survival motor neuron protein and p53 and its relationship to spinal muscular atrophy. *J Biol Chem* 277, 2852-9.

In: Cell Aging   ISBN: 978-1-61324-369-5
Editors: Jack W. Perloft and Alexander H. Wong   2012 Nova Science Publishers, Inc.

Chapter 4

AKT/PROTEIN KINASE B (PKB) IN CELL AGING

Eric R. Blough [1,2,3,4,5] *and Miaozong Wu* [1,2,3,6*]

[1] Center for Diagnostic Nanosystems
[2] Department of Biological Sciences
[3] Department of Exercise Science, Sport and Recreation, College of Education and Human Services
[4] Department of Pharmacology, Physiology and Toxicology
[5] Department of Cardiology
[6] Department of Internal Medicine,
Joan C. Edwards School of Medicine,
Marshall University, Huntington, WV, U. S.

ABSTRACT

The increased costs associated with an ever-growing aged population worldwide are expected to pose a significant burden on health care resources. From a biological standpoint, aging is an accelerated deteriorative process in tissue structure and function that is associated with higher morbidity and mortality. The Akt / protein kinase B (PKB) is a family of serine / threonine protein kinases, which play prominent roles in regulation of cellular homeostasis including cell survival, cell growth, gene expression, apoptosis, protein synthesis, and energy metabolism. It has been demonstrated that diminished Akt activity is associated with dysregulation of cellular metabolism and cell death while Akt over-activation has been linked to inappropriate cell growth and proliferation. It is likely that age-related changes in tissue structure (such as atrophy and hypertrophy) and function are related to alterations in Akt expression and Akt-dependent signaling. Although the regulation of Akt function has been well characterized *in vitro*, much less is known regarding the function of Akt *in vivo*. In this article we examine how Akt and Akt-dependent signaling may be regulated in aged cells/tissues, and how age-related alterations in Akt signaling may play a role in influencing cellular metabolism and the response of cells/tissues to environmental factors. Finally, we attempt, where possible, to highlight how such changes may have clinical relevance while also commenting on potential new directions of inquiry.

---
* Corresponding Author

1. INTRODUCTION

Population aging is becoming a worldwide issue. The number of people aged 60 and above accounted for 10% of global population in 2000, and it is projected that this will increase to 21.8% in 2050 and 32.2% in 2100 [1]. Aging exerts a significant economic burden on human society. In the United States, the health care spending per capita for the elderly population is more than 3 and 5 times higher than that for working-age person and child, respectively [2]. According to the United States Department of Health and Human Services, the national health expenditure, which accounted for 16.2% of the Gross Domestic Product in 2007, is projected to increase 6.2% per year over the next decade.

Aging is a complex biological process that is associated with decreased physiological function leading to increases in morbidity and mortality. For example, the age-related loss of skeletal muscle mass leads to decreases in muscle strength which if severe can result in disability. Similar to skeletal muscle, the cardiovascular system also experiences dimished function with age, as it exhibits diminished cardiomyocyte contractility, a prolongation of contraction and relaxation, and eventually cardiovascular failure, a major cause of death in the elderly. The aged brain also undergoes atrophy and degeneration, which if allowed to proceed unchecked can lead to neurodegenerative diseases, such as Alzheimer's. Similarly, the aged respiratory system experiences an increase in dead space and collapsed airways, both conditions which are associated with a higher prevalence of asthma and respiratory failure. The aged immune system is characterized by diminished development of functional T and B cells and antibody production which can lead to increased susceptibility to inflammation and a higher incidence of infections [3].

The underpinning mechanism(s) for increased age-associated morbidity and mortality are not entirely clear but are likely related to molecular changes within the cells. For example, cellular reactive oxygen species (ROS) are increased with age, which can induce genomic and mitochondrial DNA damage and oxidatively modified proteins and lipids. These latter events can result in cellular apoptosis, impaired protein synthesis and cell survival, and have been linked to various pathological developments, including sarcopenia, neurodegeneration and cardiovascular disease (CVD). Aging also appears to affect cellular metabolism. For example, glucose uptake into the cell is diminished with aging, which may contribute to the progressive development of age-associated hyperglycemia, insulin resistance, and cellular ROS production. The Akt / protein kinase B (PKB) is thought to play a central role in regulating cell survival, apoptosis, protein synthesis and glucose metabolism as it functions to integrate anabolic and catabolic responses by transducing extracellular (e.g. growth factors, nutrients and cytokines) and mechanical stimuli (e.g. load and contraction) to its downstream signaling cascades via the phosphorylation of numerous Akt substrates. Herein we summarize two of our recent articles (Front Biosci (Schol Ed) 2010; 2:1169-88 and J Cell Physiol 2011; 226(1):29-36), and investigate the role that Akt may play in the aging progress [3, 4].

2. AKT/PROTEIN KINASE B AND ITS ACTIVATION PATHWAY

The Akt / PKB molecules are a family of serine / threonine-specific protein kinases (EC 2.7.11.1) consisting of three different members: Akt1 (also called PKB-alpha), Akt2 (PKB-beta) and Akt3 (PKB-gamma). These isoforms are coded by three different but highly homologous genes and share significant homology in amino acid sequences. The molecular structure of Akt contains three important domains that are crucial for its kinase activity: an amino-terminal pleckstrin homology (PH) domain, which can interact with phospholipid messenger phosphoinositides-(3,4,5)-P3 (PIP3) and other molecules, a catalytic domain, and a carboxyl-terminal hydrophobic domain [5, 6]. Within these latter two domains are important serine (Ser473 in Akt1, Ser474 in Akt2 and Ser472 in Akt3) and threonine (Thr308 in Akt1, Thr309 in Akt2 and Thr305 in Akt3) residues that can undergo reversible phosphorylation that act, at least in part, to regulate the kinase activity of the Akt molecule.

Many stimuli can modulate Akt activity (Figure 1). Growth factors, such as insulin and insulin-like growth factor-1 (IGF-1), can rapidly activate Akt via binding to tyrosine kinase-type growth factor receptors (TKR) [7]. Besides TKR, Akt may also be activated by G protein-coupled receptors (GPCR). The GPCRs, a large family of seven-transmembrane receptors composed of perhaps more than 800 different subtypes, are responsive to many different hormones, neurotransmitters, calcium and chemokines which suggests that Akt may be responsive to a large number of different biological stimuli [8]. Akt is highly mechanosensitive, as such, it is readily activated by muscle loading and contraction, which leads to increased muscle glucose uptake and protein synthesis [9, 10]. Some heavy metal ions, such as Zn^{2+} and Cu^{2+}, have also been shown to activate phosphoinositide 3-kinases (PI3K) / Akt signaling possibly via inhibition of protein tyrosine phosphatases (PTPases) [11]. Finally, some Akt-binding proteins, e.g. Tcl1 and carboxyl-terminal modulator protein (CTMP), can interact with Akt which may be an additional mechanism to further modulate Akt activity [12, 13].

The sequence of events leading to the activation of Akt is highly conserved [4]. Upon binding to ligand, the conformational change of the TKR or GPCR leads to the activation/ phosphorylation of PI3K, which in turn catalyzes the synthesis of the lipid messenger PIP_3. These lipid products in turn interact with proteins via the PH domain which allows the recruitment of other signaling molecules to the cell membrane. Among the downstream effectors of PI3K, the serine / threonine kinase Akt is the most important and best studied [14]. The PIP_3 can be dephosphorylated by the phosphatase and tensin homologue deleted on chromosome 10 (PTEN), and therefore switches off the PI3K-activated pathway [15, 16]. Similar to other enzymes, Akt activity is regulated by phosphorylation / de-phosphorylation of key amino acid residues (Figure 1). Activation of PI3K elicits interaction between Akt and PIP_3 that can lead to Akt recruitment to the plasma membrane where it is activated through phosphorylation at its serine(Ser473 in Akt1, Ser474 in Akt2 and Ser472 in Akt3) and threonine residues (Thr308 in Akt1, Thr309 in Akt2 and Thr305 in Akt3) [14]. It is thought that phosphorylation of these two residues is necessary for the full activation of Akt kinase activity [14, 17]. Phosphorylation of Thr308 is mediated by 3-phosphoinositide-dependent protein kinase 1 (PDK1) which itself possesses a PH domain that is able to bind PIP_3. Binding of the Akt PH domain to PIP_3 is critical to induce a conformational change which is required for PDK1-dependent phosphorylation [14, 18]. The mechanisms regulating phosphorylation of Ser473 are not well understood. Recent data has suggested the involvement of a mTOR-containing complex (mTORC2) [19], the integrin linked kinase [20] and DNA-dependent protein kinase (DNA-PK) [17].

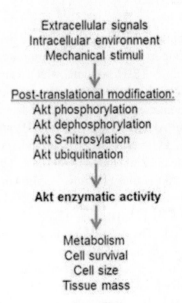

Figure 1.Regulation of Akt enzymatic activity.In response to extracellular signals (e.g. growth factors, nutrients and cytokines), intracellular environment (e.g. oxidative stress and MAPK activity) and mechanical stimuli, Akt molecules can undergo different type of post-translational modifications, including phosphorylation / dephosphorylation, S-nitrosylation and ubiquitination, which in turn lead to stimulate or inhibit Akt activity. Alteration of Akt enzymatic activity is involved in the regulation of cellular metabolism (e.g. protein synthesis and glucose uptake), cell survival (e.g. differentiation, proliferation and apoptosis) and cell size / tissue mass.Modified from [4].

In addition to changes in Akt phosphorylation, S-nitrosylation has recently begun to be appreciated as an important mechanism to negatively regulate Akt kinase activity (Figure 1)[3, 4, 21]. Nitric oxide (NO) produced by the inducible nitric oxide synthase (iNOS) or by NO donors such as GEA 5024 and sodium nitroprusside have been shown to diminish muscle glucose uptake and promote the development of insulin resistance [22-25]. Increases in muscle iNOS expression are common in several insulin resistant models, including genetic obesity, diet-induced obesity, diabetic (db/db) mice, aging and endotoxemia- or LPS-treated animal models [21-25]. The inhibitory effects of NO and iNOS on insulin signaling have been found to be mediated by S-nitrosylation of proteins that are involved in insulin signal transduction, includinginsulin receptor (IR)-beta, insulin receptor substrate-1 (IRS-1) and Akt[21-25]. In the muscle ofiNOS-knockout mice, LPS treatment failed to increase S-nitrosylation of these proteins, and more importantly, failed to produce insulin resistance [22]. We and others using animal and cell culture models have demonstrated that Akt S-nitrosylation can be decreased by acetaminophen and dithiothreitol, resulting in re-activation of Akt kinase activity [21, 25]. Importantly, this restoration of Akt kinase activity following acetaminophen intervention has been shown to decrease the amount of age-associated myocyte apoptosis and increase muscle fiber cross sectional area with this latter effect occurring most likely from an increase in the amount of myosin and actin protein [21]. Another post-translational modification, ubiquitination, recently has been shown to play a role in regulation of Akt function (Figure 1)[26-28]. This ubiquitination can regulate Akt degradation or activation depending on which residues in the Akt molecule undergo ubiquitination[28]. Whether ubiquitination of Aktmay play a role in the aging processhas not been reported.

The physiological function of Akt is mediated via phosphorylation of its downstream molecules [3]. For example, Akt increases glucose uptake via phosphorylation of Akt substrate of 160 kDa (AS160, Thr642) which functions to increase glucose transporter translocation to the cell membrane. Phosphorylation of mTOR (Ser2448) and TSC2 (Ser939 / Ser1086 / Ser1088 / Thr1462) by Akt lead to increases in protein synthesis. Phosphorylation of Bad (Ser136), Bax (Ser184), caspase-9 (Ser196) and forkhead box O-3α (FOXO3α, Thr32/Ser253/Ser315) by Akt regulate apoptotic processes and cell survival. Phosphorylation / inactivation of glycogen synthase kinase (GSK)-3α (Ser21) and GSK3β (Ser9) by Akt function to regulate glycogen synthesis and apoptosis. Phosphorylation / activation of endothelial nitric oxide synthase (eNOS, Ser1177) by Akt regulate blood pressure, vascular remodeling and angiogenesis via generation of nitric oxide (NO). Taken together, these data suggest that Akt signaling regulates diverse critical cellular processes, including cell growth, cell survival, energy metabolism, protein synthesis, and apoptosis.

3. ROLEOF AKTKINASE: TRANSGENICSTUDIES

The Akts are widely expressed in different cell types and tissues. Akt1 is ubiquitously expressed, while Akt2 is largely expressed in insulin-responsive tissues, such as skeletal muscle, adipose tissue and liver. Akt3, on the other hand, is predominately expressed in the brain, lung and kidney. The function of individual Akt molecules has been examined using gene knockout studies (Table 1). Although the knockout of Akt1 or Akt2 is not lethal, this alteration leads to growth deficiency in the transgenic animals. Knockout of Akt1 results in ~ 40% decrease in fetal survivability and a retardation of muscle growth [29, 30]. Cell culture studies have demonstrated that knockdown of Akt1 completely prevents myoblast differentiation but not proliferation, suggesting that Akt may play a critical role in the initiation / early stages of differentiation [31, 32]. This inhibitory effect of Akt1 appears to be mediated by the inhibition of MyoD transcriptional activity [32] suggesting that Akt1 is important for embryonic / myogenic development and postnatal survival. Akt1-deficient mice also exhibit a shortened life span upon exposure to genotoxic stress, a finding that may be related to increased susceptibility to apoptotic stimuli [30]. Gene silencing using small interfering RNA (siRNA) has shown that Akt1 also plays a key role in lipid metabolism, as depletion of Akt1 is associated with increased basal fatty acid uptake and beta oxidation in myotubes[33].

Like Akt1, knockout of Akt2 also results in growth retardation in transgenic mice [34]. This may be related to a deficit in cellular maturation, as Rotwein and Wilson showed that Akt2 deficiency results in delayed myotube maturation of C2 myoblasts [31]. Akt2 is also necessary for maintaining normal glucose homeostasis and insulin sensitivity. Using transgenic mice Cho and colleagues demonstrated that ablation of the Akt2 gene causes both fasting and postprandial hyperglycemia even in the presence of elevated blood insulin levels [35]. These animals also exhibited mild glucose intolerance (glucose tolerance test), impaired insulin responsiveness (insulin tolerance test), and reduced 2-deoxyglucose uptake in skeletal muscle and liver [35, 36]. *Ex vivo* muscle incubation experiments showed that the absence of

Akt2 resulted in lower insulin-stimulated glucose uptake compared to wildtype animals [37], while cell culture studies have shown that siRNA-silencing of Akt2 inhibits insulin-stimulated glucose uptake and glycogen synthesis [33]. Interestingly, Akt2 deficiency does not influence exercise-stimulated glucose uptake suggesting that Akt may play diverse roles in mediating insulin- and exercise-stimulated glucose metabolism [37].

It has been demonstrated that the loss of Akt3 gene reduces brain size in transgenic mice, mainly due to reduction of both brain cell number and size [38]. Double-knockout of Akt1 and Akt2 genes cause severe growth deficiency in mice, resulting in death shortly after birth. These mice display severe skeletal muscle atrophy (~ 50% decrease in muscle mass), mainly due to markedly decreased individual muscle fiber size [39]. Double-silencing of Akt1 and Akt2 genes by siRNA in primary human skeletal muscle myotubes reduces basal and insulin-stimulated glucose uptake and glycogen synthesis, but increases basal palmitate uptake and oxidation [33]. Similar to gene knockout studies, induction of dominant negative Akt into adult muscle fibers results in myofiber atrophy that is specific to the transfected fibers [40]. Loss of Akt1 and Akt3 caused fetal death mainly due to defects in the cardiovascular and nervous systems [41]. Animals deficient in both the Akt2 and Akt3 genes can survive but have less body weight (~25% reduction), substantial reductions in brain and testis size, and exhibit glucose and insulin intolerance [42].

Table 1. Effects of genetic manipulation ofAkt on cellular metabolism and tissue structure

Isoform	Alterations	Reference
Akt1	~ 40% decrease in fetal survivability, fetal and postnatal growth retardance, inhibition of cell differentiation, stress response, apoptosis, increased fatty acid uptake and beta oxidation.	[29-33]
Akt2	Growth retardance and impaired cellular maturation, impaired insulin sensitivity and hyperinsulinemia, impaired glucose homostatisis and hyperglycemia, decreased glycogen synthesis.	[31, 33-37, 49]
Akt3	Impaired brain development.	[38]
Akt1/Akt2	Severe fetal growth retardation, severe muscle atrophy, neonatal lethality, impaired insulin sensitivity and reduced glucose uptake, decreased glycogen synthesis, increased fatty acid uptake and beta oxidation.	[33, 39, 40]
Akt1/Akt3	Embryonic lethality (defects in the cardiovascular and nervous systems).	[41]
Akt2/Akt3	Growth retardance, smaller brain and testis size, glucose and insulin intolerance.	[42]

Modified from [3, 4].

Conversely, the overexpression of Akt has been shown to increase anabolism. Muscle-specific overexpression of constitutively active Akt1 or Akt2 results in increased p70S6 kinase (p70S6K) phosphorylation, glycogen accumulation and muscle fiber hypertrophy [43]. These effects have also been demonstrated in cell culture models. For example, Sandri and colleagues showed that overexpression of constitutively active AKT in cultured myotubes inhibits dexamethasone-induced expression of atrogin-1, an ubiquitin ligase that mediates proteolysis events during muscle atrophy [44]. Similarly, Ueki and Hajduch each demonstrated that expression of a constitutively active Akt increases basal and insulin-stimulated glucose and amino acid uptake, along with protein synthesis in L6 myotubes[45, 46]. Interestingly, the postnatal induction of Akt expression is also able to regulate muscle structure and cellular signaling. Indeed, several laboratories have demonstrated that the targeted expression of a constitutively active Akt can increase myofiber cross sectional area and muscle mass in adult muscle, and reduce body adipose tissue mass [47, 48]. In addition, transfection with constitutively active Akt in adult skeletal muscle can also prevent denervation-induced myofiber atrophy. This effect appears to be specific only to the transfected fibers, as surrounding fibers that had not taken up the Akt construct appear to be normal in appearance, suggesting an autonomous regulation of cell growth by Akt[40]. More recently, constitutively active Akt1 in skeletal muscle has been shown to produce type IIbmyofiber hypertrophy and increase muscle strength [47].

4. REGULATIONOF AKTIN SKELETAL MUSCLE PHYSIOLOGYAND PATHOLOGYOF AGING

Skeletal muscle is one of the largest tissues in the human body, composing more than 35% of total body weight, and responsible for more than one-fourth of postprandial glucose disposal. Skeletal muscle functions to generate the force needed to produce movement in the external environment. In addition, skeletal muscle is also involved in the maintenance of body posture and temperature. As we age, there is a progressive loss of skeletal muscle mass and strength known as sarcopenia. It is thought that the degree of sarcopenia is accelerated with increasing age, and clinical studies have demonstrated that the prevalence of sarcopenia increases dramatically (greater than 4-fold) from ages 70-75 to 85 and older [50]. Similar age-related changes have also been demonstrated in the laboratory rat. For example, in the aging Fischer 344 / NNIaHSD × Brown Norway / BiNia (F344BN) rat model [51], age-associated decreases in muscle mass, muscle cross-sectional area, and diminished muscle function are accelerated after 30 months of age [21, 52, 53]. Sarcopenia is an important health problem as it is directly influences the capacity of an individual to maintain quality of life. Indeed, the development of sarcopenia is associated with a loss of balance, increased muscle weakness and fatigue, increased incidence of falling and fracture, and a higher prevalence of disability [50]. Skeletal muscle exhibits a great deal of plasticity in response to changes in contractile activity (e.g. loading), circulating growth factors, cytokines, and nutrients. However, muscle from aged animals exhibits a diminished ability to adapt the stimuli compared to that observed in their younger counterparts. For example, Hwee and colleagues using the F344BN rat model demonstrated that the ability of the plantaris to undergo muscle hypertrophy in response to functional overload was decreased ~40% at the beginning of middle age (18-weeks old) to about 60% at 30-months of age [54].

It is thought that skeletal muscle mass is determined by the balance between protein synthesis and degradation. Aged muscle has a lower cellular protein-to-RNA ratio and a decreased ability to increase protein synthesis following contractile-, growth factor- or nutrient-stimulation. Among the signaling regulators governing protein synthesis in skeletal muscle, the Akt / mTOR pathway appears to play a critical role. Indeed, Akt / mTORsignaling is activated in response to anabolic stimuli where it is thought to regulate protein synthesis via the phosphorylation of various downstream regulators, including activation of S6 ribosomal protein and inhibition of eukaryotic translation initiation factor-4E (eIF4E) binding protein-1 (4EBP1). The ability of muscle to activate these signaling pathways appears to be attenuated with aging, and this may be related to the diminished capacity of aged muscle to undergo muscle growth in response to an anabolic stimulus. For example, Funai and colleagues showed that aging decreased the capacity of skeletal muscle to phosphorylate 4EBP1, an inhibitory regulator of translation initiation when non-phosphorylated, in response to high-frequency electrical stimulation [55]. Other work from our laboratory has demonstrated similar findings as aging appears to be associated with diminished phosphorylation of the S6 ribosomal protein (Ser235/236) and increased inhibition of eIF4E by binding to 4EBP1 [56], both of which are, in turn, associated with diminished muscle fiber size and decreased expression of the contractile proteins myosin and actin [21]. In addition to the effect of Akt / mTOR signaling on protein synthesis, it is likely that Akt also participates in regulating protein degradation via FOXO signaling. Indeed, the mRNA expression of the muscle-specific E3 ubiquitin ligases muscle RING-finger 1 (MuRF1) and muscle atrophy F-box (MAFbx / atrogin-1) is inhibited by the phosphorylation of FOXO transcription factors by Akt[57]. It has been reported that MuRF1 and MAFbx mRNA and proteins are increased with age in several animal models, and that the increased expression of these molecules appears to related to diminished Akt signaling [58, 59].

As one of major sources of fuel, glucose is taken up into the cell largely in response to insulin stimulation. Insulin can rapidly stimulate Akt activation / phosphorylation and GLUT4 translocation to the plasma membrane within minutes. Research using both humans and animal models have demonstrated that the incidence of insulin resistance increases with age [60, 61]. Although aged muscle retains the ability of insulin to stimulate Akt phosphorylation [21, 62], the magnitude of Akt activation and subsequent phosphorylation of its downstream regulators, such as AS160, is significantly decreased with age [60]. In addition, skeletal muscle from aged animals also exhibit a decreased ability to activate Akt in response to elevations in extracellular glucose [63]. Why aging may diminish insulin sensitivity in skeletal muscle is not well understood. Oh and colleagues found that the expression of Akt and caveolin-1, an insulin sensitivity modulator, is down-regulated in the muscles of aged C57BL/6 × DBA/2 mice (JYD) mice, and that these changes may be related to why these animals suffer from a higher incidence of type 2 diabetes [64]. Consistent with this notion, transient induction of caveolin-1 has been demonstrated able to improve insulin tolerance and muscle glucose uptake in these animals [64]. Duan and co-workers demonstrated that SH2-B, a Src homology 2 (SH2) and PH domain-containing adaptor protein, can bind to insulin receptor in response to insulin. Further, they also noted that disruption of SH2-B can result in age-dependent hyperglycemia, hyperinsulinemia, and glucose intolerance [65]. Moreno and colleagues showed that chronic 17beta-estradiol treatment is able to decrease the deleterious effect of aging on Akt phosphorylation (Ser473) and GLUT4 membrane translocation, indicating age-associated loss of gonadal function may also play a role in the development of skeletal muscle insulin resistance [66].

The ability to respond to stress is an important characteristic of muscle as it is involved in regulation of cellular proliferation, differentiation and muscle homeostasis. The mitogen activated protein kinases (MAPK) are serine / threonine protein kinases that are responsive to and activated by stress stimuli, such as proinflammatory cytokines, mechanical stimuli and oxidative stress. Given that is generally accepted that both Akt and the MAPK signaling cascades function to integrate several different signaling inputs it is likely that the interplay between these two pathways plays an important role in regulating skeletal muscle function. Indeed, during myoblast differentiation, p38 MAPK-induced myogenic differentiation appears to be mediated by the concurrent activation of Akt, while inhibition of p38 by dominant negative MAPK kinase-3 (MKK3) results in decreased Akt expression and Akt activation [67]. Conversely, in the mature myocyte or under conditions of extensive or prolonged stress, the degree of MAPK activation is typically inversely correlated to the amount of Akt activation. For example, Wu and co-workers observed that hyper-phosphorylation of p38 MAPK and ERK1/2 is accompanied by diminished Akt kinase activity in the very aged soleus, while acetaminophen-induced reductions in MAPK activity are associated with the restoration of Akt function [21, 61]. Furthermore, it is noteworthy that these changes also resulted in improvements in protein translational signaling (increased phosphorylation / activation of S6 ribosomal protein (Ser235/236) and translation initiation factor eIF4E (Ser209)), increases in the amount of muscle GLUT4, myosin and actin, along with a decrease in myocyte apoptosis and the prevention of age-associated hyperglycemia [21, 56, 61].

A review of the literature suggests that aging is typically characterized by increases in Akt expression suggesting that Akt levels may not be limiting per se in aged muscle(Table 2) [3]. Nonetheless, the literature does suggest that aging may be associated with alterations in Akt activation and its signaling. Indeed, several studies have demonstrated an age-related mismatch between the degree of Akt phosphorylation level and Akt-dependent downstream signaling. For example, Hwee and colleagues found that even though the amount of basal and overload-induced levels of Akt phosphorylation are increased with age, the ability of Akt to phosphorylate eIF2B epsilonis actually diminished [54]. Similarly, Li and colleagues reported that although IGF-1 increases Akt1 phosphorylation in muscle to a similar degree in young and aged mice the ability of IGF-1- to induce p70S6k phosphorylation appears to be diminished in aged muscle [68]. Given these apparent mismatches between Akt phosphorylation and downstream signaling, we have wondered whether these changes in Akt signaling could be ascribed to changes in the Akt molecule itself. To this end, we have shown that aged skeletal muscle exhibits an uncoupling between Akt phosphorylation and the phosphorylation of its downstream substrates GSK3α and GSK3β [21]. In addition, there is a mismatch between the enzymatic activity of Akt and the degree of Akt phosphorylation at Ser473 and Thr308, as higher Akt phosphorylation was not associated with higher Akt activity as determined by *in vitro* activity assays. Further, we also demonstrated that Akt dysfunction, at least in the skeletal muscle of the aged F344BN rat model, is related to increases in the degree of Akt S-nitrosylation, as reduction of S-nitrosylatedAkt by chronic acetaminophen administration was found to increase *ex vivo* insulin responsiveness, myocyte size, and expression of the contractile proteins myosin and actin [21].

Table 2. Expression of Akt in aged skeletal muscle

Model	Gender	Age	Expression	Reference
Human	Male	70 vs. 20 yr.	↑	[69]
F344BN rats	Male	26 vs. 10 mo.	Akt1: ND	[70]
F344BN rats	Male	30 vs. 6 mo.	↑	[71]
F344BN rats	Male	33 vs. 6 mo.	ND	[21]
Fischer344 rats	Male	24 vs. 3 mo.	ND	[60]
C57Bl/6 mice	Female	24 vs. 5 mo.	Akt1: ND	[68]
Sprague-Dawley rats	Female	30 vs. 4 mo.	↑	[72]

↑: increased; ND: no difference; Akt isoform indicated where appropriate.
Adapted from [3].

5. REGULATION OF AKT IN THE AGING OF THE CARDIOVASCULARSYSTEM

Aging-associated heart diseases are major cause of death, accounting for about 29% of total death in the elderly [73]. Although not well understood it is thought that the profound impact of age on the risk of the occurrence, severity and prognosis of cardiovascular disease is due, in part, to age-associated changes in cardiovascular structure and/or function. Aging in the healthy adult heart is characterized by increased left ventricular (LV) wall thickness, diminished cardiomyocyte contractility, alterationsin the diastolic filling pattern, impaired LV ejection, diminished reserve capacity, prolonged contraction and relaxation, altered heart rhythm, large arterythickening, changes in calcium sensitivity, increased deposition of ECM, increased sarcomere length and number, the loss of nuclear shape, lipofuscin accumulation, and cardiomyocyte hypertrophy. Aging in the vascular system is characterized by vessel elongation, torsion, enlargement of the lumen, a thickening of the vascular walls, and endothelial dysfunction [74, 75]. Under a normal lifetime of operation, the pulsatile nature of blood flow causes the vasculature to undergo a tremendous number of cyclic expansions and contractions. Given the tempestuous demands placed on the vessel walls, the contractile-elastic unit, elastin, begins to fatigue by the sixth decade of life. Eventually, this fatigue is counteracted by changes in the ECM which may include the proliferation of collagen and calcium deposition [76]. How changes in the vessel ECM may affect the ability of the vasculature to respond to mechanical stimuli is not well understood.

Both the Akt1 and Akt2 isoforms are highly expressed in the mammalian heart [77]. Similar to that observed in skeletal muscle, tyrosine kinase receptor binding of various cytokines and growth factors such as insulin and IGF-1 activate PI3K / Akt signaling in the heart. Akt is also involved in the regulation of heart development. The transcription factors FOXO1 and FOXO3 are downstream targets of PI3K / Akt signaling and regulate the expression of cyclin kinase inhibitors (e.g. Cip/Kip). An increase in FOXO1 protein is thought to inhibit myocyte proliferation which could in turn, affect heart development [78]. Akt activation also participates in several cellular processes including glucose uptake, glycogen synthesis, cellular growth and survival, and apoptosis. Akt signaling has been found to play a role in cardiac hypertrophy, apoptosis and angiogenesis. A time dependent effect of Akt activation in the heart has also been shown [79]where short term activation of Akt1 in cardiac muscle by insulin / IGF-1 or other factors results in cardiac hypertrophy, possibly through the mTOR pathway. Akt activation may also play a protective role, as the Akt-associated activation of vascular endothelial growth factor (VEGF) appears to function in preventing ischemia / reperfusion (I/R) injury. In contrast, long-term activation of Akt is associated with dysregulation of intracellular signaling which can promote cardiac hypertrophy and a concomitant inhibition of VEGF and increased myocyte injury following I/R [80]. The up-regulation of FOXO proteins by Akt activation may also serve to protect cardiac myocytes against damage due to ROS [81]. Akt expression levels are decreased in the senescent rat and in aged human hearts [79, 82]. These age-related changes in Akt expression may be related to increases in myocardial fibrosis which is a known contributor of the cardiovascular disease process [82]. A decrease in IGF-1 and downstream Akt-related signaling with aging may also desensitize the cardiomyocyte to mechanical stimuli [83]. Age-associated decreases in Akt expression and activation / phosphorylation have been found to be associated with increased cardiomyocyte apoptosis, altered composition and localization of the cardiac dystrophin glycoprotein complex (DGC), and impaired cardiac membrane integrity [79].

Aktmay also play a key role in the regulation of vascular permeability, angiogenesis and endothelial function. The activation of PI3K / Akt in the vasculature promotes macrophage survival and atherosclerosis [84]. Knockout of Akt1 enhances the high cholesterol diet induced atherosclerosis in ApoE-/- mice [85]. Likewise, insulin and IGF-1 are also implicated in atherosclerotic lesion development [86]. It is postulated that cell senescence is one of the triggering mechanisms of atherosclerotic lesion development [87]. Akt phosphorylation is increased in the balloon injured vascular wall of young rats, but not in old rats. This loss of Akt activation in the vascular wall with aging is associated with a decrease in repair of the endothelium after injury [88]. *In vitro* studies have shown that late-passage endothelial and vascular smooth muscle cells (VSMCs) exhibit less migration and Akt phosphorylation in response to shear stress than early-passage cells [89]. Other studies have begun to establish the role of Akt in cellular senescence and atherogenesis. It is thought that Akt can reduce the lifespan of cultured endothelial cells via a p53/p21 dependent pathway [87]. Similarly, oxidized low density lipoprotein (LDL) suppresses telomerase activity through inhibition of the PI3K / Akt pathway which can lead to pro-atherogenic effects on endothelial cells (ECs) [90]. This loss of telomerase may also contribute to hypertension by increasing endothelin-1 expression (ET-1) [91]. The maintenance of the EC layer is an essential defense mechanism against vascular disorders [92] and is thought to be regulated, at least in part, by endothelial progenitor cells (EPCs) [92]. A decrease in EPC telomerase activity [93] possibly occurring through impaired PI3K / Akt signaling can contribute to the senescence of EPCs and the impairment of EC maintenance [90]. Interestingly, high density lipoprotein (HDL) can increase NO release by activating eNOS which has been shown to increase PI3K / Akt signaling thereby promoting telomerase activity to slow EPCs senescence [94]. Likewise, growth hormone (GH) stimulation can upregulate IGF-1 through PI3K / Akt pathway signaling which in turn is associated with the partial prevention of EPC senescence [95].

Akt is also implicated in the regulation of thrombotic events. Plasminogen activator inhibitor-1 (PAI-1) promotes the thrombotic process by attenuating the fibrinolytic system [96]. PAI-1 is up-regulated in senescent endothelial cells in a process that is inhibited by the activation of the PI3K / Akt pathway [97, 98]. Similarly, the apoptotic effects of thromboxane A2 (TxA2) is also facilitated by the inhibition of Akt phosphorylation [99]. Changes in Akt activity may also be related to the maintenance of vaso-tone as impaired vaso-relaxation is associated with impaired activation of the Akt / eNOS pathway [100].

Aging is also associated with an increased incidence of hypertension. The characteristic pathological change of the vascular wall in aging and hypertension is a progressive increase in the thickness and rigidity of smooth muscle layer in capacitance arteries [101]. Whether Akt may play a direct role in this process is not entirely clear however Akt1 expression is up-regulated in the VSMC of capacitance arteries in hypertensive animals [102]. This finding may be of importance given that increased Akt1 levels in VSMC have been shown to increase cyclin B degradation and produce polypoid / hypertrophic changes [102]. It is thought that these processes might involve the anti-apoptotic effect of Akt1 and its regulation of the Bcl-2 family protein, Bad.

It is likely that aging also contributes to the loss of control of vasomotor function, as senescent endothelial cells exhibit decreased production of NO [103, 104]. This decrease may be related to changes in eNOS expression as the levels of this protein appear to be diminished in the aorta of aged rats, likely by a mechanism involving decreased levels of Akt[103]. Similarly, eNOS phosphorylation is directly modulated by Akt[105]. The factors that may regulate this process are not entirely understood however it is interesting that tribbles homolog (TRIB3) has been shown to impair NO production through its ability to inhibit Akt phosphorylation. As expected, the expression of TRIB3 is dramatically induced during insulin resistance and early aging [106]. Whether TRIB3 plays a direct role in regulating age-related changes in vasculature function by its affects of Akt is currently unclear.

6. REGULATION OF AKT IN THE AGING OF THE NEURONAL SYSTEM

As the neuronal system ages, it undergoes a variety of morphological, functional, and biochemical changes that include neuronal loss, degeneration of synaptic transmission, a loss of glutamate receptors, a decrease in the abundance of chemical messengers such as acetylcholine, dopamine, and norepinephrine, and an increase in oxidative stress and inflammation. These age-associated alterations, if localized to the hippocampus and cerebral cortex, can result in memory impairment and the decline of cognitive function. The formation of senile plaques and neurofibrillary tangles can lead to Alzheimer's disease, the most common neurodegenerative disease in people over 65 years of age and one of top five causes of death in the elderly [73]. Studies have shown that several signaling pathways contribute to age-related changes in neuronal systems including those associated with Akt, a finding that is none too surprising given the role of this molecule in regulating cell survival, cellular apoptotic, and glucose uptake in neuronal system.

As the brain contains diverse functional regions, the expression, activation and function of Akt may also differ. For example, the degree of Akt phosphorylation was significantly increased in the aged rat hippocampus, cerebellum, striatum, and thalamic area, even though there were not significant changes in the expression level of Akt[107]. Age-associated impairment in learning and memory from a senescence-accelerated (SA) mouse model that simulates the early onset of neurodegenerative dementia appears to be related to a reduced phosphorylation of Akt Ser473, as the level of Akt mRNA and protein do not appear to change with aging [108]. This finding has led some to postulate that age-related reductions in Akt activity may be related to neuronal death, which, in turn, could lead to learning and memory impairment [108]. It has also been shown aging may be associated with a significant increases in the number of neuronal and glial cells demonstrating Akt1 immunoreactivity in the CA1 hippocampus section ofmice [109]. However, in the hippocampus of F344BN rats Jackson and colleagues demonstrated a reduction in the amount of phosphorylated Akt Thr308 with age in the CA1 region, and increased phosphorylation in the CA3 region. Conversely, the phosphorylation level of Akt Ser473 did not change significantly with age in either the CA1 or CA3 region [110]. It has also been found that signaling through the PI3K / Akt pathway is impaired in Alzheimer's disease, and importantly, that this diminished signaling may be linked to the formation of senile plaques, neurofibrillary tangles, and neuronal loss [111, 112]. The molecular mechanism(s) responsible for these types of changes are currently unclear. Nonetheless, it is interesting to note that Akt substrates, such as GSK3α and GSK3β, can phosphorylate the tau protein [112]. This may be important given the fact that hyperphosphorylation of tau is associated with the increase of neurofibrillary tangles in Alzheimer's disease. The Aβ, a 40- or 42-amino acid product derived from the serial cleavage of the amyloid precursor protein (APP) by β-secretase and γ-secretase, is a major component of senile plaques, whose production is promoted by activated GSK3α. Attesting to the potential role of Akt in these processes is the finding that PI3K / Akt signaling appears to be inhibited in Alzheimer's disease. This reduced activation of Akt is thought to lead to decreased inhibition (i.e., decreased phosphorylation) of GSK3β, which may be associated with the accumulation of hyperphosphorylated tau protein and neuronal death [112]. Similarly, reduced activation of Akt also leads to decreased inhibition of GSK3α, which can lead to the accumulation of Aβ and further inhibition of PI3K / Akt signaling [112]. Interestingly, intracellular accumulation of Aβ42 actually can affect Akt activation. Using primary cultures of cortical cells, Querfurth and co-workers found that Aβ42 can decrease Akt phosphorylation without changes in the amount of total Akt expression [113]. This Aβ42-induced decrease in Akt phosphorylation resulted in reduced GSK3β phosphorylation [113]. Although the mechanism(s) underlying these changes are not fully understood, these data led to the postulate that increased levels of intracellular Aβ might act to inhibit the interaction between PDK1 and Akt, thus inhibiting the PDK-dependent activation of Akt[111]. Additional research to explore how aging and Alzheimer's disease may affect Akt signaling will no doubt be useful in furthering our understanding of how differences in Akt activity may be related to this neurological disease.

7. REGULATION OF AKT IN THE AGING OF OTHER PHYSIOLOGICALSYSTEMS

Respiratory diseases cause approximately 6% of the total deaths in people 65 years of age and older [73]. As we age, the lung is thought to become less elastic, and muscles of chest wall, airway and diaphragm lose mass and strength, resulting in increases in the volume of dead space, diminished peak airflow, and reduced capability of lung alveolar gas exchange. These age-related alterations in respiratory tissues are associated with an increased possibility of developing respiratory failure, pneumonia, asthma, fibrosis, chronic bronchitis, chronic obstructive pulmonary disease, emphysema, aspiration, influenza, tuberculosis, and pleurisy [3, 9]. Although the basal activity of Aktis higher in adult and aged diaphragms, mechanical stretch-induced activation of Akt and its downstream substrate I-kappa B-kinase (IKK), which is thought to lead to increased binding of the FOXO transcription factor to DNA were diminished in the diaphragms from aged mice, suggesting that aging in the diaphragm is associated with a loss of muscle mechanosensitivity[114]. *In vitro* cell culture studies also demonstrated an age-associated impairment in Akt signaling. Although the level of Akt Ser473 phosphorylation in cultured senescent WI-38 diploid fibroblast cells were not different from that in pre-senescent cells [115, 116], the capability of Akt to phosphorylate nuclear targets was reduced in the senescent cells [115].

The function of immune system gradually declines with age. The aged immune system not only produces fewer antibodies in response to vaccination, but also the produced antibodies have a lower affinity for their antigens, resulting in an impaired capability to distinguish foreign invaders from "self". These changes, together, may result in increased susceptibility to inflammation and increased risks of autoimmune disorders and other diseases, such as cancer [117, 118]. Though seldom done, studies have shown that age-associated changes in the immune system may be related to alterations in the Akt pathway. Using cells obtained from elderly humans, Larbi and colleagues found an age-associated impairment in lipid raft polarization, a process critical for assembling the T-cell receptor (TCR) signaling and T-cell activation. The impaired lipid rafts in both resting and stimulated CD4+ T cells from elderly human subjects were associated with decreased Akt phosphorylation when compared to their younger counterparts [119]. Similarly, it has been reported that a reduced phosphorylation of Akt in dendritic cells from the elderly subjects is associated with impaired phagocytosis and migration, resulting in an impaired innate immune response in the aged [120]. Additional research to explore age-associated modification in Akt signaling in the immune system is needed to better understand the mechanisms behind these changes.

Aging results in dramatic changes in renal morphology and physiology, including a loss of renal function, glomerulosclerosis, interstitial fibrosis, the infiltration of inflammatory cells and tubular casts. It have been reported that age-associated renal cortex dysfunction in rats is accompanied by a significant decrease in Akt activity, which results in the attenuation of cell growth and replacement [121]. Similarly, increased oxidative stress and its deleterious effects possibly on Akt activity may also be involved in the age-related renal disorders [122].

The percentage of body fat is accumulating with age, and more importantly the preferential accumulation of adipose tissue in the abdominal region is excessively increased. These age-associated increases in body fat, in turn, increase the risk of obesity, diabetes, cardiovascular disease and thus may accelerate the aging process. Of the three Akt isoforms, Akt2 appears to be the most abundant in adipose tissue [49]. Using Akt2-null mice model, Coleman and colleagues demonstrated that both male and female mice missing the Akt2 gene exhibited aged-dependent lipoatrophy (i.e., loss of adipose tissue)[49], suggesting that Akt2 may function in the regulation of adipose tissue size. Similar to other tissues, Akt also performs important functions in regulation of glucose metabolism. Although the basal Akt phosphorylation in epididymal adipose tissue from aged animals is higher than young counterparts, insulin fails to increase Akt phosphorylation and Akt translocation to the plasma membrane in aged fat [62, 123]. Associated with this impairment of Akt activity, the insulin-stimulated translocation of GLUT4 from intracellular compartments to the plasma membrane was also significantly decreased in aged rat adipocytes [123]. Although not fully understood, it is thought that the reduced activation of Akt and diminished GLUT4 membrane translocation with aging may be related to age-associated differences in the subcellular distribution and phosphorylation of IRS-1 and IRS-3 proteins [123].

The quality of skeletal system is also greatly affected by aging. It is thought osteoblasts function to construct bone tissue while osteoclasts maintain mineral homeostasis by reabsorbing bone. Working together, these two cell types act to maintain the balance between bone deposition and resorption[124]. Nonetheless, as we age the number of osteoblast cells tends to decrease, impairing the natural processes of bone tissue renovation and a reduction in bone mass. Left unchecked, bone loss can approach up to 70% by the age of 70 [125]. Osteoporosis, defined by compromising bone strength with an increasing risk of fracture, is perhaps the most prevalent bone disease in elderly population [126]. With increasing age, it is thought that bone exhibits diminished activation of PI3K / Akt signaling in response to IGF-1[127]. Alterations in Akt signaling may also be involved in the development of osteoarthritis, with this process occurring through the inhibition of Akt by TRIB3 protein, an inhibitor of insulin signaling [128].

8. REGULATION OF AKT IN LIFESPAN

The use of different model organisms, such as the budding yeast *Saccharomyces cerevisiae*, the nematode *Caenorhabditiselegans* and transgenic mice, has done much to help unravel the different molecular mechanisms involved in lifespan regulation. Recent findings have suggested that the silent information regulator-2 (Sir2) family of genes in *S. cerevisiae* and daf-16 gene in *C. elegans* may play a key role in regulating lifespan. These genes are conserved during evolution and homologs of sirtuins and FoxOs (daf-16 homologs) are thought to be involved in regulating the expression of anti-aging genes. For example, over-activation of insulin / IGF-1 / PI3K / Akt pathway may function as a negative regulator of lifespan via its ability to transactivate the FoxO genes. Indeed, in *C. elegans*, increased phosphorylation of forkhead / winged-helix transcription factor DAF-16 by Akt prevents the nuclear localization of DAF-16 and hence the induction of DAF-16 dependent gene expression [129]. Interestingly, *in vivo* and *in vitro* studies have shown that increased Akt phosphorylation negatively regulates longevity [87, 130-132]. On the other hand, absence of Akt can also lead to shortened life span, as Chen and colleagues have shown that Akt1-deficient mice have a shortened life span upon exposure to genotoxic stress [30].

Energy metabolism is also thought to be coupled to life span, and calorie restriction has been investigated as an approach to extend life span. Studies have shown that calorie restriction can reduce adipose tissue mass and that this change is accompanied by decreases in oxygen metabolism and the production of mitochondrial hydrogen peroxide. Akt over-activation may increase oxidative stress through inhibition of FOXO transcription factors [133]. Aging can also increases the activation of NF-kB signaling via the PI3K / Akt pathway [134], a process that can be suppressed by caloric restriction [134]. Park and colleagues studied the effect of calorie restriction (reduction by 30% for 4 months) on insulin signaling in the epididymal adipose tissue of young (7 mo.) and aged (22 mo.) male F344 rats [135]. Their findings suggested that the basal level of Akt phosphorylation in the young calorie-restricted group was increased compared to age-matched controls, and that this alteration was associated with differences in the activation of the insulin signaling pathway. Conversely, the basal level of phosphorylated Akt was decreased in the aged calorie-restricted group when compared to their younger counterparts, a finding they attributed to age-associated desensitization of insulin signaling [135].

CONCLUSIONS AND PERSPECTIVES

Age-associated alterations in tissue structure and function are postulated to play a role in the physiological impairment seen with aging. The mechanism(s) underlying these changes are not well understood given the complexity of cellular function and the difficulty of studying physiological processes in the aged tissues. Akt / protein kinase B belongs to a family of serine/threonine protein kinases, which play prominent roles in a diverse number of processes including cell survival, cell growth, gene expression, apoptosis, protein synthesis, and energy metabolism in response to extracellular stimuli (such as growth factors, mechanical loading) and intracellular interaction (such as Akt-binding proteins, post-translational modifications). Given the importance of Akt in regulating such a wide variety of cellular functions, it is likely that age-related changes in tissue structure and function performance may be related to alterations in Akt expression and Akt-dependent signaling. Recent data from a variety of different animal models and other work using *in vitro* approaches appear to support this notion. For example, the restoration of Akt function in aged animals has been shown to increase muscle mass, prevent denervation-induced myofiber atrophy and reduce body adipose tissue mass [40, 47]. Although these findings are promising, more work remains. For example, a review of the literature indicates that most of the investigations to date have focused on the role of Akt in the aging male. It is well known that sex may have an effect on the aging process while other studies have shown that gonadal hormones can affect Akt activation and other physiological processes such as glucose metabolism [53, 66]. Which residue(s) in the Akt molecule undergo S-nitrosylation*in vivo* and how might these changes contribute to the etiology of insulin resistance? Beyond S-nitrosylation and its better studied partner phosphorylation, are there other modifications involved in positively or negatively regulating Akt kinase activity? It is anticipated that additional studies will further our understanding of the role that Akt plays in the aging progress and will yield valuable clinical insight for the increasing aged population.

REFERENCES

[1]   Lutz, W., W. Sanderson, and S. Scherbov, The coming acceleration of global population ageing. *Nature*, 2008. 451(7179): p. 716-9.

[2]   Hartman, M., A. Catlin, D. Lassman, J. Cylus, et al., U.S. Health spending by age, selected years through 2004. *Health Aff* (Millwood), 2008. 27(1): p. w1-w12.

[3]   Wu, M., B. Wang, J. Fei, N. Santanam, et al., Important roles of Akt/PKB signaling in the aging process. *Front Biosci* (Schol Ed), 2010. 2: p. 1169-88.

[4]   Wu, M., M. Falasca, and E.R. Blough, Akt/protein kinase B in skeletal muscle physiology and pathology. *J Cell Physiol*, 2011. 226(1): p. 29-36.

[5]   Brazil, D.P., J. Park, and B.A. Hemmings, PKB binding proteins. *Getting in on the Akt. Cell*, 2002. 111(3): p. 293-303.

[6]   Franke, T.F., Intracellular signaling by Akt: bound to be specific. *Sci Signal*, 2008. 1(24): p. pe29.

[7]   Altomare, D.A., G.E. Lyons, Y. Mitsuuchi, J.Q. Cheng, et al., Akt2 mRNA is highly expressed in embryonic brown fat and the AKT2 kinase is activated by insulin. *Oncogene*, 1998. 16(18): p. 2407-11.

[8]   New, D.C., K. Wu, A.W. Kwok, and Y.H. Wong, G protein-coupled receptor-induced Akt activity in cellular proliferation and apoptosis. *Febs J*, 2007. 274(23): p. 6025-36.

[9]   Wu, M., J. Fannin, K.M. Rice, B. Wang, et al., Effect of aging on cellular mechanotransduction. *Ageing Res Rev*, 2001.10(1): p. 1-15.

[10] Katta, A., S. Kakarla, M. Wu, S. Paturi, et al., Altered regulation of contraction-induced Akt/mTOR/p70S6k pathway signaling in skeletal muscle of the obese Zucker rat. *Exp Diabetes Res*, 2009. 2009: p. 384683.

[11] Barthel, A., E.A. Ostrakhovitch, P.L. Walter, A. Kampkotter, et al., Stimulation of phosphoinositide 3-kinase/Akt signaling by copper and zinc ions: mechanisms and consequences. *Arch Biochem Biophys*, 2007. 463(2): p. 175-82.

[12] Noguchi, M., V. Ropars, C. Roumestand, and F. Suizu, Proto-oncogene TCL1: more than just a coactivator for Akt. *Faseb J*, 2007. 21(10): p. 2273-84.

[13] Maira, S.M., I. Galetic, D.P. Brazil, S. Kaech, et al., Carboxyl-terminal modulator protein (CTMP), a negative regulator of PKB/Akt and v-Akt at the plasma membrane. *Science*, 2001. 294(5541): p. 374-80.

[14] Alessi, D.R., S.R. James, C.P. Downes, A.B. Holmes, et al., Characterization of a 3-phosphoinositide-dependent protein kinase which phosphorylates and activates protein kinase Balpha. *Curr Biol*, 1997. 7(4): p. 261-9.

[15] Wan, X. and L.J. Helman, Levels of PTEN protein modulate Akt phosphorylation on serine 473, but not on threonine 308, in IGF-II-overexpressing rhabdomyosarcomas cells. *Oncogene*, 2003. 22(50): p. 8205-11.

[16] Leslie, N.R., I.H. Batty, H. Maccario, L. Davidson, et al., Understanding PTEN regulation: PIP2, polarity and protein stability. *Oncogene*, 2008. 27(41): p. 5464-76.

[17] Feng, J., J. Park, P. Cron, D. Hess, et al., Identification of a PKB/Akt hydrophobic motif Ser-473 kinase as DNA-dependent protein kinase. *J Biol Chem*, 2004. 279(39): p. 41189-96.

[18] Alessi, D.R., M. Deak, A. Casamayor, F.B. Caudwell, et al., 3-Phosphoinositide-dependent protein kinase-1 (PDK1): structural and functional homology with the Drosophila DSTPK61 kinase. *Curr Biol*, 1997. 7(10): p. 776-89.

[19] Sarbassov, D.D., D.A. Guertin, S.M. Ali, and D.M. Sabatini, Phosphorylation and regulation of Akt/PKB by the rictor-mTOR complex. *Science*, 2005. 307(5712): p. 1098-101.

[20] Delcommenne, M., C. Tan, V. Gray, L. Rue, et al., Phosphoinositide-3-OH kinase-dependent regulation of glycogen synthase kinase 3 and protein kinase B/AKT by the integrin-linked kinase. *Proc Natl Acad Sci U S A*, 1998. 95(19): p. 11211-6.

[21] Wu, M., A. Katta, M.K. Gadde, H. Liu, et al., Aging-associated dysfunction of akt/protein kinase B: s-nitrosylation and acetaminophen intervention. *PLoS One*, 2009. 4(7): p. e6430.

[22] Carvalho-Filho, M.A., M. Ueno, J.B. Carvalheira, L.A. Velloso, et al., Targeted disruption of iNOS prevents LPS-induced S-nitrosation of IRbeta/IRS-1 and Akt and insulin resistance in muscle of mice. *Am J Physiol Endocrinol Metab*, 2006. 291(3): p. E476-82.

[23] Carvalho-Filho, M.A., M. Ueno, S.M. Hirabara, A.B. Seabra, et al., S-nitrosation of the insulin receptor, insulin receptor substrate 1, and protein kinase B/Akt: a novel mechanism of insulin resistance. *Diabetes*, 2005. 54(4): p. 959-67.

[24] Kapur, S., S. Bedard, B. Marcotte, C.H. Cote, et al., Expression of nitric oxide synthase in skeletal muscle: a novel role for nitric oxide as a modulator of insulin action. *Diabetes*, 1997. 46(11): p. 1691-700.

[25] Yasukawa, T., E. Tokunaga, H. Ota, H. Sugita, et al., S-nitrosylation-dependent inactivation of Akt/protein kinase B in insulin resistance. *J Biol Chem*, 2005. 280(9): p. 7511-8.

[26] Medina, E.A., R.R. Afsari, T. Ravid, S.S. Castillo, et al., Tumor necrosis factor-{alpha} decreases Akt protein levels in 3T3-L1 adipocytes via the caspase-dependent ubiquitination of Akt. *Endocrinology*, 2005. 146(6): p. 2726-35.

[27] Yang, W.L., J. Wang, C.H. Chan, S.W. Lee, et al., The E3 ligase TRAF6 regulates Akt ubiquitination and activation. *Science*, 2009. 325(5944): p. 1134-8.

[28] Yang, W.L., C.Y. Wu, J. Wu, and H.K. Lin, Regulation of Akt signaling activation by ubiquitination. *Cell Cycle*, 2010. 9(3): p. 487-97.

[29] Cho, H., J.L. Thorvaldsen, Q. Chu, F. Feng, et al., Akt1/PKBalpha is required for normal growth but dispensable for maintenance of glucose homeostasis in mice. *J Biol Chem*, 2001. 276(42): p. 38349-52.

[30] Chen, W.S., P.Z. Xu, K. Gottlob, M.L. Chen, et al., Growth retardation and increased apoptosis in mice with homozygous disruption of the Akt1 gene. *Genes Dev*, 2001. 15(17): p. 2203-8.

[31] Rotwein, P. and E.M. Wilson, Distinct actions of Akt1 and Akt2 in skeletal muscle differentiation. *J Cell Physiol*, 2009. 219(2): p. 503-11.

[32] Wilson, E.M. and P. Rotwein, Selective control of skeletal muscle differentiation by Akt1. *J Biol Chem*, 2007. 282(8): p. 5106-10.

[33] Bouzakri, K., A. Zachrisson, L. Al-Khalili, B.B. Zhang, et al., siRNA-based gene silencing reveals specialized roles of IRS-1/Akt2 and IRS-2/Akt1 in glucose and lipid metabolism in human skeletal muscle. *Cell Metab*, 2006. 4(1): p. 89-96.

[34] McCurdy, C.E. and G.D. Cartee, Akt2 is essential for the full effect of calorie restriction on insulin-stimulated glucose uptake in skeletal muscle. *Diabetes*, 2005. 54(5): p. 1349-56.

[35] Cho, H., J. Mu, J.K. Kim, J.L. Thorvaldsen, et al., Insulin resistance and a diabetes mellitus-like syndrome in mice lacking the protein kinase Akt2 (PKB beta). *Science*, 2001. 292(5522): p. 1728-31.

[36] Chen, W.S., X.D. Peng, Y. Wang, P.Z. Xu, et al., Leptin deficiency and beta-cell dysfunction underlie type 2 diabetes in compound Akt knockout mice. *Mol Cell Biol*, 2009. 29(11): p. 3151-62.

[37] Sakamoto, K., D.E. Arnolds, N. Fujii, H.F. Kramer, et al., Role of Akt2 in contraction-stimulated cell signaling and glucose uptake in skeletal muscle. *Am J Physiol Endocrinol Metab*, 2006. 291(5): p. E1031-7.

[38] Easton, R.M., H. Cho, K. Roovers, D.W. Shineman, et al., Role for Akt3/protein kinase Bgamma in attainment of normal brain size. *Mol Cell Biol*, 2005. 25(5): p. 1869-78.

[39] Peng, X.D., P.Z. Xu, M.L. Chen, A. Hahn-Windgassen, et al., Dwarfism, impaired skin development, skeletal muscle atrophy, delayed bone development, and impeded adipogenesis in mice lacking Akt1 and Akt2. *Genes Dev*, 2003. 17(11): p. 1352-65.

[40] Pallafacchina, G., E. Calabria, A.L. Serrano, J.M. Kalhovde, et al., A protein kinase B-dependent and rapamycin-sensitive pathway controls skeletal muscle growth but not fiber type specification. *Proc Natl Acad Sci U S A*, 2002. 99(14): p. 9213-8.

[41] Yang, Z.Z., O. Tschopp, N. Di-Poi, E. Bruder, et al., Dosage-dependent effects of Akt1/protein kinase Balpha (PKBalpha) and Akt3/PKBgamma on thymus, skin, and

cardiovascular and nervous system development in mice. *Mol Cell Biol*, 2005. 25(23): p. 10407-18.

[42] Dummler, B., O. Tschopp, D. Hynx, Z.Z. Yang, et al., Life with a single isoform of Akt: mice lacking Akt2 and Akt3 are viable but display impaired glucose homeostasis and growth deficiencies. *Mol Cell Biol*, 2006. 26(21): p. 8042-51.

[43] Cleasby, M.E., T.A. Reinten, G.J. Cooney, D.E. James, et al., Functional studies of Akt isoform specificity in skeletal muscle in vivo; maintained insulin sensitivity despite reduced insulin receptor substrate-1 expression. *Mol Endocrinol*, 2007. 21(1): p. 215-28.

[44] Sandri, M., C. Sandri, A. Gilbert, C. Skurk, et al., Foxo transcription factors induce the atrophy-related ubiquitin ligase atrogin-1 and cause skeletal muscle atrophy. *Cell*, 2004. 117(3): p. 399-412.

[45] Ueki, K., R. Yamamoto-Honda, Y. Kaburagi, T. Yamauchi, et al., Potential role of protein kinase B in insulin-induced glucose transport, glycogen synthesis, and protein synthesis. *J Biol Chem*, 1998. 273(9): p. 5315-22.

[46] Hajduch, E., D.R. Alessi, B.A. Hemmings, and H.S. Hundal, Constitutive activation of protein kinase B alpha by membrane targeting promotes glucose and system A amino acid transport, protein synthesis, and inactivation of glycogen synthase kinase 3 in L6 muscle cells. *Diabetes*, 1998. 47(7): p. 1006-13.

[47] Izumiya, Y., T. Hopkins, C. Morris, K. Sato, et al., Fast/Glycolytic muscle fiber growth reduces fat mass and improves metabolic parameters in obese mice. *Cell Metab*, 2008. 7(2): p. 159-72.

[48] Lai, K.M., M. Gonzalez, W.T. Poueymirou, W.O. Kline, et al., Conditional activation of akt in adult skeletal muscle induces rapid hypertrophy. *Mol Cell Biol*, 2004. 24(21): p. 9295-304.

[49] Garofalo, R.S., S.J. Orena, K. Rafidi, A.J. Torchia, et al., Severe diabetes, age-dependent loss of adipose tissue, and mild growth deficiency in mice lacking Akt2/PKB beta. *J Clin Invest*, 2003. 112(2): p. 197-208.

[50] Castillo, E.M., D. Goodman-Gruen, D. Kritz-Silverstein, D.J. Morton, et al., Sarcopenia in elderly men and women: the Rancho Bernardo study. *Am J Prev Med*, 2003. 25(3): p. 226-31.

[51] Rice, K.M., M. Wu, and E.R. Blough, Aortic aging in the Fischer 344 / NNiaHSd x Brown Norway / BiNia Rat. *J Pharmacol Sci*, 2008. 108(4): p. 393-8.

[52] Lushaj, E.B., J.K. Johnson, D. McKenzie, and J.M. Aiken, Sarcopenia accelerates at advanced ages in Fisher 344xBrown Norway rats. *J Gerontol A Biol Sci Med Sci*, 2008. 63(9): p. 921-7.

[53] Paturi, S., A.K. Gutta, A. Katta, S.K. Kakarla, et al., Effects of aging and gender on muscle mass and regulation of Akt-mTOR-p70s6k related signaling in the F344BN rat model. *Mech Ageing Dev*. 131(3): p. 202-9.

[54] Hwee, D.T. and S.C. Bodine, Age-related deficit in load-induced skeletal muscle growth. *J Gerontol A Biol Sci Med Sci*, 2009. 64(6): p. 618-28.

[55] Funai, K., J.D. Parkington, S. Carambula, and R.A. Fielding, Age-associated decrease in contraction-induced activation of downstream targets of Akt/mTor signaling in skeletal muscle. *Am J Physiol Regul Integr Comp Physiol*, 2006. 290(4): p. R1080-6.

[56] Wu, M., H. Liu, J. Fannin, A. Katta, et al., Acetaminophen improves protein translational signaling in aged skeletal muscle. *Rejuvenation Res*, 2010. 13(5): p. 571-9.

[57] Stitt, T.N., D. Drujan, B.A. Clarke, F. Panaro, et al., The IGF-1/PI3K/Akt pathway prevents expression of muscle atrophy-induced ubiquitin ligases by inhibiting FOXO transcription factors. *Mol Cell*, 2004. 14(3): p. 395-403.

[58] Hepple, R.T., M. Qin, H. Nakamoto, and S. Goto, Caloric restriction optimizes the proteasome pathway with aging in rat plantaris muscle: implications for sarcopenia. *Am J Physiol Regul Integr Comp Physiol*, 2008. 295(4): p. R1231-7.

[59] Clavel, S., A.S. Coldefy, E. Kurkdjian, J. Salles, et al., Atrophy-related ubiquitin ligases, atrogin-1 and MuRF1 are up-regulated in aged rat Tibialis Anterior muscle. *Mech Ageing Dev*, 2006. 127(10): p. 794-801.

[60] Gupte, A.A., G.L. Bomhoff, and P.C. Geiger, Age-related differences in skeletal muscle insulin signaling: the role of stress kinases and heat shock proteins. *J Appl Physiol*, 2008. 105(3): p. 839-48.

[61] Wu, M., D.H. Desai, S.K. Kakarla, A. Katta, et al., Acetaminophen prevents aging-associated hyperglycemia in aged rats: effect of aging-associated hyperactivation of p38-MAPK and ERK1/2. *Diabetes Metab Res Rev*, 2009. 25(3): p. 279-286.

[62] Serrano, R., M. Villar, N. Gallardo, J.M. Carrascosa, et al., The effect of aging on insulin signalling pathway is tissue dependent: central role of adipose tissue in the insulin resistance of aging. *Mech Ageing Dev*, 2009. 130(3): p. 189-97.

[63] Park, S., T. Komatsu, H. Hayashi, H. Yamaza, et al., Calorie restriction initiated at middle age improved glucose tolerance without affecting age-related impairments of insulin signaling in rat skeletal muscle. *Exp Gerontol*, 2006. 41(9): p. 837-45.

[64] Oh, Y.S., L.Y. Khil, K.A. Cho, S.J. Ryu, et al., A potential role for skeletal muscle caveolin-1 as an insulin sensitivity modulator in ageing-dependent non-obese type 2 diabetes: studies in a new mouse model. *Diabetologia*, 2008. 51(6): p. 1025-34.

[65] Duan, C., H. Yang, M.F. White, and L. Rui, Disruption of the SH2-B gene causes age-dependent insulin resistance and glucose intolerance. *Mol Cell Biol*, 2004. 24(17): p. 7435-43.

[66] Moreno, M., P. Ordonez, A. Alonso, F. Diaz, et al., Chronic 17beta-estradiol treatment improves skeletal muscle insulin signaling pathway components in insulin resistance associated with aging. *Age* (Dordr), 2010. 32(1): p. 1-13.

[67] Cabane, C., A.S. Coldefy, K. Yeow, and B. Derijard, The p38 pathway regulates Akt both at the protein and transcriptional activation levels during myogenesis. *Cell Signal*, 2004. 16(12): p. 1405-15.

[68] Li, M., C. Li, and W.S. Parkhouse, Age-related differences in the des IGF-I-mediated activation of Akt-1 and p70 S6K in mouse skeletal muscle. *Mech Ageing Dev*, 2003. 124(7): p. 771-8.

[69] Leger, B., W. Derave, K. De Bock, P. Hespel, et al., Human sarcopenia reveals an increase in SOCS-3 and myostatin and a reduced efficiency of Akt phosphorylation. *Rejuvenation Res*, 2008. 11(1): p. 163-175B.

[70] Arias, E.B., L.E. Gosselin, and G.D. Cartee, Exercise training eliminates age-related differences in skeletal muscle insulin receptor and IRS-1 abundance in rats. *J Gerontol A Biol Sci Med Sci*, 2001. 56(10): p. B449-55.

[71] Haddad, F. and G.R. Adams, Aging-sensitive cellular and molecular mechanisms associated with skeletal muscle hypertrophy. *J Appl Physiol*, 2006. 100(4): p. 1188-203.

[72] Edstrom, E., M. Altun, M. Hagglund, and B. Ulfhake, Atrogin-1/MAFbx and MuRF1 are downregulated in aging-related loss of skeletal muscle. *J Gerontol A Biol Sci Med Sci*, 2006. 61(7): p. 663-74.

[73] Heron, M.P., D. Hoyert, L, J. Xu, C. Scott, et al., Deaths: preliminary data for 2006. *Natl Vital Stat Rep*, 2008. 56(16): p. 31.

[74] Ferrari, A.U., A. Radaelli, and M. Centola, Invited review: aging and the cardiovascular system. *J Appl Physiol*, 2003. 95(6): p. 2591-7.

[75] Lakatta, E.G., Aging and cardiovascular structure and function in healthy sedentary humans. *Aging* (Milano), 1998. 10(2): p. 162-4.

[76] Franklin, S.S., Hypertension in older people: part 1. *J Clin Hypertens* (Greenwich), 2006. 8(6): p. 444-9.

[77] Muslin, A.J. and B. DeBosch, Role of Akt in cardiac growth and metabolism. *Novartis Found Symp*, 2006. 274: p. 118-26; discussion 126-31, 152-5, 272-6.

[78] Evans-Anderson, H.J., C.M. Alfieri, and K.E. Yutzey, Regulation of cardiomyocyte proliferation and myocardial growth during development by FOXO transcription factors. *Circ Res*, 2008. 102(6): p. 686-94.

[79] Kakarla, S.K., K.M. Rice, A. Katta, S. Paturi, et al., Possible molecular mechanisms underlying age-related cardiomyocyte apoptosis in the F344XBN rat heart. *J Gerontol A Biol Sci Med Sci*. 65(2): p. 147-55.

[80] O'Neill, B.T. and E.D. Abel, Akt1 in the cardiovascular system: friend or foe? *J Clin Invest*, 2005. 115(8): p. 2059-64.

[81] Pham, F.H., P.H. Sugden, and A. Clerk, Regulation of protein kinase B and 4E-BP1 by oxidative stress in cardiac myocytes. Circ Res, 2000. 86(12): p. 1252-8.

[82] Diez, C., M. Nestler, U. Friedrich, M. Vieth, et al., Down-regulation of Akt/PKB in senescent cardiac fibroblasts impairs PDGF-induced cell proliferation. *Cardiovasc Res*, 2001. 49(4): p. 731-40.

[83] Li, Q., A.F. Ceylan-Isik, J. Li, and J. Ren, Deficiency of insulin-like growth factor 1 reduces sensitivity to aging-associated cardiomyocyte dysfunction. *Rejuvenation Res*, 2008. 11(4): p. 725-33.

[84] Dzau, V.J., R.C. Braun-Dullaeus, and D.G. Sedding, Vascular proliferation and atherosclerosis: new perspectives and therapeutic strategies. *Nat Med*, 2002. 8(11): p. 1249-56.

[85] Fernandez-Hernando, C., E. Ackah, J. Yu, Y. Suarez, et al., Loss of Akt1 leads to severe atherosclerosis and occlusive coronary artery disease. *Cell Metab*, 2007. 6(6): p. 446-57.

[86] Li, M., J.F. Chiu, J. Gagne, and N.K. Fukagawa, Age-related differences in insulin-like growth factor-1 receptor signaling regulates Akt/FOXO3a and ERK/Fos pathways in vascular smooth muscle cells. *J Cell Physiol*, 2008. 217(2): p. 377-87.

[87] Miyauchi, H., T. Minamino, K. Tateno, T. Kunieda, et al., Akt negatively regulates the in vitro lifespan of human endothelial cells via a p53/p21-dependent pathway. *Embo J*, 2004. 23(1): p. 212-20.

[88] Torella, D., D. Leosco, C. Indolfi, A. Curcio, et al., Aging exacerbates negative remodeling and impairs endothelial regeneration after balloon injury. *Am J Physiol Heart Circ Physiol*, 2004. 287(6): p. H2850-60.

[89] Kudo, F.A., B. Warycha, P.J. Juran, H. Asada, et al., Differential responsiveness of early- and late-passage endothelial cells to shear stress. *Am J Surg*, 2005. 190(5): p. 763-9.

[90] Breitschopf, K., A.M. Zeiher, and S. Dimmeler, Pro-atherogenic factors induce telomerase inactivation in endothelial cells through an Akt-dependent mechanism. *FEBS Lett*, 2001. 493(1): p. 21-5.

[91] Perez-Rivero, G., M.P. Ruiz-Torres, J.V. Rivas-Elena, M. Jerkic, et al., Mice deficient in telomerase activity develop hypertension because of an excess of endothelin production. *Circulation*, 2006. 114(4): p. 309-17.

[92] Op den Buijs, J., M. Musters, T. Verrips, J.A. Post, et al., Mathematical modeling of vascular endothelial layer maintenance: the role of endothelial cell division, progenitor cell homing, and telomere shortening. *Am J Physiol Heart Circ Physiol*, 2004. 287(6): p. H2651-8.

[93] Imanishi, T., K. Kobayashi, T. Hano, and I. Nishio, Effect of estrogen on differentiation and senescence in endothelial progenitor cells derived from bone marrow in spontaneously hypertensive rats. *Hypertens Res*, 2005. 28(9): p. 763-72.

[94] Pu, D.R. and L. Liu, HDL slowing down endothelial progenitor cells senescence: a novel anti-atherogenic property of HDL. *Med Hypotheses*, 2008. 70(2): p. 338-42.

[95] Thum, T., S. Hoeber, S. Froese, I. Klink, et al., Age-dependent impairment of endothelial progenitor cells is corrected by growth-hormone-mediated increase of insulin-like growth-factor-1. *Circ Res*, 2007. 100(3): p. 434-43.

[96] Nordt, T.K., K. Peter, J. Ruef, W. Kubler, et al., Plasminogen activator inhibitor type-1 (PAI-1) and its role in cardiovascular disease. *Thromb Haemost*, 1999. 82 Suppl 1: p. 14-8.

[97] Mukai, Y., C.Y. Wang, Y. Rikitake, and J.K. Liao, Phosphatidylinositol 3-kinase/protein kinase Akt negatively regulates plasminogen activator inhibitor type 1 expression in vascular endothelial cells. *Am J Physiol Heart Circ Physiol*, 2007. 292(4): p. H1937-42.

[98] Comi, P., R. Chiaramonte, and J.A. Maier, Senescence-dependent regulation of type 1 plasminogen activator inhibitor in human vascular endothelial cells. *Exp Cell Res*, 1995. 219(1): p. 304-8.

[99] Gao, Y., R. Yokota, S. Tang, A.W. Ashton, et al., Reversal of angiogenesis in vitro, induction of apoptosis, and inhibition of AKT phosphorylation in endothelial cells by thromboxane A(2). *Circ Res*, 2000. 87(9): p. 739-45.

[100] Schulman, I.H., M.S. Zhou, E.A. Jaimes, and L. Raij, Dissociation between metabolic and vascular insulin resistance in aging. *Am J Physiol Heart Circ Physiol*, 2007. 293(1): p. H853-9.

[101] Avolio, A.P., S.G. Chen, R.P. Wang, C.L. Zhang, et al., Effects of aging on changing arterial compliance and left ventricular load in a northern Chinese urban community. *Circulation*, 1983. 68(1): p. 50-8.

[102] Hixon, M.L., C. Muro-Cacho, M.W. Wagner, C. Obejero-Paz, et al., Akt1/PKB upregulation leads to vascular smooth muscle cell hypertrophy and polyploidization. *J Clin Invest*, 2000. 106(8): p. 1011-20.

[103] Smith, A.R. and T.M. Hagen, Vascular endothelial dysfunction in aging: loss of Akt-dependent endothelial nitric oxide synthase phosphorylation and partial restoration by (R)-alpha-lipoic acid. *Biochem Soc Trans*, 2003. 31(Pt 6): p. 1447-9.

[104] Payne, J.A., J.F. Reckelhoff, and R.A. Khalil, Role of oxidative stress in age-related reduction of NO-cGMP-mediated vascular relaxation in SHR. *Am J Physiol Regul Integr Comp Physiol*, 2003. 285(3): p. R542-51.

[105] Dimmeler, S., I. Fleming, B. Fisslthaler, C. Hermann, et al., Activation of nitric oxide synthase in endothelial cells by Akt-dependent phosphorylation. *Nature*, 1999. 399(6736): p. 601-5.

[106] Andreozzi, F., G. Formoso, S. Prudente, M.L. Hribal, et al., TRIB3 R84 variant is associated with impaired insulin-mediated nitric oxide production in human endothelial cells. *Arterioscler Thromb Vasc Biol*, 2008. 28(7): p. 1355-60.

[107] Song, G.Y., J.S. Kang, S.Y. Lee, and C.S. Myung, Region-specific reduction of Gbeta4 expression and induction of the phosphorylation of PKB/Akt and ERK1/2 by aging in rat brain. *Pharmacol Res*, 2007. 56(4): p. 295-302.

[108] Nie, K., J.C. Yu, Y. Fu, H.Y. Cheng, et al., Age-related decrease in constructive activation of Akt/PKB in SAMP10 hippocampus. *Biochem Biophys Res Commun*, 2009. 378(1): p. 103-7.

[109] Eto, R., M. Abe, N. Hayakawa, H. Kato, et al., Age-related changes of calcineurin and Akt1/protein kinase Balpha (Akt1/PKBalpha) immunoreactivity in the mouse hippocampal CA1 sector: an immunohistochemical study. *Metab Brain Dis*, 2008. 23(4): p. 399-409.

[110] Jackson, T.C., A. Rani, A. Kumar, and T.C. Foster, Regional hippocampal differences in AKT survival signaling across the lifespan: implications for CA1 vulnerability with aging. *Cell Death Differ*, 2009. 16(3): p. 439-48.

[111] Lee, H.K., P. Kumar, Q. Fu, K.M. Rosen, et al., The insulin/Akt signaling pathway is targeted by intracellular beta-amyloid. *Mol Biol Cell*, 2009. 20(5): p. 1533-44.

[112] Takashima, A., GSK-3 is essential in the pathogenesis of Alzheimer's disease. *J Alzheimers Dis*, 2006. 9(3 Suppl): p. 309-17.

[113] Magrane, J., K.M. Rosen, R.C. Smith, K. Walsh, et al., Intraneuronal beta-amyloid expression downregulates the Akt survival pathway and blunts the stress response. *J Neurosci*, 2005. 25(47): p. 10960-9.

[114] Pardo, P.S., M.A. Lopez, and A.M. Boriek, FOXO transcription factors are mechanosensitive and their regulation is altered with aging in the respiratory pump. *Am J Physiol Cell Physiol*, 2008. 294(4): p. C1056-66.

[115] Lorenzini, A., M. Tresini, M. Mawal-Dewan, L. Frisoni, et al., Role of the Raf/MEK/ERK and the PI3K/Akt(PKB) pathways in fibroblast senescence. *Exp Gerontol*, 2002. 37(10-11): p. 1149-56.

[116] Bartling, B., G. Rehbein, R.E. Silber, and A. Simm, Senescent fibroblasts induce moderate stress in lung epithelial cells in vitro. *Exp Gerontol*, 2006. 41(5): p. 532-9.

[117] Grubeck-Loebenstein, B., S. Della Bella, A.M. Iorio, J.P. Michel, et al., Immunosenescence and vaccine failure in the elderly. *Aging Clin Exp Res*, 2009. 21(3): p. 201-9.

[118] Weiskopf, D., B. Weinberger, and B. Grubeck-Loebenstein, The aging of the immune system. *Transpl Int*, 2009. 22(11): p. 1041-50.

[119] Larbi, A., G. Dupuis, A. Khalil, N. Douziech, et al., Differential role of lipid rafts in the functions of CD4+ and CD8+ human T lymphocytes with aging. *Cell Signal*, 2006. 18(7): p. 1017-30.

[120] Agrawal, A., S. Agrawal, J.N. Cao, H. Su, et al., Altered innate immune functioning of dendritic cells in elderly humans: a role of phosphoinositide 3-kinase-signaling pathway. *J Immunol*, 2007. 178(11): p. 6912-22.

[121] Parekh, V.V., J.C. Falcone, L.A. Wills-Frank, I.G. Joshua, et al., Protein kinase B, P34cdc2 kinase, and p21 ras GTP-binding in kidneys of aging rats. *Exp Biol Med* (Maywood), 2004. 229(8): p. 850-6.

[122] Percy, C.J., L. Brown, D.A. Power, D.W. Johnson, et al., Obesity and hypertension have differing oxidant handling molecular pathways in age-related chronic kidney disease. *Mech Ageing Dev*, 2009. 130(3): p. 129-38.

[123] Villar, M., R. Serrano, N. Gallardo, J.M. Carrascosa, et al., Altered subcellular distribution of IRS-1 and IRS-3 is associated with defective Akt activation and GLUT4 translocation in insulin-resistant old rat adipocytes. *Biochim Biophys Acta*, 2006. 1763(2): p. 197-206.

[124] Zaidi, M., Skeletal remodeling in health and disease. *Nat Med*, 2007. 13(7): p. 791-801.

[125] Frost, H.M., On our age-related bone loss: insights from a new paradigm. *J Bone Miner Res*, 1997. 12(10): p. 1539-46.

[126] Kanis, J.A. and J.Y. Reginster, European guidance for the diagnosis and management of osteoporosis in postmenopausal women--what is the current message for clinical practice? *Pol Arch Med Wewn*, 2008. 118(10): p. 538-40.

[127] Cao, J.J., P. Kurimoto, B. Boudignon, C. Rosen, et al., Aging impairs IGF-I receptor activation and induces skeletal resistance to IGF-I. *J Bone Miner Res*, 2007. 22(8): p. 1271-9.

[128] Cravero, J.D., C.S. Carlson, H.J. Im, R.R. Yammani, et al., Increased expression of the Akt/PKB inhibitor TRB3 in osteoarthritic chondrocytes inhibits insulin-like growth factor 1-mediated cell survival and proteoglycan synthesis. *Arthritis Rheum*, 2009. 60(2): p. 492-500.

[129] Lin, K., H. Hsin, N. Libina, and C. Kenyon, Regulation of the Caenorhabditis elegans longevity protein DAF-16 by insulin/IGF-1 and germline signaling. *Nat Genet*, 2001. 28(2): p. 139-45.

[130] Moore, T., L. Beltran, S. Carbajal, S. Strom, et al., Dietary energy balance modulates signaling through the Akt/mammalian target of rapamycin pathways in multiple epithelial tissues. *Cancer Prev Res* (Phila Pa), 2008. 1(1): p. 65-76.

[131] Bonkowski, M.S., F.P. Dominici, O. Arum, J.S. Rocha, et al., Disruption of growth hormone receptor prevents calorie restriction from improving insulin action and longevity. *PLoS One*, 2009. 4(2): p. e4567.

[132] Hsieh, C.C. and J. Papaconstantinou, Akt/PKB and p38 MAPK signaling, translational initiation and longevity in Snell dwarf mouse livers. *Mech Ageing Dev*, 2004. 125(10-11): p. 785-98.

[133] Sedding, D.G., FoxO transcription factors in oxidative stress response and ageing--a new fork on the way to longevity? Biol Chem, 2008. 389(3): p. 279-83.

[134] Kim, D.H., J.Y. Kim, B.P. Yu, and H.Y. Chung, The activation of NF-kappaB through Akt-induced FOXO1 phosphorylation during aging and its modulation by calorie restriction. *Biogerontology*, 2008. 9(1): p. 33-47.

[135] Park, S., T. Komatsu, H. Hayashi, H. Yamaza, et al., Calorie restriction initiated at a young age activates the Akt/PKC zeta/lambda-Glut4 pathway in rat white adipose tissue in an insulin-independent manner. *Age* (Dordr), 2008. 30(4): p. 293-302.

In: Cell Aging   ISBN: 978-1-61324-369-5
Editors: Jack W. Perloft and Alexander H. Wong   2012 Nova Science Publishers, Inc.

Chapter 5

AGELESSNESS OF MESENCHYMAL STEM CELLS

*Regina Brunauer, Stephan Reitinger and Günter Lepperdinger**
Institute for Biomedical Aging Research, Austrian Academy of Sciences,
Innsbruck, Austria

ABSTRACT

Multipotential mesenchymal stem cells (MSC) are present as a rare subpopulation within any type of stroma in the body of higher organisms. They are deliberately considered of being capable to terminally differentiating into a broad variety of tissue cell types. When required, MSC are activated accordingly and deployed for tissue remodeling and regeneration after injury. In this vein, MSC are not active only during early development and growth, yet also in repair and regeneration throughout adulthood up to highly advanced ages. Besides replenishing mesenchymal tissues, MSC also modulate hematopoiesis as well as immune response.

During aging, tissue function declines, as does hematopoietic activity, or immune function. Consequences are bone loss, fat redistribution, cartilage defects, autoimmune diseases, and increased incidence of sarcomas. While numerous studies have unveiled the multipotentiality of these cells and demonstrated their capacity to regenerate and repair tissues, only recently the concept of stem cells to encounter aging has been acknowledged as an increasingly important field to appropriately tackle these unanswered questions.

Scientists are just at the beginning to conceive the complexity and importance of mechanisms that undermine and attenuate MSC function as these cells age. Notably, we and others have found that during aging, overall MSC number and their proliferative potential are hardly ever affected. Yet in aging donors, we could reveal distinct MSC subpopulations which express marked levels of vascular cell-adhesion molecule 1 (VCAM1/CD106) or leptin receptor (LEPR/CD295). MSC show such changes as a response to extrinsic factors such as decreased oxygen supply, extracellular matrix aging, e.g. collagen crosslinking, or when getting in close contact with infiltrating lymphocytes which establish a chronic proinflammatory milieu (inflamm-aging), as well as in respect of age-associated intrinsic changes with regard to altered DNA methylation, and the shifted expression of miRNAs.

---

* Corresponding author: Phone: +43 512 5839 1940; Fax +43 512 5839 198;
Email guenter.lepperdinger@oeaw.ac.at

Herein, we review the current perception regarding possible causes of MSC aging, and consequences thereof.

Keywords: self-renewal, differentiation, cellular aging, inflammation, epigenetics, pathology.

INTRODUCTION

Aging is paralleled by the loss of regenerative capacity together with a steady decline in repair and remodeling of worn-out tissues. Normally, tissue homeostasis, regeneration and repair involve the emergence and integration of new parenchymal cells descending from undifferentiated precursors. The latter are resident in many, albeit not all body organs and tissues. By this token, a decline of stem cell function and activity with advancing age has its share in delaying the replacement and turnover of damaged cells in compromised renewable tissues. Together, this eventually leads up to dysfunction of distinct parts of the body. As stem cells are perceived the single most important starting point for tissue regeneration, fitness of this specialized cell type is considered pivot, not only in the developing and growing organism, but seemingly decisive also later during adult life.

EXTRAORDINARY ITEMS WITH SPECIAL RESERVE BURIED IN STROMA: MESENCHYMAL STEM CELLS

Multipotent stromal progenitor cells also known as Mesenchymal Stem Cells (MSC) are pertinent tissue-specific stem cells in adult beings (Caplan, 1991). The MSC appears to be particularly interesting given that it can bring forth a large spectrum of cell types as diverse as bone, cartilage, tendon, or fat precursor cells. Besides these descendants, also those cells which are considered most widespread throughout many tissues in the body, namely interstitial stromal cells and fibroblasts, are thought direct offspring of MSC. Actually most work in this context has focused on isolated MSC in culture. Presently, MSC are being isolated and cultured following different methods and protocols, and common characteristics of MSC could only be insufficiently specified, first and foremost because the cell isolates are heterogeneous and comprise more than one mesenchymal cell type. It is therefore difficult to resolve how results regarding ex vivo propagated cells relate to each other.

A large body of information is available on many aspects of cultivated MSC properties. They firmly adhere to cell culture plastic, which is a distinctive functional criterion for selection during isolation. They exhibit clonogenic growth, which is a further distinctive selection criterion during expansion after low density seeding. Unfortunately there is still little consensus on their ex vivo antigenic definition. Early on, CD105 and CD73 have been reported indispensable markers (Caplan, 1991), furthermore CD29, CD44, and CD90 are pertinent determinants (Pittenger et al., 2000). The STRO-1 antibody identifies an immature population of mesenchymal cells (Gronthos et al., 1994), and many other determinants have been studied as well (Horwitz and Keating, 2000). Given these observations, the following minimum specifications have been proposed only recently by a group of peers in the field in order to generally define MSC: (i) plastic-adherence when maintained in standard culture conditions, (ii) expression of surface molecules CD105, CD73 and CD90, and lack of expression of CD45, CD34, CD14 or CD11b, CD79alpha or CD19 and HLA-DR, as determined by the Mesenchymal and Tissue Stem Cell Committee of the International Society for Cellular Therapy, and last but not least (iii) differentiation potential towards osteogenic, adipogenic and chondrogenic cell types in vitro (Dominici et al., 2006). As there is no universal antigenic definition of MSC, analogous to CD34 for hematopoietic stem cells, and there is also no generally accepted assay, analogous to hematopoietic repopulation assays, there is only little data available from in vivo experimentation, in particular with the intention to comparing MSC obtained from different tissues regarding differentiation potential and proliferation capacity (Caplan, 2007).

In order to ensure mesenchymal tissue maintenance, every single step of the complex processes of stem cell self renewal, transient amplification and progenitor differentiation has to be tightly regulated. Now with advancing age, deterioration of tight control appears to increasingly emerge, and it is reasoned that this ascending deficit may lead to derailed guidance of cell differentiation, which in turn is growing up to adverse effects such as the accumulation of fat deposits in bone and muscles, or impaired healing and fibrosis after severe injury, yet also altered hematopoiesis and autoimmunity.

LIVE-IN REFUGE, OR, CONTROL AND DISTRIBUTING CENTER: THE MSC NICHE

Besides intrinsic stem cell factors, it is generally believed that the respective stem cell niche provides cell-cell and cell-matrix interactions, which duly control tissue-specific lineage commitment. Upon stimulation (e.g. after injury or toxic insults), a stem cell that nests well protected in its niche starts to grow and subsequently divides, thereby giving rise to one stem cell (self-renewal) and one progenitor cell. The latter is thought to exit the niche, where upon it continues amplification to yield the necessary number of cells needed to regain functional mesenchymal tissue before this offspring turns into appropriately matured tissue cells. There are many good reasons to believe that it is the niche which provides means and measures for the delicate balance between self-renewal and generation of progeny through asymmetric cell division; said that, it is also assumed that a proper niche comforts and protects stem cells from adverse influences throughout life.

MSC are part of a highly specialized microenvironment and thus generally believed to participate in the regulation of hematopoietic precursor cell differentiation into mature progeny (Mendez-Ferrer et al., 2010; Muguruma et al., 2006). They reside in a complex three-dimensional network, which comprises a plethora of other cell types such as, in the case of bone marrow, HSC, adipocytes, and endothelial cells, supported by an extracellular matrix consisting mainly of fibronectin, collagen I and IV, heparin sulphate, chondroitin sulphate, and hyaluronan (Dorshkind, 1990). Hence, there are lots of opportunities for MSC to interact at both the cellular and matrix level, and all of these elements may modify MSC behavior, in particular in respect to pathological alterations, traumatic insults, as well as organismic or cellular aging (Fehrer et al., 2007).

To fully understand the effects of physiological niches, it is important to consider injury-derived changes to its microenvironment such as hypoxia. Anoxic conditions may occur after injury and during wound healing (Jauniaux et al., 1999). For instance, after bone fracture, O_2 levels are known to transiently drop to 0-2% at the centre of the wound (Heppenstall et al., 1975). To date it is technically unfeasible to assess spatial differences in pO_2 within cancellous bone and marrow. It is known that blood in human bone marrow contains about 7% O_2 (Ishikawa and Ito, 1988), from which O_2 gradients in surrounding tissues have been further extrapolated applying mathematical modeling. Thereby O_2 gradients have been calculated in bone marrow to be highest (~5%) near the sinuses and lowest (~1%) at the inner surface of cortical bone (Antoniou et al., 2004).

Explanted mesenchymal progenitors display the following characteristic variations when being incubated at physiological O_2 levels during ex vivo cultures (Fehrer et al., 2007): (i) investigation of the transcriptional profile of mesenchymal progenitors revealed particular changes in the expression level of a subset of functionally interesting genes. However when culturing MSC at 3% O_2, hypoxia-induced transcriptional pathways are not activated; (ii) with regard to replication potential and in vitro aging, 3% O_2 is beneficial for cultured MSC; (iii) in comparison to traditional culture conditions, in vitro differentiation into adipogenic progeny was found to be markedly reduced, when O_2 content is lowered to 3% during induction. No osteogenic differentiation was evident at 3% O_2. This suppression of differentiation can however be overcome by increasing the level of O_2 to ambient air condition. Still in most cases MSC are expanded under ambient atmospheric conditions of ~ 20% O_2 for later use in biomedicine and tissue engineering. These findings greatly suggest that oxygen is an important determinant of the MSC niche. In turn, occlusion and/or degeneration of the microvascular plexus within mesenchymal tissue as an accessory implication of aging may have a profound impact on the basic properties of tissue-borne MSC.

Only recently, data became available showing the existence of mesenchymal stem and progenitor cells in perivascular locations, and therefore these cells have often been called pericytes (Crisan et al., 2008). In view of these data, MSC, which as already mentioned above contribute to tissue homeostasis by e.g. supporting bone regeneration, or fulfilling permissive action on hematopoiesis, appear to exert yet another function, namely maintaining blood vessel integrity (da Silva Meirelles et al., 2008). It can further be envisaged that upon tissue damage and injury, MSC/pericytes are either being activated or released from the perivascular niche, and thus support wound healing and tissue regeneration.

PAST THE PRIME OF TIME: AGE-ASSOCIATED CHANGES OF MSC

To date it is generally accepted in the field that stem cells undergo donor specific age-associated changes. One major feature of aging is an overall decline in regenerative vigor in many parts of the body. Similarly to somatic cells, stem cells are exposed to threads such as reactive oxygen species, harmful chemical agents or physical stressors, which altogether may lead to premature senescence of individual cells, or is provocative to accelerate cell death, or may put cells at risk to cellular transformation. In this context, it is also worthwhile to assume that stem cells exhibit enhanced cellular repair capacities, or other still unveiled protective measures, which allow them to efficiently restore otherwise irreparable damage.

While extensive research regarding this particular question has been undertaken with HSC, and distinct mechanisms, which trigger age-related changes therein, could be deciphered, detailed knowledge about MSC aging is still scarce. Most of the available literature is concerned with bone marrow-derived MSC. Published work regarding this particular topic often yielded greatly divergent results (Fehrer and Lepperdinger, 2005; Sethe et al., 2006). For instance, there are conflicting results regarding the question whether MSC numbers change during life span. MSC can be easily selected within low-density mononuclear cell isolates based on their characteristic of tightly adhering to the plastic surface of a culture dish. Due to their inherent clonogenic growth properties, fibroblastic colonies are formed after extended periods of cultivation, and the respective number, also called colony-forming unit-fibroblasts (CFU-f) can be reliably estimated. Yet some laboratories report that total CFU-f decrease with age (Baxter et al., 2004; Caplan, 2007; Caplan, 2009; Fan et al., 2010; Majors et al., 1997; Muschler et al., 2001; Nishida et al., 1999; Stolzing et al., 2008a; Stolzing et al., 2008b; Tokalov et al., 2007) while we and others find only a minor or no significant decline (D'Ippolito et al., 1999; Glowacki, 1995; Justesen et al., 2002; Laschober et al. 2011a; Oreffo et al., 1998; Stenderup et al., 2001).

As already pointed out, one particular problem with this approach is that there is no agreement on a single, standardized protocol for MSC isolation as well as how to proceed with further analysis of the primary cell isolates. Another issue is that the primary cell isolates are, though at a varying extent, contaminated by other cells, in particular by macrophages and hematopoietic cells. Lastly, single clones within the heterogeneous cultures exhibit greatly varying differentiation potential, which was demonstrated by in vitro cell cloning of human stromal cultures: only around 30% of the CFU-f are multipotential and can thus be considered true MSC (Kuznetsov et al., 1997). In addition to the aforementioned observations, Muraglia and colleagues reported that the number of bi- or tri-potential colonies declined with age (Muraglia et al., 2000).

Considering the proliferative potential with age, the published data are less conflicting because most authors report declining performances in long-term culture (Kretlow et al., 2008; Laschober et al., 2011a; Stenderup et al., 2003; Stolzing et al., 2008b). Much of the here mentioned data are regarding MSC which were isolated from bone and bone marrow. Most conspicuously, there are many other sites containing mesenchymal progenitor cells such as the muscle (Bosch et al., 2000), vessel-associated pericytes (Collett and Canfield, 2005a), or blood (Eghbali-Fatourechi et al., 2005), and evidently the aforementioned questions also apply here: as of now, no stringent conclusions can be drawn at this point as most questions in this particular context are still unanswered.

Given the ubiquitous presence of MSC, it is apparent that this particular cell type subsists in greatly variant anatomical sites. As a consequence thereof, individual MSC may escape later in life from common fate control and become diverted to bring forth unwanted progeny. In the case of a vascular mural location, it is conceivable that MSC-like cells may become involved in ectopic calcification of arteries, skeletal muscles and heart valves, which often occurs in the course of diabetes, renal failure or atherosclerosis. In the latter condition, plaques are most often found in combination with intimal neovascularisation, and interestingly enough it is only those vessels that exhibit neovascularisation, which also exhibit considerable calcification (Kamat et al., 1987). Also within heart valves with progressive calcification, cartilaginous nodules and mature lamellar bone with hemopoietic elements had been recognized (Mohler et al., 2001). Although vascular calcification appears to be triggered by molecular signaling events, such as BMP2 and TGFß, which are also known to be active during endochondral bone formation and repair (Collett and Canfield, 2005b; Neven et al., 2007; Vattikuti and Towler, 2004), it has been unveiled that pericytes are undergoing osteogenic differentiation also when facing inflammatory conditions. In addition to that, and consistent with the situation described above, it is anti-angiogenic factors which are capable of inhibiting progression of calcification (Collett and Canfield, 2005a). These results suggest that MSC within an aged vascular wall show the propensity to develop along the osteogenic lineage rather than turning into cell types that support vascular architecture, be it smooth muscle cells located within the tunica media or interstitial fibroblasts of the tunica adventitia.

Ectopic bone formation is considered a good example for a well established theory in aging biology named antagonistic pleiotropy (Rose and Graves, 1989). Most notably, in osseous tissues during advanced aging, both mass and mineral density of cortical and cancellous bone steadily decreases while at the same time more and more fat cells do emerge within bone marrow. The process underlying this phenomenon is referred to the term "adipogenic switch" (Meunier et al., 1971; Tokalov et al., 2007). The critical determinants of this switch are unclear. The mutual exclusivity of the osteoblast and adipocyte cell fates are thought to be determined by signals such as the CCAAT/enhancer binding protein (C/EBP), a trigger for adipogenesis, PPARγ, which promotes adipocyte maturation, or Runx2, an osteoblastic transcriptional mediator. Impaired PPARγ signaling shifts the fate of MSC toward the osteoblast lineage. The wnt pathway can suppress PPARγ, favoring MSC differentiation to osteoblasts. Other transcriptional mediators associated with osteoblast/adipocyte specification include ΔFosB, TAZ, Esr1, Msx2, C/EBPβ, and Id4, though their specificity as determinants of the age-related switch in bone is unclear (McCauley, 2010). Deviations of the MSC's multipotential differentiation capacity was reported by many laboratories inasmuch osteogenic potential of MSC isolated from aged donors declined while the respective adipogenic differentiation performance remained unchanged, or more than that, was found to be enhanced (Moerman et al., 2004; Roura et al., 2006; Stenderup et al., 2003; Stolzing et al., 2008b). This subtle fate change is however considered but one of many explanations, since only a few reports documented an overall decline in multilineage differentiation (Justesen et al., 2002; Kretlow et al., 2008; Muraglia et al., 2000; Tokalov et al., 2007), while still others failed to find respective changes confirming this assumption (Justesen et al., 2002; Tokalov et al., 2007).

Several functional studies have also tackled the question of whether age-associated changes would impinge on MSC properties with respect to their inherent regenerative potential. In an immune suppressed rat model for ischemic cardiomyopathy, human MSC from aged donors performed worse (Fan et al., 2010). In an attempt to use MSC in cell replacement therapies for neurologic disorders, the neuroectodermal differentiation potential, which can be provoked in vitro in MSC derived from young donors, was completely lost in MSC from old donors (Hermann et al., 2010). In several studies addressing the pivotal role of MSC supporting hematopoietic stem cells and their progeny, cells derived from young bone marrow were superior over the aged ones (Mauch et al., 1982; Schraml et al., 2008; Wagner et al., 2008; Walenda et al., 2010). As MSC are clinically applied to suppress graft-versus-host-disease, it would be interesting to uncover whether age-associated changes also skews

MSC in this regard. To our knowledge no such attempts have been undertaken, or experimental results addressing these questions have been published to date.

Provided those reports, which indicate that upon aging primary properties of MSC may become corrupted, molecular pathways deviating from the norm and becoming manifest during MSC aging were investigated in great detail. In vitro, the proliferative capacity of primary mammalian cells is finite. Propagation of somatic cells for prolonged periods in culture flaws and injures cells, leading to accumulation of damage and the cells' capacity to efficiently repair lesions seizes. Hence during prolonged periods of culture, cells emerge which are metabolically active yet are permanently stalled in cell cycle. Unlike quiescence, this apparent change is unresponsive to growth factors and paralleled by suppression of apoptosis. After extensive expansion, also cultured MSC cease growth and enter a state of irreversible growth arrest, referred to as 'replicative senescence' (Beausejour, 2007; Fehrer and Lepperdinger, 2005; Sethe et al., 2006). The cells however remain alive and stay metabolically active. Lifelong accumulation of damage, such as genomic mutations as a consequence of oxidative stress, or aberrant intercellular signaling due to critical shortening of telomeres may contribute to the induction of this terminal state (Campisi and d'Adda di Fagagna, 2007). As one particular role model for cellular aging, replicative senescence paved the way to discover major insights about terminal changes of aged cells. By this token, cellular senescence is thought to be a strict program to limit the proliferative potential of premalignant cells, hence not narrowing lifespan but actually resulting in a controlled extension by ultimately safeguarding the organism's longevity.

When growing MSC in long-term culture, we recognized, similarly to what has been reported by others only recently (Shibata et al., 2007; Stolzing et al., 2008b; Wagner et al., 2008), that the number of population doublings (PD) accumulated by MSC lines derived from individual donors varies considerably (Laschober et al., 2011a). Interestingly, a good correlation between donor age and the respective PD became evident. At the first glance, this circumstance indicated that MSC may age *in vivo* in a similar fashion as they do in vitro. Although it is rather unclear to date what are the distinct implications of the in vivo niche with respect to cellular aging of MSC, replicative senescence as a model system was deliberately employed to sort out molecular cues of how MSC potentially age in vivo (Baxter et al., 2004; Bork et al., 2010; Wagner et al., 2009). However in fact, there are good reasons to believe that studying MSC senescence acquired in long-term culture might be misleading. In culture, MSC show telomere attrition at high passages, and this type of genotoxic stress eventually contributes to the limited replicative life-span. This phenomenon can indeed be overcome by forced expression of telomerase (Simonsen et al., 2002). Yet many laboratories could show that the length of telomeric ends, although being significantly higher in children (Choumerianou et al., 2010; Mareschi et al., 2006) is maintained at a considerable long length in adult age (Laschober et al., 2011a; Lund et al., 2011; Wagner et al., 2009). This suggests that expression of telomerase takes place in MSC in vivo be it at a very low constitutive level, or in a transient fashion thereby maintaining the proper structure of chromosome ends. Hence, changes occurring in MSC in vivo are only insufficiently described by patchy examinations of homogenous cultures of replicatively aging cells.

Senescent cells can be distinguished based on the accumulation of senescence-associated beta-galactosidase (SA-β-gal)(Dimri et al., 1995). This lysosomal protein is predominantly active in senescent fibroblasts and also, albeit at much lower extent, in MSC (Fehrer et al., 2007). Despite the limitations in quantification and prospective analysis of MSC, SA-β-gal is a widely used biomarker for senescent cells. Senescent cells were found in vivo in various tissues derived from individuals of advanced age, e.g. skin, eye, prostate, arteries or liver (Erusalimsky and Kurz, 2005). In lieu of specific markers which allow to distinguish MSC in their in vivo niche, it is even harder to fancy a way how to detect a senescent MSC in vivo, moreover as there is also experimental evidence from a mouse model that emerging senescent cells can be recognized and subsequently cleared by immune surveillance mechanisms (Xue et al., 2007). Whether a presenescent cell is tagged and in subsequence withdrawn from its niche by immune surveillance measures, remains to be determined though.

As highlighted above, long-term culture results in cellular senescence. Yet a fit cell can withstand accrued damage and compounding of irreversible molecular changes longer than cells that have prior to that been aged by whatever means. As differential expression analysis is easy to apprehend nowadays, characterization of dynamic aberrations before the cell slithers into the final state of senescence could be meaningful. In this vein, specific molecular markers that prospectively reflect the degree of cellular aging in MSC could be helpful. In this context, we were able to sort out CD295 (leptin receptor) to increase as a function of intrinsic cellular aging, (Laschober et al., 2009). Flow cytometric analysis of this surface marker further discriminated a distinct CD295-bright subpopulation, which interestingly enough also stained positive for annexin V. Enhanced CD295 expression therefore marks apoptotic cells, and this rate steadily raises with increasing cellular age.

It might therefore be more valuable to distinguish phenotypic appearances in MSC which have been isolated from differently aged healthy individuals in an attempt to understand those mechanisms that actually bring forth naturally occurring alterations instead of studying in vitro MSC senescence. In an aging-associated context, gross evaluations regarding the increased production of reactive oxygen species (Stolzing et al., 2008b), deviating SOD activity (Kemp et al., 2010), regulation of apoptosis rate (Laschober et al., 2009) whole genome gene expression profiling (Fehrer et al., 2007; Laschober et al., 2011a; Laschober et al., 2011b; Scaffidi and Misteli, 2008; Wagner et al., 2009), and epigenetic signatures (Bork et al., 2010) have been documented.

STONES OF FALLING AND ROCKS OF TROUBLE: CAUSES OF MSC AGING

The efficiency of tissue regeneration processes is influenced by stem cell-intrinsic factors, yet certainly is also geared by extrinsic determinants and signals, both of which are subject to change with advancing age. As a result thereof, biological communication systems and interconnected physiological networks may become derailed (Figure 1).

Most organs consist of cells that simultaneously sense and communicate with each other, thereby being engulfed in extracellular matrix, further instructed by hormones, which are distributed through the circulation after being released into the system elsewhere in the body, nourished by blood constituents, yet subject to strict scrutiny by the immune system, and lastly erected and formed in well assembled three dimensional structures as the sole basis for their proper function. These subunits, be it one alone or in combinations, are all subject to change along with aging. Malfunction of a single part may impinge on other albeit not damaged parts. Age-related impacts of systemic factors on stem cell functions have been clearly shown, e.g. applying parabiosis, in which two circulatory systems were surgically connected in mice (Conboy et al., 2005), or interchange of differently aged hematopoietic stem cells in heterochronic recipients (Waterstrat and Van Zant, 2009).

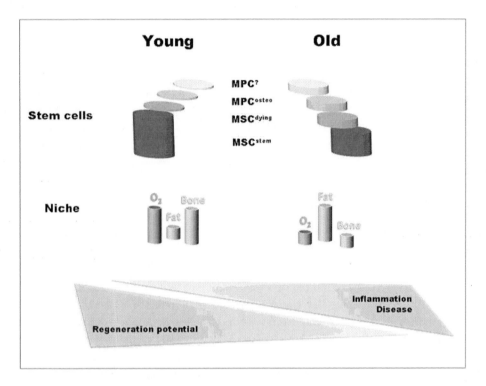

Figure 1. During aging, the number of mesenchymal stem cells (MSC) in bone is by and large maintained; however, subpopulations MSC^{dying}, MPC^{osteo}, and mesenchymal progenitor cells (MPC) arise due to increased levels of pro-inflammatory cytokines and decreased oxygen levels. The consequence is increased fat mass and decreased bone mass.

Extracellular Cues

It is generally accepted, yet far too often also omitted, that the role of extracellular matrix (ECM) is key for proper tissue function, in particular, in the context of regulation of cell fate, growth and function within the tissue. Matrix mediates the effects of signals for cell proliferation and differentiation, and plays a decisive role to control apoptosis, as cell adhesion to ECM, may, in and of itself, help preventing apoptosis. It is widely believed that ECM integrity and composition influences angiogenesis. During aging, blood vessels, nerves and skin are altered both in structure and function. It is the stem cell that may play a role in maintaining cell numbers and function and in healing of aging tissues. The elderly frequently experience problems associated with connective tissue that affect their quality of life. In some aged tissues, increased turnover of extracellular matrix as well as changes in composition of the matrix have been observed.

Paradoxically, the matrix can become increasingly insoluble, less digestible and more cross-linked with age. Products of advanced protein glycosylation (advanced glycation end products - AGEs) accumulate in tissues over time only depending on elevated levels of free sugars. The formation of AGEs is brought forward by non-enzymatic means. The modification is irreversible, and protein-cross links that evolve from the Maillard reaction are protease-resistant (Munch et al., 1997). As a matter of fact, proteins and peptides modified in this way accumulate in cells and extracellular matrix in vivo particularly at old age (Grillo and Colombatto, 2008). Especially proteins with remarkably long life spans such as collagen or lens crystalline serve as accumulators. AGEs not only induce permanent abnormalities in ECM components, but also directly stimulate a proinflammatory cellular response, as NF-κB signaling is sustainingly initiated (Sparvero et al., 2009). Subsequently, reactive oxygen species production is triggered, which in due course adversely modifies intracellular proteins (Ramasamy et al., 2005). The effect of AGEs leading to increased oxidative stress and, as a consequence thereof, activation of pro-apoptotic cytokines have been investigated in several cell types, amongst them retinal pericytes and renal mesangial cells (Alikhani et al., 2005). Also MSC are disturbed by the action of AGE derivatives, as they decrease their proliferation index and at the same time exhibit elevated apoptosis rates. Also differentiation potential of MSC is greatly perturbed inasmuch that osteogenesis, adipogenesis and chondrogenesis are effectively blocked (Kume et al., 2005).

Fibrosis is yet another example of an extracellular environment that may perturb control of stem cell fate and may be triggered by derailed balance of stem cell proliferation, differentiation and apoptosis. Successful wound healing requires a number of overlapping events including leukocyte recruitment, matrix deposition, epithelialization, and ultimately resolution of inflammation with the formation of a mature scar. Age-related delayed wound healing was found to result from impaired cytokine signal transduction, unchecked inflammation, an altered balance of protein synthesis and degradation, and subsequent downstream effects on the rate and quality of the wound healing response (Ashcroft et al., 2002). As a consequence thereof, stromal fibroblasts commence to transdifferentiate into myofibroblasts resulting in pronounced contraction and enhanced cellular migration. Myofibroblasts are known to distinctly secrete cytokines which stimulate an inflammatory response that complicates appropriate regeneration of the tissue and cell layers in the case of a persisting activity of myofibroblasts. Eventually fibrotic scars and tissues are formed.

Inflammatory Cues

Levels of pro-inflammatory cytokines such as tumor necrosis factor alpha (TNFα) increase in older people (de Gonzalo-Calvo et al., 2010). In excess of inflammation and/or a chronic manifestation thereof it has been proposed long time ago that tumor formation and propagation is instigated (Dolberg et al., 1985; Dvorak, 1986), and indeed there exists well documented evidence that chronic inflammation may for instance cause colon cancer (Hahn et al., 2008; Khatami, 2009).

MSC treated with increasing concentrations of interferon gamma (IFNγ) or TNFα activate CD106/Vascular Cell Adhesion Molecule 1 (VCAM1) expression on the surface of MSC (Laschober et al., 2011a). We also found this marker differentially expressed in primary MSC derived from bone of differently aged donors. Interestingly however, proliferation and self-renewal assays as well as quantitative assessment of the levels of various transcription factors involved in the regulation and control of stemness and differentiation indicated that regardless of CD106 expression, MSC exhibited no apparent difference. In addition to that, we were also able to show that IFNγ effectively modulates the commitment of MSC with regard to their respective differentiation capacity while at the same time not disturbing their proliferative potential. Hence according to defined environmental specifications, VCAM1/CD106 is upregulated in MSC which commit themselves to lineage-specific differentiation. It has been appreciated though that well adjusted, elevated levels of inflammatory stimuli go along with wound and bone healing and are thus supporting osteoblastogenesis (Duque et al., 2008). However, lack of this type of instructive signal would rather favor adipogenic differentiation. This is in line with observations that bone loss is associated, if not caused, by inflammatory disorders (Hardy and Cooper, 2009). Thus, dominant aberrations within the MSC microenvironment may arise from systemic chronic inflammation, which as mentioned above occurs regularly in elderly persons, or unbalanced inflammatory and anti-inflammatory networks as a consequence to life-long antigenic burden or age-related diseases (Franceschi et al., 2007). Such cues are likely leading to a deviation from properly guiding molecular signaling mechanisms during regeneration, repair and tissue remodelling, and may thus gear a pronounced impetus for adipogenic upgrowth, to name only one prominent example, bone marrow adipositas.

Key sources of proinflammatory cytokines such as IFNγ are T lymphocytes and natural killer (NK) cells. In the last few years, it became clear that MSC exhibit immune regulatory properties (Chen et al., 2006; Dazzi et al., 2006; Nauta and Fibbe, 2007). It could be further shown that the suppressive activity of MSC requires the presence of IFNγ produced by T cells and NK cells (Krampera et al 2006). This observation indicates that MSC not only passively withstand inflammatory stimuli, but more than that, MSC decisively respond to inflammatory cues, e.g. by promoting the expression of indoleamine 2,3-dioxygenase, which leads to tryptophan depletion in lymphocytes, and may by the same token modulate lymphocyte activity and proliferation (Frumento et al., 2002; Meisel et al., 2004). Taken together, fate and function of both cell classes are tightly interconnected and age-related changes appear to be compensated to a certain degree by complementary modes (Tokoyoda et al., 2010).

Epigenetic Cues

Chromatin comprises DNA and DNA-bound proteins such as histones. Both, DNA and histones, can be modified, resulting in a looser (euchromatin) or denser (heterochromatin) packing of DNA, in consequence leading to programmed gene activation or gene repression. It has also been acknowledged that non-coding RNA contributes to heterochromatin formation. Major covalent modifications of the genome include methylation of CpG nucleotides, especially CpG islands at promoters of repressed genes, and a broad range of histone tail modifications such as methylation, acetylation, phosphorylation, ubiquitinylation, and sumoylation. DNA methylation is established by DNA methyltransferases (DNMT) 3a and 3b and is maintained during cell division by DNMT1. DNA methylation is relatively stable; no definitive DNA de-methylating enzyme was identified so far. Histone modifications are more reversible. Methylations are established by histone methyltransferases (HMTs) such as the Drosophila trithorax (Trx) and its mammalian homolog mixed lineage leukaemia (MLL), and proteins of the polycomb group (PcG), which form the complexes PRC1 and 2 (polycomb repressive complex).

There are many reasons to believe that in course of aging, specific chromatin patterns are altered. Aged DNA of human and murine tissues appear globally hypomethylated, whereas specific regions are hypermethylated (Maegawa et al., 2010). Interestingly, similar changes can be observed in tumor cells, and it has further been demonstrated that hypomethylation activates oncogenes such as cyclin D2 and r-ras, while at the same time de-represses retrotransposons, leading to genomic instability. Contrastingly, within the number of hypermethylated genes, well-known tumor suppressors such as p16 and pRB have been singled out (Cruickshanks and Adams, 2010; Siddiqi et al., 2010).

Being highly plastic in terms of differentiation stem cells are seemingly articulated in terms of chromatin state. Plasticity is thought to emerge through the action of so-called "poised" or "bivalent" chromatin at loci harboring early differentiation genes. A poised state is accomplished by co-existence of repressive histone marks (H3K27me), active histone marks (H3K4me) and lack of DNA methylation. By this means, a stem cell is able to respond to differentiation cues quickly solely by activation of early differentiation genes (Lunyak and Rosenfeld, 2008; Wu and Sun, 2006). The respective histone marks are established and maintained by members of PRC1 and 2, such as Suz12, Eed and EZH2 (for H3K27me), as well as trithorax proteins such as MLL (for H3K4me). Interestingly, a high percentage of genes repressed by transcription factors which are potent inducers of pluripotency, such as OCT4, SOX2 and NANOG, are actually targets of polycomb group-mediated repression in embryonic stem cells (Boyer et al., 2005). Lastly, there are good evidences to believe that the state of chromatin is critical to assign, direct and guide DNA repair (Benetti et al., 2008; Braconi et al., 2010; Juan et al., 2009; Lee et al., 2010; Pegoraro et al., 2009; Pegoraro and Misteli, 2009), which is considered integral to preserve the naïve and unchanged state of stem cells.

Comparable to somatic cells, stem cells display variant epigenetic modes during aging. Due to their peculiar chromatin state, they further bear the risk of transformation, in particular because pluripotency genes are often actively transcribed in this class of cells. Epigenetic changes have also been investigated in aging MSC, however most results available to date are picturing deviations after long-term culture when cells turned replicatively senescent (Alessio et al., 2010; Jung et al., 2010; Napolitano et al., 2007). Viewed from a similar angle, an increase in EZH2 expression was noticed in adipose-derived MSC during successive steps of long-term culture. In parallel, these cells lost differentiation potential, which correlated with stable histone methylation and hypoacetylation on lineage-specific promoters (Noer et al., 2009). The global methylation pattern of MSC from differently aged donors appear to remain unchanged with age; yet distinct exceptions manifest and stand out such as differential methylation of particular genes, such as Hoxa11, which discriminates craniofacial MSC from skeletal ones (Leucht et al., 2008), or the tumor suppressor p15 (Bork et al., 2010).

A class of short non-coding RNAs, miRNAs, is regarded potent epigenetic regulators of gene expression as they are capable of degrading target mRNAs or otherwise inhibiting translation thereof. In many studies, miRNAs have been reported to target oncogenes, tumor suppressors as well as differentiation markers, which typically should be kept silenced in stem cells in order to prevent deliberate differentiation instead of committing themselves to self-renewing (Imam et al., 2010; Urbich et al., 2008; Wang et al., 2011). Individual miRNAs can also regulate epigenetic regulators such as DNMT, HDAC, high-mobility group AT-hook 2

(HMGA2) and PcG (Benetti et al., 2008; Braconi et al., 2010; Juan et al., 2009; Lee et al., 2010). Only recently, data from global analysis of miRNA expression profiles in MSC derived from bone marrow of old rhesus monkeys became available. Thereby, down-regulation of miR-let-7f, miR-23a, miR-132, miR-125b, miR-199-3p, miR-221, miR-222, miR-365, and miR-720 and an up-regulation of miR-291a-5p, miR-558, miR-466, miR-466-3p as well as miR-766 became apparent (Yu et al., 2011). Down to the present day, the full activity of most of the identified miRNAs is not well defined. Besides miR-221 and miR-222 (Lambeth et al., 2009; Lorimer, 2009), which appear to be involved in regulation of genes relevant to tumorigenesis and oncogenic activity, many of the others are clustered together in miR-17-92, a cluster which was found to be differentially regulated in many cellular aging models (Grillari et al., 2010). Notably, the down-regulation of miR-365 was noticed to be associated with growth arrest (Maes et al., 2009).

Although evident that stemness and differentiation potential of MSC is varying over lifetime, and the inherent dynamics are superimposed to variant epigenetic manifestations, the detailed functions of primary epigenetic regulatory factors such as DNMT, HDAC, PcG and miRNAs still remain to be elucidated, mainly due to the fact that their actions greatly overlap and these factor interplay and cross-regulate each other.

DISPARATE IMPACTS: CONSEQUENCES OF MSC AGING

Fracture Healing

With increasing age, not only the fracture incidence is increased, but also the capacity of bone to regenerate slows down with advancing age (Gruber et al., 2006). Age-related changes include the detention in the onset of the periosteal reaction, delays in cell differentiation, decreased bone formation, retarded angiogenic invasion of cartilage, a protracted period of endochondral ossification and impaired bone remodeling (Lu et al., 2005). As osteogenic differentiation is highly associated with blood vessel formation, which actually is greatly retarded in the elderly and thus the prime cause for slow fracture healing, critical contribution of debilitated MSC remains unclear.

Wound Repair

Normal wound repair is often decelerated in elderly persons or profoundly impacted in several age-related disease which display states of non-healing or poor healing of lesions. As MSC can be isolated from virtually all tissue types, it appears likely that they are involved in normal wound repair. It is not known whether MSC simply divide and increase their representation in the wound bed, but it is tempting to fancy that these cells communicate with overlying epithelial stem cells and the remainder of the wound-associated mesenchyme. MSC either can provide daughter cells that differentiate and then directly participate in the structural repair of a wound, or can supply secreted factors that support wound repair and modulate the immune system (Wu et al., 2007). In lieu of basic knowledge regarding their actual involvement in this process, it is furthermore highly speculative whether aged MSC are confounding denominators of age-associated impairment in wound healing.

Loss of Active Bone Marrow

Both mass and mineral density of cortical and cancellous bone decreases with aging, while at the same time fat cells are enriched in bone marrow at the expense of hematopoietically active regions. Recently, it was shown that adipose tissue is not a passive space filler, but influences hematopoietic repopulation activity negatively, possibly by secretion of cytokines such as TNFα (Naveiras et al., 2009). Therefore it is likely that the enhanced propensity of aged MSC to differentiate into adipocytes rather than osteoblasts may have a detrimental effect on hematopoiesis in the elderly.

Age-Associated Diseases

Evidence that MSC may be dysfunctional in age-associated disease is controversial. At least two reports (Dalle Carbonare et al., 2009; Rodriguez et al., 1999) suggest that abnormal osteogenic potential contributes to osteoporosis and that MSC growth, proliferative response, and osteogenic differentiation may be affected in this disease.

MSC may also play a decisive role in the immunopathology of rheumatoid arthritis (Marinova-Mutafchieva et al., 2002). It could be shown in the TNFα transgenic mouse model that MSC can migrate to the joint space before the onset of acute inflammation, while anti-TNFα therapy could actually inhibit local migration of MSC and prevent the clinical signs of arthritis. Yet paradoxically, injection of MSC into the diseased joints prevented the occurrence of severe arthritis, most likely because production of pro-inflammatory cytokines was effectively blocked. However on the long run this treatment seems to yield no curative effect. Whether it is aged MSC which are all-dominantly involved in disease manifestation remains to be determined.

Cancer

There is good evidence that analogous to hematopoietic stem / progenitor cells which are originators of leukemia stem cells (Passegué et al., 2003), MSC and mesenchymal progenitors are trailheads of sarcomas, as sarcoma formation could potentially originate from aberrantly differentiating MSC (Tang et al., 2008). In line with this assumption, gene expression signatures of cells derived from malignant fibrous histiocytoma (MFH), a highly undifferentiated pleomorphic sarcoma with an age onset of 50-70 years (Matushansky et al., 2008), as well as that of Ewing's sarcoma cells, which in turn is most frequent in male teenagers (Tirode et al., 2007) was found to resemble genome-wide expression of MSC. Apparent in these cases are signaling pathways which are involved in guiding MSC differentiation and are seemingly activated in a comparable fashion in the pathological situation. Prominent examples elucidated in this way are wnt/beta-catenin targets and activation of c-myc, which are known to be positive regulators of osteogenic differentiation

(Krishnan et al., 2006) while at the same time inhibiting adipogenic differentiation (Ross et al., 2000). In a comparable fashion, also the etiology of liposarcomas could be linked to certain stages in MSC adipogenic differentiation (Matushansky et al., 2008). However, the fact that the prevalence of sarcomas remains low late in life greatly suggests that alterations of MSC corresponding with advancing age are not excessive enough to promote tumorigenesis. Despite this observation, the concept of cancer originating from cancer stem cells (CSC) is also of interest in the context of MSC biogerontology. Needless to say that there are no experimental results available to date, which proof whether CSC exhibit quasi multipotential differentiation potential. Also no in vivo lineage tracing could be established which clarifies a direct interrelationship of tissue-borne MSC with CSC.

Although proofs for tumor initiation by aged MSC is vague, there are by far better evidences that MSC are involved in the progression of carcinogenesis, as MSC appear to migrate towards primary tumors and metastatic sites (Lazennec and Jorgensen, 2008). It is highly likely that determining chemokines, such as CCL5 or SDF-1 alpha, which are being secreted by attracted MSC, actually promote emergence of metastases as shown in the case of breast cancer (Karnoub et al., 2007), or facilitate the recruitment of breast cancer cells to the bone marrow which was shown to be mediated through the action of CXCR4 (Corcoran et al., 2008). Provided these evidences, it would be tempting to examine whether chemokine expression is indeed changed in aging MSC.

In contrast to the aforementioned information, also an anti-proliferative effect of MSC could be distinguished in a model of colon carcinogenesis in rats. Here, tumor initiating cells co-injected with MSC exhibited significant growth inhibition (Ohlsson et al., 2003). Also in the case of an experimental model for Kaposi's sarcoma, MSC exerted a potent tumor-suppressive effect *in vivo*, through direct cell contact (Khakoo et al., 2006). Yet in another model, MSC migrated to and inhibited proliferation of glioma cells (Nakamura et al., 2004). Taken together, it appears likely that MSC may induce apoptosis of cancer cells albeit this fuzzy skill appears to greatly depend on the origin of the tumor and the ratio of MSC and tumor cells, and again it is not clear to date whether there are age-associated changes to this particular property.

EPILOGUE

It is tempting to see that over the recent years, increasingly more information became available concerning the alterations of stem cells for tissue repair as they age, and it is thus considered an emerging field of interest in the proliferating discipline of Aging Research. Understanding of how stem cells are able to continuously self-renew may lead to a more complete understanding of how the body ages, and subsequently will nourish ideas on how to slow down or even cease the aging of tissues. Moreover, this particular theme will also become increasingly important as the number of MSC therapies steadily rises, while pertinent questions about MSC aging are still unanswered.

ACKNOWLEDGMENTS

Our work was previously funded by the Austrian Science Fund (NFN Project S93) and the Jubilee Fund of the Austrian National Bank. Currently our research is aided by grants gifted by the EC (FP7 - Health VascuBone), the Austrian Research Agency (Laura Bassi Centre for Excellence - DIALIFE) and The Tyrolean Future Fund (Translational Project - Smart Implants).

DISCLOSURE OF POTENTIAL CONFLICTS OF INTEREST

No potential conflicts of interest were disclosed.

REFERENCES

Alessio, N., Squillaro, T., Cipollaro, M., Bagella, L., Giordano, A., and Galderisi, U. (2010). The BRG1 ATPase of chromatin remodeling complexes is involved in modulation of mesenchymal stem cell senescence through RB-P53 pathways. *Oncogene*. 29, 5452-5463.

Alikhani, Z., Alikhani, M., Boyd, C.M., Nagao, K., Trackman, P.C., and Graves, D.T. (2005). Advanced glycation end products enhance expression of pro-apoptotic genes and stimulate fibroblast apoptosis through cytoplasmic and mitochondrial pathways. *J Biol Chem*. 280, 12087-12095.

Antoniou, E.S., Sund, S., Homsi, E.N., Challenger, L.F., and Rameshwar, P. (2004). A theoretical simulation of hematopoietic stem cells during oxygen fluctuations: prediction of bone marrow responses during hemorrhagic shock. *Shock*, 22, 415-422.

Ashcroft, G.S., Mills, S.J., and Ashworth, J.J. (2002). Ageing and wound healing. *Biogerontology*, 3, 337-345.

Baxter, M.A., Wynn, R.F., Jowitt, S.N., Wraith, J.E., Fairbairn, L.J., and Bellantuono, I. (2004). Study of telomere length reveals rapid aging of human marrow stromal cells following in vitro expansion. *Stem Cells*, 22, 675-682.

Beausejour, C. (2007). Bone marrow-derived cells: the influence of aging and cellular senescence. *Handbook of experimental pharmacology*, 67-88.

Benetti, R., Gonzalo, S., Jaco, I., Munoz, P., Gonzalez, S., Schoeftner, S., Murchison, E., Andl, T., Chen, T., Klatt, P., et al. (2008). A mammalian microRNA cluster controls DNA methylation and telomere recombination via Rbl2-dependent regulation of DNA methyltransferases. *Nat Struct Mol Biol*. 15, 998.

Bork, S., Pfister, S., Witt, H., Horn, P., Korn, B., Ho, A.D., and Wagner, W. (2010). DNA methylation pattern changes upon long-term culture and aging of human mesenchymal stromal cells. *Aging Cell*, 9, 54-63.

Bosch, P., Musgrave, D.S., Lee, J.Y., Cummins, J., Shuler, T., Ghivizzani, T.C., Evans, T., Robbins, T.D., and Huard (2000). Osteoprogenitor cells within skeletal muscle. *J Orthop Res*. 18, 933-944.

Boyer, L.A., Lee, T.I., Cole, M.F., Johnstone, S.E., Levine, S.S., Zucker, J.P., Guenther, M.G., Kumar, R.M., Murray, H.L., Jenner, R.G., et al. (2005). Core transcriptional regulatory circuitry in human embryonic stem cells. *Cell*, 122, 947-956.

Braconi, C., Huang, N., and Patel, T. (2010). MicroRNA-dependent regulation of DNA methyltransferase-1 and tumor suppressor gene expression by interleukin-6 in human malignant cholangiocytes. *Hepatology*, 51, 881-890.

Campisi, J., and d'Adda di Fagagna, F. (2007). Cellular senescence: when bad things happen to good cells. *Nat Rev Mol Cell Biol*. 8, 729-740.

Caplan, A.I. (1991). Mesenchymal stem cells. J Orthop Res 9, 641-650.

Caplan, A.I. (2007). Adult mesenchymal stem cells for tissue engineering versus regenerative medicine. *J Cell Physiol*. 213, 341-347.

Caplan, A.I. (2009). Why are MSCs therapeutic? New data: new insight. *The Journal of pathology*, 217, 318-324.

Chen, X., Armstrong, M.A., and Li, G. (2006). Mesenchymal stem cells in immunoregulation. *Immunology and cell biology*, 84, 413-421.

Choumerianou, D.M., Martimianaki, G., Stiakaki, E., Kalmanti, L., Kalmanti, M., and Dimitriou, H. (2010). Comparative study of stemness characteristics of mesenchymal cells from bone marrow of children and adults. *Cytotherapy*, 12, 881-887.

Collett, G.D., and Canfield, A.E. (2005a). Angiogenesis and pericytes in the initiation of ectopic calcification. *Circ Res*. 96, 930-938.

Collett, G.D.M., and Canfield, A.E. (2005b). Angiogenesis and pericytes in the initiation of ectopic calcification. *Circulation research*, 96, 930-938.

Conboy, I.M., Conboy, M.J., Wagers, A.J., Girma, E.R., Weissman, I.L., and Rando, T.A. (2005). Rejuvenation of aged progenitor cells by exposure to a young systemic environment. *Nature*, 433, 760-764.

Corcoran, K.E., Trzaska, K.A., Fernandes, H., Bryan, M., Taborga, M., Srinivas, V., Packman, K., Patel, P.S., and Rameshwar, P. (2008). Mesenchymal stem cells in early entry of breast cancer into bone marrow. *PloS One*, 3, e2563.

Crisan, M., Yap, S., Casteilla, L., Chen, C.W., Corselli, M., Park, T.S., Andriolo, G., Sun, B., Zheng, B., Zhang, L., et al. (2008). A perivascular origin for mesenchymal stem cells in multiple human organs. *Cell Stem Cell*, 3, 301-313.

Cruickshanks, H.A., and Adams, P.D. (2010). Chromatin: a molecular interface between cancer and aging. *Current Opin Gen Dev*.

D'Ippolito, G., Schiller, P.C., Ricordi, C., Roos, B.A., and Howard, G.A. (1999). Age-related osteogenic potential of mesenchymal stromal stem cells from human vertebral bone marrow. *J Bone Miner Res*. 14, 1115-1122.

da Silva Meirelles, L., Caplan, A.I., and Nardi, N.B. (2008). In search of the in vivo identity of mesenchymal stem cells. *Stem Cells*, 26, 2287-2299.

Dalle Carbonare, L., Valenti, M.T., Zanatta, M., Donatelli, L., and Lo Cascio, V. (2009). Circulating mesenchymal stem cells with abnormal osteogenic differentiation in patients with osteoporosis. *Arthritis Rheum*. 60, 3356-3365.

Dazzi, F., Ramasamy, R., Glennie, S., Jones, S.P., and Roberts, I. (2006). The role of mesenchymal stem cells in haemopoiesis. *Blood reviews*, 20, 161-171.

de Gonzalo-Calvo, D., Neitzert, K., Fernández, M., Vega-Naredo, I., Caballero, B., García-Macía, M., SuÃ¡rez, F.M., Rodríguez-Colunga, M.J., Solano, J.J., and Coto-Montes, A.

(2010). Differential inflammatory responses in aging and disease: TNF-alpha and IL-6 as possible biomarkers. *Free Radical Biol Med*. 49, 733-737.

Dimri, G.P., Lee, X., Basile, G., Acosta, M., Scott, G., Roskelley, C., Medrano, E.E., Linskens, M., Rubelj, I., Pereira-Smith, O., et al. (1995). A biomarker that identifies senescent human cells in culture and in aging skin in vivo. *Proc Natl Acad Sci U S A*, 92, 9363-9367.

Dolberg, D.S., Hollingsworth, R., Hertle, M., and Bissell, M.J. (1985). Wounding and its role in RSV-mediated tumor formation. *Science* (New York, NY) 230, 676-678.

Dominici, M., Le Blanc, K., Mueller, I., Slaper-Cortenbach, I., Marini, F., Krause, D., Deans, R., Keating, A., Prockop, D., and Horwitz, E. (2006). Minimal criteria for defining multipotent mesenchymal stromal cells. The International Society for Cellular Therapy position statement. *Cytotherapy*, 8, 315-317.

Dorshkind, K. (1990). Regulation of hemopoiesis by bone marrow stromal cells and their products. *Annu Rev Immunol*. 8, 111-137.

Duque, G., Huang, D.C., Macoritto, M., Rivas, D., Yang, X.F., Ste-Marie, L.G., and Kremer, R. (2008). Autocrine Regulation of Interferon {gamma} in Mesenchymal Stem Cells plays a Role in Early Osteoblastogenesis. Stem Cells.

Dvorak, H.F. (1986). Tumors: wounds that do not heal. Similarities between tumor stroma generation and wound healing. *N Engl J Med*. 315, 1650-1659.

Eghbali-Fatourechi, G.Z., Lamsam, J., Fraser, D., Nagel, D., Riggs, B.L., and Khosla, S. (2005). Circulating osteoblast-lineage cells in humans. *N Engl J Med*. 352, 1959-1966.

Erusalimsky, J.D., and Kurz, D.J. (2005). Cellular senescence in vivo: its relevance in ageing and cardiovascular disease. *Exp Gerontol*. 40, 634-642.

Fan, M., Chen, W., Liu, W., Du, G.-Q., Jiang, S.-L., Tian, W.-C., Sun, L., Li, R.-K., and Tian, H. (2010). The Effect of Age on the Efficacy of Human Mesenchymal Stem Cell Transplantation after a Myocardial Infarction. Rejuvenation research.

Fehrer, C., Brunauer, R., Laschober, G., Unterluggauer, H., Reitinger, S., Kloss, F., Gülly, C., Gassner, R., and Lepperdinger, G. (2007). Reduced oxygen tension attenuates differentiation capacity of human mesenchymal stem cells and prolongs their life span. *Aging Cell*, 6, 745-757.

Fehrer, C., and Lepperdinger, G. (2005). Mesenchymal stem cell aging. *Exp Gerontol*. 40, 926-930.

Franceschi, C., Capri, M., Monti, D., Giunta, S., Olivieri, F., Sevini, F., Panourgia, M.P., Invidia, L., Celani, L., Scurti, M., et al. (2007). Inflammaging and anti-inflammaging: a systemic perspective on aging and longevity emerged from studies in humans. *Mech Ageing Dev*. 128, 92-105.

Frumento, G., Rotondo, R., Tonetti, M., Damonte, G., Benatti, U., and Ferrara, G.B. (2002). Tryptophan-derived catabolites are responsible for inhibition of T and natural killer cell proliferation induced by indoleamine 2,3-dioxygenase. *The Journal of experimental medicine*, 196, 459-468.

Glowacki, J. (1995). Influence of age on human marrow. *Calcif Tissue Int*. 56 Suppl 1, S50-51.

Grillari, J., Hackl, M., and Grillari-Voglauer, R. (2010). miR-17–92 cluster: ups and downs in cancer and aging. *Biogerontology*, 11, 501-506.

Grillo, M.A., and Colombatto, S. (2008). Advanced glycation end-products (AGEs): involvement in aging and in neurodegenerative diseases. Amino Acids 35, 29-36.

Gronthos, S., Graves, S.E., Ohta, S., and Simmons, P.J. (1994). The STRO-1+ fraction of adult human bone marrow contains the osteogenic precursors. *Blood*, 84, 4164-4173.

Gruber, R., Koch, H., Doll, B.A., Tegtmeier, F., Einhorn, T.A., and Hollinger, J.O. (2006). Fracture healing in the elderly patient. *Exp Gerontol*. 41, 1080-1093.

Hahn, M., Hahn, T., Lee, D.-H., Esworthy, R.S., Kim, B.-w., Riggs, A.D., Chu, F.-F., and Pfeifer, G.P. (2008). Methylation of Polycomb Target Genes in Intestinal Cancer Is Mediated by Inflammation. *Cancer Research*, 68, 10280-10289.

Hardy, R., and Cooper, M.S. (2009). Bone loss in inflammatory disorders. *J Endocrinol*. 201, 309-320.

Heppenstall, R.B., Grislis, G., and Hunt, T.K. (1975). Tissue gas tensions and oxygen consumption in healing bone defects. *Clin Orthop Relat Res*, 357-365.

Hermann, A., List, C., Habisch, H.-J., Vukicevic, V., Ehrhart-Bornstein, M., Brenner, R., Bernstein, P., Fickert, S., and Storch, A. (2010). Age-dependent neuroectodermal differentiation capacity of human mesenchymal stromal cells: limitations for autologous cell replacement strategies. *Cytotherapy*, 12, 17-30.

Horwitz, E.M., and Keating, A. (2000). Nonhematopoietic mesenchymal stem cells: what are they? *Cytotherapy*, 2, 387-388.

Imam, J.S., Buddavarapu, K., Lee-Chang, J.S., Ganapathy, S., Camosy, C., Chen, Y., and Rao, M.K. (2010). MicroRNA-185 suppresses tumor growth and progression by targeting the Six1 oncogene in human cancers. *Oncogene*. 29, 4971-4979.

Ishikawa, Y., and Ito, T. (1988). Kinetics of hemopoietic stem cells in a hypoxic culture. *Eur J Haematol*. 40, 126-129.

Jauniaux, E., Watson, A., Ozturk, O., Quick, D., and Burton, G. (1999). In-vivo measurement of intrauterine gases and acid-base values early in human pregnancy. Hum Reprod. 14, 2901-2904.

Juan, A.H., Kumar, R.M., Marx, J.G., Young, R.A., and Sartorelli, V. (2009). Mir-214-dependent regulation of the polycomb protein Ezh2 in skeletal muscle and embryonic stem cells. *Mol Cell*, 36, 61-74.

Jung, J.-W., Lee, S., Seo, M.-S., Park, S.-B., Kurtz, A., Kang, S.-K., and Kang, K.-S. (2010). Histone deacetylase controls adult stem cell aging by balancing the expression of polycomb genes and jumonji domain containing 3. *Cell Mol Life Sci*. 67, 1165-1176.

Justesen, J., Stenderup, K., Eriksen, E.F., and Kassem, M. (2002). Maintenance of osteoblastic and adipocytic differentiation potential with age and osteoporosis in human marrow stromal cell cultures. *Calcif Tissue Int*. 71, 36-44.

Kamat, B.R., Galli, S.J., Barger, A.C., Lainey, L.L., and Silverman, K.J. (1987). Neovascularization and coronary atherosclerotic plaque: cinematographic localization and quantitative histologic analysis. *Human pathology*, 18, 1036-1042.

Karnoub, A.E., Dash, A.B., Vo, A.P., Sullivan, A., Brooks, M.W., Bell, G.W., Richardson, A.L., Polyak, K., Tubo, R., and Weinberg, R.A. (2007). Mesenchymal stem cells within tumour stroma promote breast cancer metastasis. *Nature*, 449, 557-563.

Kemp, K., Gray, E., Mallam, E., Scolding, N., and Wilkins, A. (2010). Inflammatory cytokine induced regulation of superoxide dismutase 3 expression by human mesenchymal stem cells. *Stem Cell Rev*. 6, 548-559.

Khakoo, A.Y., Pati, S., Anderson, S.A., Reid, W., Elshal, M.F., Rovira, II, Nguyen, A.T., Malide, D., Combs, C.A., Hall, G., et al. (2006). Human mesenchymal stem cells exert potent antitumorigenic effects in a model of Kaposi's sarcoma. *The Journal of experimental medicine*, 203, 1235-1247.

Khatami, M. (2009). Inflammation, aging, and cancer: tumoricidal versus tumorigenesis of immunity: a common denominator mapping chronic diseases. *Cell Biochem Biophys*. 55, 55-79.

Krampera M et al (2006): Role for Interferon-☐ in the Immunomodulatory Activity of Human Bone Marrow Mesenchymal Stem Cells. *Stem Cells*, 24, 386-398

Kretlow, J.D., Jin, Y.-Q., Liu, W., Zhang, W.J., Hong, T.-H., Zhou, G., Baggett, L.S., Mikos, A.G., and Cao, Y. (2008). Donor age and cell passage affects differentiation potential of murine bone marrow-derived stem cells. *BMC cell biology*, 9, 60.

Krishnan, V., Bryant, H.U., and Macdougald, O.A. (2006). Regulation of bone mass by Wnt signaling. *The Journal of clinical investigation*, 116, 1202-1209.

Kume, S., Kato, S., Yamagishi, S., Inagaki, Y., Ueda, S., Arima, N., Okawa, T., Kojiro, M., and Nagata, K. (2005). Advanced glycation end-products attenuate human mesenchymal stem cells and prevent cognate differentiation into adipose tissue, cartilage, and bone. *J Bone Miner Res*. 20, 1647-1658.

Kuznetsov, S.A., Krebsbach, P.H., Satomura, K., Kerr, J., Riminucci, M., Benayahu, D., and Robey, P.G. (1997). Single-colony derived strains of human marrow stromal fibroblasts form bone after transplantation in vivo. *J Bone Miner Res*. 12, 1335-1347.

Lambeth, L.S., Yao, Y., Smith, L.P., Zhao, Y., and Nair, V. (2009). MicroRNAs 221 and 222 target p27Kip1 in Marek's disease virus-transformed tumour cell line MSB-1. *J Gen Virol* 90, 1164-1171.

Laschober, G.T., Brunauer, R., Jamnig, A., Fehrer, C., Greiderer, B., and Lepperdinger, G. (2009). Leptin receptor/CD295 is upregulated on primary human mesenchymal stem cells of advancing biological age and distinctly marks the subpopulation of dying cells. *Exp Gerontol*. 44, 57-62.

Laschober, G.T., Brunauer, R., Jamnig, A., Singh, S., Hafen, U., Fehrer, C., Kloss, F., Gassner, R., and Lepperdinger, G. (2011a). Age-Specific Changes of Mesenchymal Stem Cells Are Paralleled by Upregulation of CD106 Expression As a Response to an Inflammatory Environment. Rejuvenation Res in press.

Laschober, G.T., Ruli, D., Hofer, E., Muck, C., Carmona-Gutierrez, D., Ring, J., Hutter, E., Ruckenstuhl, C., Micutkova, L., Brunauer, R., et al. (2011b). Identification of evolutionarily conserved genetic regulators of cellular aging. *Aging Cell*, 9, 1084-1097.

Lazennec, G., and Jorgensen, C. (2008). Concise review: adult multipotent stromal cells and cancer: risk or benefit? Stem Cells 26, 1387-1394.

Lee, S., Jung, J.W., Park, S.B., Roh, K., Lee, S.Y., Kim, J.H., Kang, S.K., and Kang, K.S. (2010). Histone deacetylase regulates high mobility group A2-targeting microRNAs in human cord blood-derived multipotent stem cell aging. Cell Mol Life Sci 68, 325-336.

Leucht, P., Kim, J.-B., Amasha, R., James, A.W., Girod, S., and Helms, J.A. (2008). Embryonic origin and Hox status determine progenitor cell fate during adult bone regeneration. *Development* (Cambridge, England) 135, 2845-2854.

Lorimer, I.A. (2009). Regulation of p27Kip1 by miRNA 221/222 in glioblastoma. *Cell Cycle*, 8, 2685.

Lu, C., Miclau, T., Hu, D., Hansen, E., Tsui, K., Puttlitz, C., and Marcucio, R.S. (2005). Cellular basis for age-related changes in fracture repair. *J Orthop Res*. 23, 1300-1307.

Lund, T.C., Kobs, A., Blazar, B.R., and Tolar, J. (2011). Mesenchymal stromal cells from donors varying widely in age are of equal cellular fitness after in vitro expansion under hypoxic conditions. *Cytotherapy*, 12, 971-981.

Lunyak, V.V., and Rosenfeld, M.G. (2008). Epigenetic regulation of stem cell fate. *Hum Mol Gen.* 17, R28-36.

Maegawa, S., Hinkal, G., Kim, H.S., Shen, L., Zhang, L., Zhang, J., Zhang, N., Liang, S., Donehower, L.A., and Issa, J.-P. (2010). Widespread and tissue specific age-related DNA methylation changes in mice. *Genome Res.* 20, 332-340.

Maes, O.C., Sarojini, H., and Wang, E. (2009). Stepwise up-regulation of MicroRNA expression levels from replicating to reversible and irreversible growth arrest states in WI-38 human fibroblasts. *Journal of Cellular Physiology*, 221, 109-119.

Majors, A.K., Boehm, C.A., Nitto, H., Midura, R.J., and Muschler, G.F. (1997). Characterization of human bone marrow stromal cells with respect to osteoblastic differentiation. *J Orthop Res.* 15, 546-557.

Mareschi, K., Ferrero, I., Rustichelli, D., Aschero, S., Gammaitoni, L., Aglietta, M., Madon, E., and Fagioli, F. (2006). Expansion of mesenchymal stem cells isolated from pediatric and adult donor bone marrow. *J Cell Biochem.* 97, 744-754.

Marinova-Mutafchieva, L., Williams, R.O., Funa, K., Maini, R.N., and Zvaifler, N.J. (2002). Inflammation is preceded by tumor necrosis factor-dependent infiltration of mesenchymal cells in experimental arthritis. Arthritis Rheum 46, 507-513.

Matushansky, I., Hernando, E., Socci, N.D., Matos, T., Mills, J., Edgar, M.A., Schwartz, G.K., Singer, S., Cordon-Cardo, C., and Maki, R.G. (2008). A developmental model of sarcomagenesis defines a differentiation-based classification for liposarcomas. *Am J Pathol.* 172, 1069-1080.

Mauch, P., Botnick, L.E., Hannon, E.C., Obbagy, J., and Hellman, S. (1982). Decline in bone marrow proliferative capacity as a function of age. Blood 60, 245-252.

McCauley, L.K. (2010). c-Maf and you won't see fat. *J Clin Invest.* 120, 3440-3442.

Meisel, R., Zibert, A., Laryea, M., Gobel, U., Daubener, W., and Dilloo, D. (2004). Human bone marrow stromal cells inhibit allogeneic T-cell responses by indoleamine 2,3-dioxygenase-mediated tryptophan degradation. *Blood*, 103, 4619-4621.

Mendez-Ferrer, S., Michurina, T.V., Ferraro, F., Mazloom, A.R., Macarthur, B.D., Lira, S.A., Scadden, D.T., Ma'ayan, A., Enikolopov, G.N., and Frenette, P.S. (2010). Mesenchymal and haematopoietic stem cells form a unique bone marrow niche. *Nature*, 466, 829-834.

Meunier, P., Aaron, J., Edouard, C., and Vignon, G. (1971). Osteoporosis and the replacement of cell populations of the marrow by adipose tissue. A quantitative study of 84 iliac bone biopsies. *Clin Orthop Relat Res.* 80, 147-154.

Moerman, E.J., Teng, K., Lipschitz, D.a., and Lecka-Czernik, B. (2004). Aging activates adipogenic and suppresses osteogenic programs in mesenchymal marrow stroma/stem cells: the role of PPAR-gamma2 transcription factor and TGF-beta/BMP signaling pathways. *Aging cell*, 3, 379-389.

Mohler, E.R., Gannon, F., Reynolds, C., Zimmerman, R., Keane, M.G., and Kaplan, F.S. (2001). Bone formation and inflammation in cardiac valves. *Circulation*, 103, 1522-1528.

Muguruma, Y., Yahata, T., Miyatake, H., Sato, T., Uno, T., Itoh, J., Kato, S., Ito, M., Hotta, T., and Ando, K. (2006). Reconstitution of the functional human hematopoietic microenvironment derived from human mesenchymal stem cells in the murine bone marrow compartment. *Blood*, 107, 1878-1887.

Munch, G., Thome, J., Foley, P., Schinzel, R., and Riederer, P. (1997). Advanced glycation endproducts in ageing and Alzheimer's disease. *Brain Res Brain Res Rev*. 23, 134-143.

Muraglia, A., Cancedda, R., and Quarto, R. (2000). Clonal mesenchymal progenitors from human bone marrow differentiate in vitro according to a hierarchical model. *J Cell Sci*. 113 (Pt 7), 1161-1166.

Muschler, G.F., Nitto, H., Boehm, C.A., and Easley, K.A. (2001). Age- and gender-related changes in the cellularity of human bone marrow and the prevalence of osteoblastic progenitors. *J Orthop Res*. 19, 117-125.

Nakamura, K., Ito, Y., Kawano, Y., Kurozumi, K., Kobune, M., Tsuda, H., Bizen, A., Honmou, O., Niitsu, Y., and Hamada, H. (2004). Antitumor effect of genetically engineered mesenchymal stem cells in a rat glioma model. *Gene Ther*. 11, 1155-1164.

Napolitano, M.A., Cipollaro, M., Cascino, A., Melone, M.A.B., Giordano, A., and Galderisi, U. (2007). Brg1 chromatin remodeling factor is involved in cell growth arrest, apoptosis and senescence of rat mesenchymal stem cells. *Journal of cell science*, 120, 2904-2911.

Nauta, A.J., and Fibbe, W.E. (2007). Immunomodulatory properties of mesenchymal stromal cells. *Blood*, 110, 3499-3506.

Naveiras, O., Nardi, V., Wenzel, P.L., Hauschka, P.V., Fahey, F., and Daley, G.Q. (2009). Bone-marrow adipocytes as negative regulators of the haematopoietic microenvironment. *Nature*, 460, 259-263.

Neven, E., Dauwe, S., {De Broe}, M.E., D'Haese, P.C., and Persy, V. (2007). Endochondral bone formation is involved in media calcification in rats and in men. *Kidney international*, 72, 574-581.

Nishida, S., Endo, N., Yamagiwa, H., Tanizawa, T., and Takahashi, H.E. (1999). Number of osteoprogenitor cells in human bone marrow markedly decreases after skeletal maturation. *J Bone Miner Metab*. 17, 171-177.

Noer, A., Lindeman, L.C., and Collas, P. (2009). Histone H3 modifications associated with differentiation and long-term culture of mesenchymal adipose stem cells. *Stem cells and development*, 18, 725-736.

Ohlsson, L.B., Varas, L., Kjellman, C., Edvardsen, K., and Lindvall, M. (2003). Mesenchymal progenitor cell-mediated inhibition of tumor growth in vivo and in vitro in gelatin matrix. *Exp Mol Pathol*. 75, 248-255.

Oreffo, R.O., Bennett, A., Carr, A.J., and Triffitt, J.T. (1998). Patients with primary osteoarthritis show no change with ageing in the number of osteogenic precursors. *Scand J Rheumatol*. 27, 415-424.

Passegué, E., Jamieson, C.H.M., Ailles, L.E., and Weissman, I.L. (2003). Normal and leukemic hematopoiesis: are leukemias a stem cell disorder or a reacquisition of stem cell characteristics? *PNAS*, 100 Suppl, 11842-11849.

Pegoraro, G., Kubben, N., Wickert, U., GÃ¶hler, H., Hoffmann, K., and Misteli, T. (2009). Ageing-related chromatin defects through loss of the NURD complex. *Nature cell biology*, 11, 1261-1267.

Pegoraro, G., and Misteli, T. (2009). The central role of chromatin maintenance in aging. *Aging*, 1, 1017-1022.

Pittenger, M.F., Mosca, J.D., and McIntosh, K.R. (2000). Human mesenchymal stem cells: progenitor cells for cartilage, bone, fat and stroma. *Curr Top Microbiol Immunol*. 251, 3-11.

Ramasamy, R., Vannucci, S.J., Yan, S.S., Herold, K., Yan, S.F., and Schmidt, A.M. (2005). Advanced glycation end products and RAGE: a common thread in aging, diabetes, neurodegeneration, and inflammation. *Glycobiology*, 15, 16R-28R.

Rodriguez, J.P., Garat, S., Gajardo, H., Pino, A.M., and Seitz, G. (1999). Abnormal osteogenesis in osteoporotic patients is reflected by altered mesenchymal stem cells dynamics. *J Cell Biochem*. 75, 414-423.

Rose, M.R., and Graves, J.L., Jr. (1989). What evolutionary biology can do for gerontology. *J Gerontol*. 44, B27-29.

Ross, S.E., Hemati, N., Longo, K.A., Bennett, C.N., Lucas, P.C., Erickson, R.L., and MacDougald, O.A. (2000). Inhibition of adipogenesis by Wnt signaling. Science (New York, NY) 289, 950-953.

Roura, S., FarrÃ©, J., Soler-Botija, C., Llach, A., Hove-Madsen, L., CairÃ³, J.J., GÃ²dia, F., Cinca, J., and Bayes-Genis, A. (2006). Effect of aging on the pluripotential capacity of human CD105+ mesenchymal stem cells. European journal of heart failure : *journal of the Working Group on Heart Failure of the European Society of Cardiology*, 8, 555-563.

Scaffidi, P., and Misteli, T. (2008). Lamin A-dependent misregulation of adult stem cells associated with accelerated ageing. *Nat Cell Biol*. 10, 452-459.

Schraml, E., Fehrer, C., Brunauer, R., Lepperdinger, G., Chesnokova, V., and Schauenstein, K. (2008). lin-Sca-1+ cells and age-dependent changes of their proliferation potential are reliant on mesenchymal stromal cells and are leukemia inhibitory factor dependent. *Gerontology*, 54, 312-323.

Sethe, S., Scutt, A., and Stolzing, A. (2006). Aging of mesenchymal stem cells. *Ageing research reviews*, 5, 91-116.

Shibata, K.R., Aoyama, T., Shima, Y., Fukiage, K., Otsuka, S., Furu, M., Kohno, Y., Ito, K., Fujibayashi, S., Neo, M., et al. (2007). Expression of the p16INK4A gene is associated closely with senescence of human mesenchymal stem cells and is potentially silenced by DNA methylation during in vitro expansion. *Stem Cells*, 25, 2371-2382.

Siddiqi, S., Mills, J., and Matushansky, I. (2010). Epigenetic remodeling of chromatin architecture: exploring tumor differentiation therapies in mesenchymal stem cells and sarcomas. *Current stem cell research & therapy*, 5, 63-73.

Simonsen, J.L., Rosada, C., Serakinci, N., Justesen, J., Stenderup, K., Rattan, S.I., Jensen, T.G., and Kassem, M. (2002). Telomerase expression extends the proliferative life-span and maintains the osteogenic potential of human bone marrow stromal cells. *Nature biotechnology*, 20, 592-596.

Sparvero, L.J., Asafu-Adjei, D., Kang, R., Tang, D., Amin, N., Im, J., Rutledge, R., Lin, B., Amoscato, A.A., Zeh, H.J., et al. (2009). RAGE (Receptor for Advanced Glycation Endproducts), RAGE ligands, and their role in cancer and inflammation. *J Transl Med*. 7, 17.

Stenderup, K., Justesen, J., Clausen, C., and Kassem, M. (2003). Aging is associated with decreased maximal life span and accelerated senescence of bone marrow stromal cells. *Bone*, 33, 919-926.

Stenderup, K., Justesen, J., Eriksen, E.F., Rattan, S.I., and Kassem, M. (2001). Number and proliferative capacity of osteogenic stem cells are maintained during aging and in patients with osteoporosis. *J Bone Miner Res*. 16, 1120-1129.

Stolzing, A., Jones, E., McGonagle, D., and Scutt, A. (2008a). Age-related changes in human bone marrow-derived mesenchymal stem cells: consequences for cell therapies. *Mech Ageing Dev.* 129, 163-173.

Stolzing, A., Jones, E., McGonagle, D., and Scutt, A. (2008b). Age-related changes in human bone marrow-derived mesenchymal stem cells: consequences for cell therapies. *Mechanisms of ageing and development*, 129, 163-173.

Tang, N., Song, W.-X., Luo, J., Haydon, R.C., and He, T.-C. (2008). Osteosarcoma development and stem cell differentiation. *Clinical orthopaedics and related research*, 466, 2114-2130.

Tirode, F., Laud-Duval, K., Prieur, A., Delorme, B., Charbord, P., and Delattre, O. (2007). Mesenchymal stem cell features of Ewing tumors. *Cancer cell*, 11, 421-429.

Tokalov, S.V., GrÃ¼ner, S., Schindler, S., Iagunov, A.S., Baumann, M., and Abolmaali, N.D. (2007). Molecules and A Number of Bone Marrow Mesenchymal Stem Cells but Neither Phenotype Nor Differentiation Capacities Changes with Age of Rats. Molecules and Cells 24, 255-260 L251 - file:///C:/Dokumente und Einstellungen/rbrun/Eigene Dateien/Mendeley Desktop/Tokalov et al. - 2007 - Molecules and A Number of Bone Marrow Mesenchymal Stem Cells but Neither Phenotype Nor Differentiation Capacities Changes with Age of Rats.pdf.

Tokoyoda, K., Hauser, A.E., Nakayama, T., and Radbruch, A. (2010). Organization of immunological memory by bone marrow stroma. *Nature reviews Immunology*, 10, 193-200.

Urbich, C., Kuehbacher, A., and Dimmeler, S. (2008). Role of microRNAs in vascular diseases, inflammation, and angiogenesis. *Cardiovasc Res.* 79, 581-588.

Vattikuti, R., and Towler, D.A. (2004). Osteogenic regulation of vascular calcification: an early perspective. American journal of physiology *Endocrinology and metabolism*, 286, E686-696.

Wagner, W., Bork, S., Horn, P., Krunic, D., Walenda, T., Diehlmann, A., Benes, V., Blake, J., Huber, F.-X., Eckstein, V., et al. (2009). Aging and replicative senescence have related effects on human stem and progenitor cells. *PloS one*, 4, e5846.

Wagner, W., Horn, P., Bork, S., and Ho, A.D. (2008). Aging of hematopoietic stem cells is regulated by the stem cell niche. *Experimental gerontology*, 43, 974-980.

Walenda, T., Bork, S., Horn, P., Wein, F., Saffrich, R., Diehlmann, A., Eckstein, V., Ho, A.D., and Wagner, W. (2010). Co-culture with mesenchymal stromal cells increases proliferation and maintenance of haematopoietic progenitor cells. *J Cell Mol Med.* 14, 337-350.

Wang, H.J., Ruan, H.J., He, X.J., Ma, Y.Y., Jiang, X.T., Xia, Y.J., Ye, Z.Y., and Tao, H.Q. (2011). MicroRNA-101 is down-regulated in gastric cancer and involved in cell migration and invasion. *Eur J Cancer*, 46, 2295-2303.

Waterstrat, A., and Van Zant, G. (2009). Effects of aging on hematopoietic stem and progenitor cells. *Curr Opin Immunol.* 21, 408-413.

Wu, H., and Sun, Y.E. (2006). Epigenetic regulation of stem cell differentiation. *Pediatric research*, 59, 21R-25R.

Wu, Y., Chen, L., Scott, P.G., and Tredget, E.E. (2007). Mesenchymal stem cells enhance wound healing through differentiation and angiogenesis. *Stem Cells*, 25, 2648-2659.

Xue, W., Zender, L., Miething, C., Dickins, R.A., Hernando, E., Krizhanovsky, V., Cordon-Cardo, C., and Lowe, S.W. (2007). Senescence and tumour clearance is triggered by p53 restoration in murine liver carcinomas. *Nature*, 445, 656-660.

Yu, J.M., Wu, X., Gimble, J.M., Guan, X., Freitas, M.A., and Bunnell, B.A. (2011). Age-related changes in mesenchymal stem cells derived from rhesus macaque bone marrow. *Aging Cell*, 10, 66-79.

In: Cell Aging
Editors: Jack W. Perloft and Alexander H. Wong

ISBN: 978-1-61324-369-5
2012 Nova Science Publishers, Inc.

Chapter 6

ERYTHROCYTE SENESCENCE

Angel Rucci, María Alejandra Ensinck, Liliana Racca, Silvia García Borrás, Claudia Biondi, Carlos Cotorruelo and Amelia Racca[*]

Laboratorio de Inmunohematología e Inmunogenética. Facultad de Ciencias Bioquímicas y Farmacéuticas. Universidad Nacional de Rosario. CONICET, Argentina

ABSTRACT

Mature red blood cells (RBCs) lack protein synthesis and are unable to restore inactivated enzymes or damaged cytoskeleton and membrane proteins. An oxidation breakdown of band 3 is probably part of the mechanism leading to the generation of a senescent cell antigen (SCA). It serves as a specific signal for the clearance of these cells by inducing the binding of autologous IgG and phagocytosis.

Whole blood samples from volunteer donors were processed. Senescent (Se) RBCs and Young (Y) RBCs were obtained by self-formed Percoll gradients. The separation of both populations was demonstrated by statistically significant changes in hematological parameters and creatine concentration. The antioxidant response in RBCs of different ages was studied. Activities of glucose-6-phosphate dehydrogenase (G6PD), soluble NADH-cytochrome b5 reductase (b5Rs) and membrane-bound b5R (b5Rm) were determined spectrophotometrically. The G6PD and b5Rm activities in SeRBCs were significantly lower than that observed in YRBCs. The decline in those activities would indicate a decrease in the antioxidant response associated to RBC aging.

Membrane proteins modifications in RBCs of different ages were assessed. Membrane proteins were analyzed by SDS-PAGE, band 3 by immunoblotting, and

---

[*] Correspondence:
Dra. Amelia Racca.
Laboratorio de Inmunohematología e Inmunogenética.
Facultad de Ciencias Bioquímicas y Farmacéuticas.
Universidad Nacional de Rosario.
Suipacha 531. 2000 Rosario. Argentina.
Tel: 54 341 4804592   Fax: 54 341 4804598
Email: aracca@fbioyf.unr.edu.ar

protein oxidation by measuring the carbonyl groups. Densitometric analysis showed no differences between mean percentage values obtained from the major bands of SeRBCs and YRBCs membrane proteins. On the contrary an increase in band 3 and its degradation products were found by immunoblotting in SeRBCs. A higher protein oxidation level was also encountered in this population. These results provide experimental evidence about protein modifications occurring during the RBC lifespan.

Then, considering that the accumulation of autologous IgG on RBCs membrane provides a direct mechanism for the removal of SeRBCs, the IgG content of intact RBCs was measured using an enzyme linked anti-immunoglobulin test. In addition, the presence of bound IgG was observed by confocal microscopy. It was shown that the amount of IgG bound to SeRBCs was significantly higher than that observed for YRBCs.

The interaction between different RBCs populations (SeRBCs, YRBCs and desialiniysed RBCs) and peripheral blood monocytes was further investigated through a functional assay. The increase observed in the percentage of erythrophagocytosis with SeRBCs confirmed the involvement of autologous IgG in the selective removal of erythrocytes. Also, the percentage of monocytes with phagocytosed desialiniysed RBCs was higher than that obtained with YRBCs. This finding suggests that a decrease in sialic acid content of RBCs may be involved in the physiological erythrophagocytosis.

In addition, cells of different ages in whole blood were characterized using light scatter, binding of autologous IgG and externalization of phosphatidylserine measurements. Dot-plot analysis based on the forward scatter versus side scatter parameters showed two RBC populations of different sizes and density. RBCs were further incubated with FITC-conjugated mouse anti-human IgG o PE-annexin-V. Binding of IgG to RBCs was analyzed by mean fluorescence intensity. The percentage of IgG positive cells was significantly higher in SeRBCs. The fraction of annexin-V positive RBCs was also larger in SeRBCs. These results indicate that flow cytometry permits differentiating erythrocyte populations of different ages. This methodology could be an alternative tool to study erythrocyte aging.

Taken together, these findings will contribute to a better understanding of the process and mechanisms involved in the erythrocyte senescence.

INTRODUCTION

The physicochemical changes associated with red blood cells (RBCs) aging have been of interest for several decades, since understanding senescence of RBCs should provide insight into cellular aging and the oldest cell clearance. In the particular case of RBCs, the choice of the term senescent was made because, by most criteria, senescent (Se) RBCs do show the greatest number of time-dependent changes. In other words, SeRBCs represent the erythrocytes population which has been condemned to death [1, 2].

Aging, a multifactorial process of enormous complexity is characterized by impairment of physicochemical and biological aspects of cellular functions. After a life span of 120 days in the bloodstream, human RBCs expose removal markers that account for their selective recognition by macrophages and clearance from circulation [3, 4].

One of earliest erythrocyte parameters that appeared to correlate with age was an increase in cell density, resulting from the shedding of vesicles with minimal loss of hemoglobin (Hb). In this sense, density gradient separation methods have allowed the collection of the densest human RBCs that would be destined for the immediate clearance [1, 3].

A mechanism for senescence that appears to be currently arousing substantial interest is that of oxidative damage to erythrocytes. Mammalian RBCs are particularly susceptible to oxidative damage because (i) being an oxygen carrier, they are exposed uninterruptedly to high oxygen tension; (ii) they lack protein synthesis and are unable to restore inactivated enzymes or damaged cytoskeleton and membrane proteins; and (iii) the Hb is susceptible to autoxidation and their membrane components to lipid peroxidation. To cope with these inabilities, RBCs are equipped with high activity of the cellular antioxidant defense systems fully adequated to sustain even excessive oxidative stress for limited time periods [5, 6, 7]. While the systemic oxidative status was identified as an important contributor to age-related processes, the modification of RBCs antioxidant defense during aging is still controversial.

Because mature RBCs are powerless to synthesise new proteins, the events that trigger removal of SeRBCs from circulation must derive from alterations in pre-existing proteins that lead to recognisable changes at the membrane surface. Reactive oxygen species (ROS) modify amino acid side chains (mostly lysine, arginine, proline and histidine) resulting in the generation of free carbonyls that are not present on nonoxidized protein. Protein carbonyl content is actually the major general indicator and by far the most commonly used marker for protein oxidation [8]. The accumulation of protein carbonyls has been observed in aging and several human diseases including Alzheimer's disease, diabetes, inflammatory bowel disease, and arthritis [9, 10].

Molecular recognition of SeRBCs would involve oxidation of membrane biomolecules leading to generation of a neoantigen and binding of physiologic natural antibodies (NAbs), which increases their avidity [11, 12, 13]. These autoantibodies induce RBCs clearance not just by binding, but also by binding and recruiting C3b deposition that trigger macrophage removal of the cell at a time when the membrane is still grossly intact [3, 14]. An oxidation breakdown of band 3 in its membrane-spanning region is probably part of the mechanism leading to the generation of a senescent cell antigen (SCA) *in vivo*. It serves as a specific signal for the clearance of these cells by inducing the binding of NAbs and phagocytosis [2, 15, 16].

Several investigations have suggested that the senescent marker derives from desialyzation of sialoglycoconjugates. Sialic acid is lost from high density RBCs reducing the total surface negative charge. This observation gave rise to the hypothesis that a loss of sialic acid from membrane glycoproteins would reduce the repulsive interaction between erythrocytes and phagocytes, thus permitting ingestion of the altered RBCs [17, 18, 19]. Therefore, the desialyzation could be correlated with the physiological mechanisms of erythrocyte aging and could be another responsible signal for the removal of the oldest RBCs from circulation by an alternative pathway, which is immunoglobulin-independent.

Recently RBC senescence has been associated with the apoptosis. Apoptosis is a well known mechanism for the clearance of the nucleated cell but its role in the removal of erythrocytes from the circulation is not well understood. Phosphatidylserine (PS) externalization is an early event associated with apoptosis. The role of the PS exposure during erythrocyte aging is still a subject of discussion. Studies in humans, rabbits and mice showed a positive correlation between the PS externalization and SeRBCs suggesting that aged erythrocytes could be eliminated from blood through a process resembling apoptosis. [20, 21, 22]. However, some authors observed that in the fraction containing the largest proportion of old cell there was no increase in the number of erythrocytes exposing PS compared with other fractions [23].

HUMAN RBCs SEPARATION

The analysis of the events that take place during erythrocyte aging is still very much hindered by the fact that it is very difficult to obtain homogeneous fractions that contain RBCs of the same age and the probability of introducing artifacts when manipulating erythrocytes *in vitro*. Density separation is the technique that has been used by the vast majority of authors [3, 5, 19, 24].

We have performed preliminary experiments based on well-established parameters in order to validate RBCs separation by preformed Percoll gradients [25]. Briefly, blood samples were centrifuged to remove the buffy coat, RBCs were washed in phosphate buffer saline (PBS) and concentrated to a hematocrit of 80%. The self-formed gradients were loaded with RBCs suspensions and centrifuged a low speed to separate them in lightest and densest populations that represent Young (Y) RBCs (Y) RBCs) and SeRBCs, respectively [26].

Characterization of the different erythrocyte suspensions obtained was performed using a SYSMEX SF3000 counter to determine the Mean Corpuscular Volume (MCV). Reticulocytes (R) percentages were calculated following the staining method with brilliant cresil blue. Erythrocyte creatine was assayed by a spectrophotometric method [27]. Statistical comparisons were performed by means of the t-test for paired samples. The results obtained showed that SeRBCs were significantly smaller than YRBCs, as determined by MCV (SeRBC: 85.80 ± 2.91 *vs* YRBC: 91.40 ± 3.36 fL, $p < 0.001$). These findings can be attributed to the fact that at the whole-cell level, SeRBCs have lost membrane in form of microvesicles, have extruded ions and cell water, and are thus smaller, denser and less deformable [3, 12]. In addition, vesicle formation has been described as an integral part of the erythrocyte aging process and could be responsible for the loss of cell volume [7]. On the other hand, the percentage of R was significantly higher in the YRBCs fractions (SeRBCs: 0.50 ± 0.18 *vs* YRBCs: 2.19 ± 0.26 %, $p < 0.001$), an expectable result since they represent the RBCs of minor age [28]. Moreover, the values of the erythrocyte creatine were statistically higher in the YRBCs suspensions (SeRBCs: 2.76 ± 0.76 *vs* YRBCs: 4.15 ± 1.12 µg/mg Hb, $p < 0.0001$). These results are due to the fact that during aging there is a decrease in phosphorylated intermediates of energy metabolism [29, 30].

These findings confirm that the separation in preformed Percoll gradients allows obtaining RBCs populations of different ages efficiently. In addition, the separating conditions of this method caused minimal cell disturbance and avoided aggregation, thus leading to a reliable correlation between cell age and density.

ANTIOXIDANT RESPONSE

Erythrocytes are a powerful antioxidant protection system. However, at a high concentration of ROS or insufficiency of the primary antioxidant protection, membrane damage may occur [6, 7, 31]. Glutathione (GSH) plays an important role in the detoxification of ROS with the production of GSH disulfide (GSSG) that is removed via the activity of glutathione reductase, which regenerates GSH in a reaction that is absolutely dependent upon NADPH. The enzyme glucose-6-phosphate deshidrogenase (G6PD) (EC 1.1.1.49) catalyzes the first reaction of pentose phosphate pathway in which glucose-6-phosphate (G6P) is oxidized and NADP is reduced resulting in NADPH production. Direct demonstrations that G6PD can modulate GSH levels were obtained in other studies [32, 33, 34]. The age-dependent behavior of G6PD activity is an area of debate. For example, meanwhile some authors [2, 35] have reported the decline of G6PD activity in SeRBCs; Beutler [36] suggested that no significant activity differences were present in YRBCs and old RBCs.

RBCs under oxidative stress present a considerable rise in the level of methemoglobin (MHb). It is not clear whether the ultimate accumulation of oxidized Hb is due to limited capacity for its proteolytic removal or an increase in the rate of its formation [31, 37, 38]. The principal enzymatic system for the reduction of MHb in RBCs is composed by a NADH-cytochrome b5 reductase (b5R) (EC 1.6.2.2) and an intermediate electron carrier, cytochrome b5.

Given this background, we investigated the antioxidant response during RBCs aging by studying the G6PD and the b5R activities. G6PD activity was evaluated spectrophotometrically [36]. This quantitative method determines enzyme activity by measuring the rate of absorbance change at 340 nm due to reduction of NADP to NADPH after addition of G6P and the activity is expressed as International Units per gram of Hb (IU/g Hb). The enzyme G6PD catalyzes the first reaction in the pathway leading to the production of pentose phosphates and reducing power in the form of NADPH for reductive biosynthesis and maintenance of the redox state of the cell. NADPH, utilized in cellular biosynthetic reductive reactions, is found to be the most significant pathway in mature erythrocytes for the production of GSH that regulates the oxidative stress within RBCs. We observed that the G6PD activity in SeRBCs was significantly lower than that observed in YRBCs (SeRBCs: 5.01±1.99 *vs* YRBCs: 7.78±1.45 IU/g Hb; p<0.005).

The separation between the soluble b5R fraction (b5Rs) and membrane bound (b5Rm) was done according to Choury [39]. Both fractions activities were determined by following the oxidation of NADH to NAD at 340 nm using potassium ferricyanide as electron acceptors [40]. b5Rs activity was expressed as Units per gram of Hb (U/g Hb) and b5Rm activity as Units per mg of protein (U/mg protein). Proteins were estimated by a modification of the Lowry procedure to simplify protein determination in membrane and lipoprotein samples [41]. In erythrocytes under oxidative stress, there is a considerable rise in the level of MHb. For converting MHb into oxyhemoglobin, there are two enzymatic systems in RBCs, one of which is related to glycolysis and the other is associated with the pentose phosphate pathway. Correspondingly, two types of enzymes function in erythrocytes: b5R and NADPH-MHb reductase, both having cytoplasmic and membrane-bound forms. It is generally assumed that in the reduction of oxidized Hb, the participation of b5R is of major physiological significance [6, 20, 39]. We observed that the activity of b5Rm in SeRBCs was significantly lower than that found in YRBCs (SeRBCs: 450.25±32.39 *vs* YRBCs: 488.02±42.39 U/mg protein; p<0.001). On the other hand, no differences were found in b5Rs of both erythrocytes populations (SeRBCs: 20.96±1.85 vs YRBCs: 21.09±1.86 U/g Hb; p>0.05).

These results provide experimental evidence that the increase in MHb with aging would be correlated with the decrease in the b5Rm activity. Our findings indicate that the antioxidant capacity during RBCs aging is decreased due to a lower production of NADPH by G6PD. We also observed a decrease in the activity of the b5Rm which would indicate a loss in the antioxidant process during the cell life span, impairing the regeneration of Hb.

Taken together, the results of this study point out a loss in the antioxidant capacity. The decreased enzymatic activity could lead to oxidative membrane changes during the life span of the RBCs. These modifications might be relevant in the process of recognition and removal of SeRBCs from the circulation. Finally, an unequivocal description of the physiological erythrocyte aging pathway will be invaluable in understanding the fate of RBCs in pathological circumstances and the survival of donor RBCs after blood transfusion.

MODIFICATIONS OF BAND 3 AND OXIDATION LEVEL OF MEMBRANE PROTEINS

During RBC aging, changes in proteins occur that alter their function and render them immunogenic. These "neoantigens" called SCA are recognized by physiologic autoantibodies initiating the removal of SeRBCs by macrophages [4, 13, 20]. However, the molecular identity of the SCA and thus the nature of these changes, have not been unambiguously identified.

Modifications of band 3 protein, the major erythrocyte membrane protein, are probably part of the mechanism leading to the generation of SCA during RBCs aging. The discussion focuses on aggregation versus breakdown of band 3 as the ultimate steps in SCA formation. In both scenarios there is a central role for oxidation as a causative event, being through hemichrome generation in the oligomerization hypothesis or through inducing an increased sensitivity of band 3 to proteolysis in the breakdown hypothesis [11, 14, 42].

We evaluated the modifications of band 3 and measured the oxidation level of membrane proteins during the RBC aging process. SeRBCs and YRBCs membrane proteins were prepared from washed cells by lysis with 5 mM sodium phosphate buffer (pH 8.0) supplemented with 1 mM phenylmethylsulfonyl fluoride. Pelleted membranes were washed several times to obtain white ghosts. Protein concentrations were determined by the method developed by Markwell [41]. Ghost membranes were analysed by SDS-polyacrylamide gel electrophoresis (SDS-PAGE) under reducing conditions. Equal amounts (30 µg) of protein were loaded per track of each gel. One gel was stained with Coomassie blue. Densitometric analyses were performed using the GelPro Analyser programme. Each band was expressed as a percentage of the total of main bands upon densitometric scanning. The other gel was transferred to a nitrocellulose membrane. Immunoblot assays were performed using a monoclonal anti-human band 3 antibody.

As erythrocytes do not exhibit protein synthesis, alterations in pre-existing proteins could result in recognizable changes at the membrane surface [1, 2, 22]. In this work, the densitometric analysis of the SDS-PAGE stained with Coomassie blue showed no differences (p>0.05) between the mean percentage values of the major bands between SeRBCs and YRBCs membrane proteins (Table 1). These findings could be explained by the fact that this approach can only identify dramatic changes in the pattern of protein spots [4, 7, 43].

Erythrocyte aging is associated with band 3 modifications although there is no consensus on whether the crucial changes leading to the SCA formation are caused by degradation or cross-linking of band 3. Also, it has been described that in some pathological conditions, increased erythrocyte removal may be associated with altered band 3 [44, 45, 46]. Band 3 immunoblotting and degradation products are shown in the Figure 1.

Table 1. Mean percentages of the major membrane protein bands with the corresponding standard deviations

	spectrins	band 3	band 4.1	band 4.2	band 5	band 7
SeRBCs	26.58	23.40	5.13	5.38	7.18	6.44
	±5.89	±4.34	±1.31	±1.53	±3.82	±4.76
YRBCs	26.10	22.20	5.08	5.07	5.75	7.28
	±3.62	±6.22	±1.75	±1.83	±3.58	±5.19

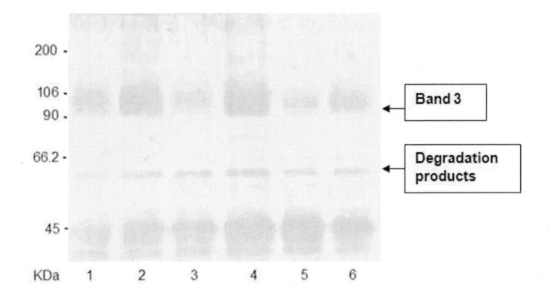

Figure 1. Immunoblotting analysis of band 3 protein in three different RBCs preparations: Lanes 1, 3 and 5: YRBCs. Lanes 2, 4 and 6: SeRBCs.

The antibody used in this analysis recognize an epitope located on the cytoplasmic side of the human RBC membrane and carried by band 3 protein and several lower molecular-mass-peptides migrating in SDS-PAGE. Densitometric analyses showed differences in the band 3 protein of 90-100 kDa (SeRBCs: 88.3 ± 15.4 *vs* YRBCs: 58.2 ± 10.2, $p<0.05$) and its degradation products of 55-60 kDa (SeRBCs: 22.1 ± 5.8 *vs* YRBCs: 17.0 ± 5.2, $p<0.05$).

These findings could be explained by retention of band 3 in the SeRBCs membrane. Vesicles formation has been described as an integral part of the erythrocyte aging process, providing an important means whereby protein sorting can occur, allowing the selective retention of certain proteins in the plasma membrane and the removal of others. Kriebardis noted a partial depletion of band 3 in vesicles in combination with a total absence of 4.2 protein that is required for normal band 3-cytoskeleton [7]. Also, Knowles observed a lower concentration of band 3 and glycophorin A in RBC-derived vesicles using mechanically induced membrane deformation [47]. The high concentration of band 3 in the erythrocyte membrane found in our work using *in vivo* aged RBCs would correlate with the decrease observed by these authors in vesicles obtained *in vitro*.

The carbonyl groups production on amino acid side chains is a common phenomenon of protein oxidation [2, 48]. Considering that the protein carbonyl groups have been extensively used as biomarkers for protein oxidative stress and aging, we measured protein carbonyls using an ELISA method [49]. We observed that the carbonyl content was significantly higher in SeRBCs populations (SeRBCs: 0.704 ± 0.041 *vs* YRBCs: 0.602 ± 0.087 nanomole per milligram protein, $p<0.005$). It has been postulated that the formation of the SCA is accompanied by oxidative damage of membrane proteins leading to an increase in protein carbonyls as well as to the appearance of new electrophoretic bands, many of which likely originate from the fragmentation of protein band 3 [4,11].

Several authors have proposed mechanisms for recognition and phagocytic removal of senescent or damaged erythrocytes based on oxidative clustering of band 3 as the starting event with subsequent opsonization and phagocytic removal by circulating or resident phagocytes [1, 13, 26]. They suggest that the progressive accumulation of Hb oxidation products or hemichromes in the cytoplasmic domain of band 3 causes it to cluster in the plane of the membrane. We observed an increased proportion of band 3 and its breakdown products and a higher oxidation level of membrane proteins in aged RBCs under physiological conditions. These findings provide new experimental evidence about the modifications that occur during the RBC lifespan that eventually transform a viable RBC into a cell destined to phagocytic removal.

MEASUREMENT OF RBC-BOUND IgG

After a life span of 120 days, SeRBCs expose removal markers that account for their selective recognition by macrophages and clearance from circulation. Although the identity of those mechanisms is still a subject of discussion, there are some evidences that the RBC aging leads to the binding of autologous IgG followed by recognition and phagocytosis. These natural autoantibodies (Nabs) are involved in the clearance of RBCs at the end of their lifespan as well as in the removal of RBCs in different hereditary hemolytic disorders and in malaria [2, 12, 14, 15]. In all cited situations, RBCs undergo oxidative stress and hemichromes are formed. Hemichromes possess a strong affinity for band 3 cytoplasmic domain and, following their binding, lead to band 3 oxidation and clusterisation. Those band 3 clusters show increased affinity for NAbs which activate complement and finally trigger the phagocytosis of altered RBCs [42, 50].

Therefore, we investigated the RBC-bound IgG in SeRBCs and YRBCs suspensions using enzyme-linked antiglobulin test. This method is based on the consumption of anti-human IgG by IgG-containing cells. RBCs were incubated with goat anti-human IgG alkaline phosphatase conjugate in a 96-wells microtiter plate. Aliquots were then transferred to new wells coated with human IgG. After washing to remove cell-bound antibody, phosphatase substrate was added and the remaining amount of antibody was measured by spectrophotometric method [51]. A standard curve was prepared using normal IgG instead of RBCs. The results of RBC-bound IgG shown that SeRBCs contain larger amount of IgG (SeRBCs: $13.30\pm1.57 \times 10^{-4}$ *vs* YRBCs: $3.35\pm1.40 \times 10^{-4}$ $\mu g/\mu L$, $p<0.0001$).

In addition, we investigated the RBC-bound IgG by confocal microscopy. Different erythrocyte suspensions: SeRBCs, YRBCs, and sensitized RBCs with IgG anti-D (SRBCs) were incubated with Alexa 488 anti-human IgG. Images were collected using a confocal microscope (Nikon C1) and processed using EZ-C1.gold version 3.70. We observed that IgG is readily detectable on SeRBCs and SRBCs but not on YRBCs. (Figure 2).

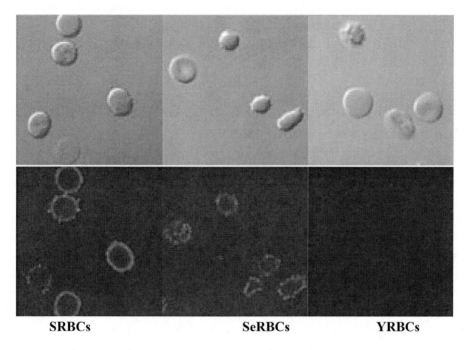

SRBCs   **SeRBCs**   **YRBCs**

Figure 2. Images of IgG bound to RBC membrane obtained by confocal microscopy.

A current hypothesis states that an immunological mediated pathway is one of the mechanisms involved in the clearance of SeRBCs. [1, 2, 12]. The denatured Hb can tightly bind to the band 3 resulting in a topographic redistribution. These clustered sites are rapidly recognised by the immune system and opsonised by NAbs and the C3 fraction of complement [13, 14, 15]. We measured the IgG content of intact RBCs using an enzyme linked anti-immunoglobulin test finding that the amount of IgG bound to SeRBCs was significantly higher than that observed for YRBCs. This method designed in our laboratory is reproducible and robust [52]. On the other hand, the images obtained with a confocal microscope showed that the IgG is detectable on SeRBCs. Taken together, these findings confirm that one of the mechanisms involved in the clearance of SeRBCs is mediated by the binding of NAbs to the erythrocyte surface. Upon deposition of sufficient antibodies and complement, the SeRBCs are removed from circulation by cells of the mononuclear phagocytic system.

INTERACTION BETWEEN PERIPHERAL BLOOD MONOCYTES AND DIFFERENT RBCS

During aging, RBCs expose the SCA that is recognized by NAbs initiating the removal of SeRBCs by circulating or resident phagocytes. Several authors have proposed that these mechanisms are based on oxidative clustering of band 3 as the starting event with subsequent opsonization and phagocytic removal by cells of the mononuclear phagocytic system [15, 46, 53].

In addition, during microcirculation, aging and storage, RBCs lose sialic acid and biomaterials suffering damages that alter cell structures, some properties and functions. Such cell damages very likely underlie the serious adverse effects of blood transfusion that are not clear yet [7, 18, 42]. Some authors have suggested that aged RBCs decrease the electrostatic charge density, which is a barrier to phagocytosis by monocytes [3, 54, 55]. However, the controversy whether the loss of sialic acid causes a decrease in the charge density and its possible correlation with the selective removal of SeRBCs remains.

Therefore, we investigated through a functional assay the interaction between SeRBCs and monocytes using erythrophagocytosis assay. Peripheral blood monocytes were obtained through their glass adhering property and incubated with different erythrocyte suspensions. The cells were fixed with methanol, stained by the May Grünwald Giemsa method and observed under the light microscope. The percentage of active monocytes (AM) with phagocytosed and adherent RBCs was determined. Non sensitized RBCs (NRBCs) were used as negative controls. Positive controls were SRBCs [57]. On the other hand, we also applied this assay to study the phagocytosis of desialysed RBCs (with neuraminidase from *Clostridium perfringens*) [17]. Differences in the percentage of AM among SeRBCs, YRBCs, desialysated RBCs, SRBCs and NRBCs were assessed by one way analysis of variance (ANOVA) and multiple comparisons. It was found that the interaction of SeRBCs with peripheral blood monocytes reflected an increase in the percentage of AM when compared to YRBCs (SeRBCs: 17.94 ± 2.24 *vs* YRBCs: 3.33 ± 0.77, $p < 0.001$). The values of AM with neuraminidase treated RBCs (10.72 ± 1.41) were lower than those obtained with SeRBCs and SRBCs but higher than those observed with YRBCs and NSRBCs ($p < 0.001$).

The involvement of autologous IgG in phagocytosis of SeRBCs was first reported by Kay [15]. Autologus IgG increased *in vitro* phagocytosis of SeRBCs, but not of YRBCs. Human IgG contains several types of NAbs that react with RBC membrane proteins. In this study, we have examined the mechanisms of SeRBCs removal by the erythrophagocytosis assay. This functional cellular assay developed by us is performed in a relatively simple way by incubating RBCs with peripheral blood monocytes and assessing different stages of the interaction such as phagocytosis and adherence [56]. Although still quite time-consuming, we think that this method is reproducible and robust. The increased rate of erythrophagocytosis of SeRBCs and SRBCs showed the involvement of autologous IgG in the selective removal of erythrocytes. Similar IgG-dependent recognition mechanism was noted in *in vitro* oxidised RBCs, enzyme-deficient RBCs, malaria-infected RBCs and in RBCs with hereditary Hb abnormalities and with hereditary spherocytosis [57, 58, 59].

Some authors have also proposed that a decrease of negative charge on the erythrocyte membrane would significantly reduce the repulsive interaction between RBCs and cells of the mononuclear phagocytic system and thus facilitate phagocytosis [2, 5, 19]. We showed that the desialysation by neuraminidase to obtain *in vitro* models of aged RBCs increased their chance of being phagocytised. The reason for this is that neuraminidase-treated RBCs have indeed a lower negative surface zeta potential than normal RBC and can approach phagocytic cells more closely. The loss of sialic acid residues from RBCs occurring *in vivo* could be due to either a low pH or, alternatively, to specific enzymatic cleavage [18, 55, 56]. Our observations give rise to the hypothesis that the decrease in the sialic acid content could be correlated with the physiological mechanism of erythrocyte aging and is one of the signals responsible for the removal of the oldest RBCs from circulation. Taken together, the results of this study showed that there are at least two erythrophagocytosis pathway for the removal of SeRBCs, one that is immunoglobulin-dependent and another involving the loss of sialic acid.

FLOW CYTOMETRIC ANALYSIS OF RBCs POPULATIONS

Erythrocyte senescence is associated with cell shrinkage, plasma membrane vesiculization, progressive shape change from discocytes to spherocytes, cytoskeletal and membrane protein alteration, and loss of plasma membrane phospholipids asymmetry leading to the externalization of PS that may represent one of the signals allowing macrophages to ingest the SeRBCs [20, 42, 60].

The major problem in studying properties of SeRBC is the lack of a reliable method to isolate RBCs populations. Density gradient separation methods have allowed the collection of the densest human RBCs that would be destined for immediate clearance [3, 5, 19, 24]. However, the separation methods used are laborious and require a considerable amount of sample.

Therefore, we characterized cells of different ages: SeRBCs and YRBCs using light scatter, binding of autologous IgG and externalization of PS measurements. Blood samples were washed in PBS, concentrated to 0.2% and labeled with FITC-conjugated mouse anti-human IgG o PE-annexin-V. Flow cytometry analysis was performed on a FACSAria II (Becton Dickinson) and analyzed using FACSDiva.

Dot-plot analysis based on the forward scatter (FSC) versus side scatter (SSC) parameters showed two RBCs populations of different size and density (Figure 3). Events that correlated with the binding of IgG to RBCs were analyzed based on the mean fluorescence intensity (MIF) (SeRBCs: 748±108 *vs* YRBCs: 47±15, p<0.001). The percentage of IgG positive cells was higher in SeRBCs (SeRBCs: 1.5 *vs* YRBCs: 0.16, p<0.01). The MIF for bound IgG and the percentage of IgG positive cells were significantly different in the two regions: low FSC and high SSC (SeRBCs) over the area of highest FSC and lowest SSC (YRBCs). The fraction of annexin-V positive RBCs was higher in SeRBCs (SeRBCs: 0.98±0.12% *vs* YRBCs: 0.15±0.05%, p<0.01) [61].

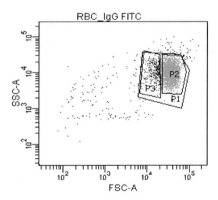

Figure 3. Flow cytometric analysis of human RBCs populations. P1 corresponds to total RBCs, P2 represents YRBCs and P3 corresponds to SeRBCs.

During *in vivo* and *in vitro* aging, as well as in a variety of pathological conditions, RBCs display molecules that lead to recognition and removal of SeRBCs and damaged cells by the immune system. The presently available data from numerous investigations have led to the identification of the SCA on band 3 and an increase of PS on the outer membrane of RBCs as the main removal signals [2, 23, 60, 62]. Subsequently, phagocytosis is triggered by NAbs binding to a SCA and/or PS binding to scavenger receptors on macrophages.

Our findings indicate that flow cytometry permit differentiating RBCs populations of different sizes and density. The aging markers studied allowed characterize cells of different ages: SeRBCs and YRBCs. This approach is capable of measuring the binding of autologous IgG and PS externalization in less than 0.1% of the total RBCs. Therefore, this methodology could be an alternative tool to study erythrocyte aging using small sample quantities.

CONCLUSION

In the absence of a classical apoptotic pathway, RBCs must possess a specific mechanism which triggers their recognition by macrophages at a defined stage of their life. It is clear that the molecular and cellular mechanisms of the senescence process are not yet solved and that this enigma represents a fascinating and exciting challenge.

Considering the results obtained, we may conclude:

- Preformed Percoll gradients allowed obtaining homogeneous RBCs populations of different ages. The conditions of RBCs separation by this method caused minimal cell disturbance and avoided aggregation, thus leading to a reliable correlation between cell age and density.
- High-density erythrocytes showed a decrease in the G6PD and b5Rm activities. The loss of the antioxidant capacity during RBCs aging could lead to oxidative membrane changes that might be relevant in the generation of the SCA involved in clearance of SeRBCs.
- The increase in the concentration of band 3 protein and its degradation products indicate protein modifications in SeRBCs. A higher protein oxidation level in this population. highlights the involvement of oxidative changes of erythrocyte membrane proteins during the RBC lifespan.
- A higher IgG content in SeRBCs corroborate that one of the mechanisms involved in the clearance of SeRBC is mediated by the binding of Nabs to the RBC surface.
- The increase in phagocytosis observed with SeRBCs and RBCs desialiniysed populations clearly show that there are at least two phagocytosis pathways for the removal of old RBCs, one that is immunoglobulin-dependent and another involving the loss of sialic acid.
- Flow cytometry permitted differentiating RBCs populations of different ages. Considering that this technique has a high sensitivity and requires small amounts of sample without using separation methods, it could be a useful tool for studying the mechanisms involved in erythrocyte senescence.

The identification of the process involved in RBCs aging constitutes a major challenge in current medicine research and transfusion because it will provide physiologically relevant insights for assessing and predicting RBCs homeostasis *in vitro* and *in vivo*, and thereby contribute to the development of rational transfusion protocols for various patient categories.

REFERENCES

[1] Bratosin, D; Mazurier, J; Tissier, J; Estaquier, J; Huart, J; Ameisen, J; Aminoff, D; Montreuil, J. Cellular and molecular mechanisms of senescent erythrocyte phagocytosis by macrophages. A review. *Biochimie,* 1998, 80:173–195.

[2] Lutz, H. Innate immune and non-immune mediators of erythrocyte clearance. *Cell Mol Biol,* 2004, 50:107-116.

[3] Piomelli, S; Seaman, C. Mechanism of red blood cell aging: Relationship of cell density and cell age. *Am J Hematol,* 1993, 42:46-52.

[4] Miki, Y; Tazawa, T; Hirano, K; Matsushima, H; Kumamoto, S; Hamasaki, N; Yamaguchi, T; Beppu, M. Clearance of oxidized erythrocytes by macrophages: involvement of caspases in the generation of clearance signal at band 3 glycoprotein. *Biochem Biophys Res Commun,* 2007, 363:57-62.

[5] Bull B. Red cell senescence. *Blood Cells,* 1994, 4:1-3.

[6] Kiefer C, Snyder L. Oxidation and erythrocyte senescence. *Curr Opin Hematol,* 2000, 7:113–116.

[7] Kriebardis A, Antonelou M, Stamoulis K, Wcibiniy-Petersen E, Margaritis L, Papassideri I. RBC-derived vesicles during storage: ultrastructure, protein composition, oxidation and signaling components. *Transfusion,* 2008, 48:1943-1953.

[8] Levine, C. Carbonyl modified proteins in cellular regulation, aging, and disease. *Free Radical Biology & Medicine* 2002, 32:790-796.

[9] Buss, H; Chan, C; Sluis, T; Domigan, N; Winterbourn, C. Protein carbonyl measurement by a sensitive ELISA method. *Free Radical Biology & Medicine,* 1997, 23:361–366.

[10] Dalle-Donne, I; Rossi, R; Giustarini, D; Gagliano, N; Di Simplicio, P; Colombo; R; Milzani; A. Methionine oxidation as a major cause of the functional impairment of oxidized actin. *Free Radical Biology & Medicine,* 2002, 32:927–937.

[11] Arese, P; Turrini, C; Schwarzer, E. Band 3/complement mediated recognition and removal of normally senescent and pathological human erythrocytes. *Cell Physiol Bioch,* 2005, 16:133-146.

[12] Berkman, P; Vardinon, N; Yust, I. Antibody dependent cell mediated cytotoxicity and phagocytosis of senescent erythrocytes by autologous peripheral blood mononuclear cells. *Autoinmunity,* 2002, 35:415-419.

[13] Kay, M. Immunoregulation of cellular life span. *Annals N Y Acad Sci,* 2005, 1057:85-111.

[14] Paleari, R; Ceriotti, F; Azzario, F; Galanello, R; Mosca, A. Experiences in the measurement of RBC-bound IgG as markers of cell age. *Bioelectrochemistry,* 2003, 62:175-179.

[15] Kay, M. Generation of senescent cell antigen on old cells initials IgG binding to a neoantigen. *Cell Mol Biol*, 1993, 39:131-153.

[16] Biondi, C; Cotorruelo, C; Ensinck, MA; García Borrás, S; Racca, A. Senescent erythrocytes: factors affecting the aging of red blood cell. *Immunol Invest*, 2002, 31:41-50.

[17] Hajoui, O; Martin, S; Alvarez, F. Study of antigenic sites on the asialoglycoprotein receptor recognized by autoantibodies. *Clin Exp Immunol,* 1998, 113:339-345.

[18] Bratosin, D; Mazurier, J; Tissier, J; Estaquier, J; Huart, J; Ameisen, J; Aminoff, D; Krierbardis, AG; Antonelou, MH; Stamoulis, KE; Economou-Petersen, E; Margaritis, LH; Papassideri, IS. Progressive oxidation of cytoskeletal proteins and accumulation of denatured hemoglobin in stored red cells. *J Cell Mol Med*, 2007, 11:148–155.

[19] Clark, M. Senescence of red blood cells: progress and problems. *Physiol reviews,* 1988, 68:503-553.

[20] Bosman, G; Willekens, F; Were, J. Erythrocyte aging: a more than superficial resemblance to apoptosis?, *Cell Physiol Biochem*, 2005, 16:1-8.

[21] Kuypers, F; Jong, K. The role of phosphatidylserine in recognition and removal of erythrocytes. *Cell Mol Biol*, 2004, 50:147-158.

[22] Bratosin, D; Estaquier, J; Petit, F; Arnoult, D; Quatannens, B; Tissier, JP;

[23] Slomianny, C; Sartiaux, C; Alonso, C; Huart, JJ; Montreuil, J; Ameisen, JC. Programmed cell death in mature erythrocytes: a model for investigating death effector pathways operating in the absence of mitochondria. *Cell Death Differ*, 2001, 8:1143–1156.

[24] Willekens, F; Were, J; Groenen-Do¨pp, Y; Roerdinkholder-Stoelwinder, B; Bosman, G. Erythrocyte vesiculation: a self-protective mechanism? *Br J Haematol,* 2008, 141, 549–556.

[25] Gifford, SC; Derganc, J; Shevkoplyas; SS; Yoshida, T; Bitensky, MW. A detailed study of time-dependent changes in human red blood cells: from reticulocyte maturation to erythrocyte senescence. *Br J Haematol,* 2006, 135:395-404.

[26] Cordero, J; Rodríguez, P; Romero, P. Differences in intramembrane particle distribution in young and old human erythrocytes. *Cell Biol Intern*, 2004, 28:423-431.

[27] Luján Brajovich, M; Rucci, A; Acosta, I; Cotorruelo, C; García Borrás, S; Racca, L; Biondi, C; Racca, A. Effects of aging on antioxidant response and phagocytosis in senescent erythrocytes. *Immun Inv*, 2009, 38:551-559.

[28] Li, P; Lee, J; Li, C; Deshpande, G. Improved method for determining erythrocyte creatine by the diacetyl α-naphthol reaction: Elimination of endogenous glutathione interference. *Clin Chem*, 1982, 28:92-96.

[29] Beutler, E; Lichtman, M; Coller, B, Kipps, T. *Williams Hematology.* International edition, Mc Graw Hill. 1995.

[30] Griffiths, WJ; Fitzpatrick, M. The effect of age of the creatine in red cells. *Br J Haematol* 1967, 13:175-180.

[31] Boches, FS; Goldberg, AL. Role for the adenosine triphosphate proteolytic pathway in reticulocyte maturation. *Science,* 1982, 215:1107-1111.

[32] Murakami, K; Mawatari, S. Oxidation of haemoglobin to methemoglobin in intact erythrocyte by a hydroperoxide induces formation of glutathionyl haemoglobin and binding of *a*-hemoglobin to membrane. *Arch Bioch Biophys*, 2003, 417:244–250.

[33] Piccinini, G; Minetti, G; Balduini, C; Brovelli, A. Oxidation state of glutathione and membrane proteins in human red cells of different age. *Mech Ageing Dev*, 1995, 78:15–26.

[34] Salvemini, F; Franze, A; Iervolino, A; Filosa, S; Salzano, S; Ursini, M. Enhance Glutathione levels and oxidoresistance mediated by increased glucose-6-phosphate dehydrogenase expression. *J Biol Chem*, 1999, 274:2750–2757.

[35] Efferth, T; Schwarzl, SM; Smith, J; Osieka, R. Role of glucose-6-phosphate dehydrogenase for oxidative stress and apoptosis. *Cell Death Diff*, 2006, 13:527–528.

[36] Rettig, M; Low, P; Gimm, A; Mohandas, N; Wang, J; Christian, J. Evaluation of biochemical changes during in vivo erythrocytes senescence in the dog. *Blood*, 1999, 1:376–384.

[37] Beutler, E. G6PD deficiency. *Blood*, 1994, 84:3613–3636.

[38] Board, PG. Brief technical note NADH-ferricyanide reductase, a convenient approach to the evaluation of NADH-methaemoglobin reductase in human erythrocytes. *Clin Chim Acta*, 1981, 109:233–237.

[39] Olek, R; Antosiewicz, J; Caulini, GC; Falcioni, G. Effect of NADH on the redox state of human hemoglobin. *Clin Chim Acta*, 2002, 324:129–134.

[40] Choury, D; Leroux, A; Kaplan, JC. Membrane-bound cytochrome b_5 reductase (methemoglobin reductase) in human erythrocytes. *J Clin Invest*, 1981, 67:149-155.

[41] Rockwood, G; Armstrong, K; Bassin, S. Species comparison of Methemoglobin Reductase. *Exp Biol Med*, 2003, 228:79–83.

[42] Markwell, M; Haas, S; Bieber, L; Tolbert, N. A modification of the Lowry procedure to simplify protein determination in membrane and lipoprotein samples. *Anal Bioch*, 1978, 87:206–210.

[43] Bossman, G; Were, J; Willekens, F; Novotny, V. Erythrocyte ageing in vivo and in vitro: structural aspects and implications for transfusion. *Transfusion Med*, 2008, 18:335-347.

[44] Mohandas, N; Gallagher, PG. Red cell membrane: Past, present, and future. *Blood*, 2008, 112:3939–3948.

[45] Binder, C; Horkko, S; Dewan, A; Chan, MK; Kieu, EP, Goodyear, C; Shaw, P; Polinski, W; Witztum, J; Silverman, G. Pneumococcal vaccination decreases atherosclerotic lesion formation: molecular mimicry between *Streptococcus pneumoniae* and oxidized LDL. *Nat Med*, 2003, 9:736–743.

[46] Kay, M; Goodman, J. Mapping of senescent cell antigen on brain anion exchanger (AE) protein isoforms using fast atom bombardment ionization mass spectrometry. *J Mol Recogn*, 2004, 17:33–40.

[47] Kay, M. Band 3 and its alterations in health and disease. *Cell Mol Biol*, 2004, 50:117-138.

[48] Knowles, D; Tilley, L; Mohandas, N; Chasis, J. Erythrocyte membrane vesiculation: Model for the molecular mechanism of protein sorting. *Proc Nat Acad Sci*, 1997, 94:12969-12974.

[49] Stadtman, E; Berlett, B. Reactive oxygen-mediated protein oxidation in aging and disease. *Chem Res Toxicol*, 1997, 10:485-494.

[50] Alamdari, D; Kostidou, E; Paletas, K; Sarigianni, M; Konstas, A; Karapiperdou, A; Koliakos, G. High sensitivity enzyme-linked immunosorbent assay (ELISA) method for

measuring protein carbonyl in samples with low amounts of protein. *Free Rad Biol Med,* 2005, 39:1362-1367.

[51] Pantaleo, A; Giribaldi, G; Mannu, F; Arese, P; Turrini, F. Naturally occurring anti-band 3 antibodies and red blood cell removal under physiological and pathological conditions. *Autoimm Rev*, 2008, 7: 457–462.

[52] Bosman, G; Bartholomeus, I; De Man, A; Van Kalmthout, P; De Grip, W. Erythrocyte membrane characteristics indicate abnormal cellular aging in patients with Alzheimer's disease. *Neurobiol Aging,* 1991, 12:13–18.

[53] Ensinck, MA; Biondi, C; Marini, A; García Borrás, S; Racca, L; Cotorruelo, C; Racca, A. Effect of membrana-bound IgG and desialysation in the interaction of monocytes with senescent erythrocytes. *Clin Exp Med*, 2006; 6: 138-142.

[54] Bruce, L; Beckman, R; Anstee, DJ; Tanner, J. A band 3- based macrocomplex of integral and peripheral proteins in the RBC membrane. *Blood*, 2003, 101:4180–4188.

[55] Mehrishi, JN; Bauer, J. Electrophoresis of cells and the biological relevance of surface charge. *Electrophoresis,* 2002; 23: 1984-1994.

[56] Chen, XY; Huang, YX; Liu, WJ; Yuan, ZJ. Membrane surface charge and morphological and mechanical properties of young and old erythrocytes. *Current Appl Phys*, 2007; 7:94-96.

[57] Biondi, C; Cotorruelo, C; Ensinck, MA; Racca, L; Racca, A. Use of the erythrophagocytosis assay for predicting the clinical consequences of immune blood cell destruction. *Clin Lab*, 2004, 50: 265-270.

[58] Cappellini, M; Gavazzi, D; Turrini, F; Arese, P; Fiorelli, G. Metabolic indicators of oxidative stress correlate with haemichrome attachment to membrane, band 3 aggregation and erythrophagocytosis in β-thalassemia intermedia. *Br J Haematol,* 1999, 104:504–512.

[59] Garibaldi, G; Ulliers, D; Mannu, F; Arese, P; Turrini, F. Growth of Plasmodium falciparum induces stage-dependent haemichrome formation, oxidative aggregation of band 3, membrane deposition of complement and antibodies, and phagocytosis of parasitized erythrocytes. *Br J Haematol*, 2001, 113:492–499.

[60] Dhermy, D; Bournier, O; Bourgeois, M; Grandchamp, B. The red blood cell band 3 variant associated with dominant hereditary spherocytosis causes defective membrane targeting of the molecule and a dominant negative effect. *Mol Membr Biol*, 1999, 16:305–312

[61] Bratosin, D; Tcacenco, L; Sidoroff, M; Cotoraci, C; Slomianny, C; Estaquier, J; Montreuil, J. Active caspases-8 and -3 in circulating human erythrocytes purified on immobilized Annexin-V: A cytometric demonstration. *Cytometry*, 2009, Part A 75A:236-244.

[62] Ensinck, MA; Fiorenza, G; Rucci, A; Biondi, C; Racca, A. Flow cytometric analysis of erytrhocytees subpopulations using light scatter measurements. *Vox Sanguinis,* 2010; 99: 327.

[63] Bratosin, D; Tissier, J-P; Lapillonne, H; Hermine, O; De Villemeur, TB; Cotoraci, C; Montreuil, J; Mignot, C. A cytometric study of the red blood cells in Gaucher disease reveals their abnormal shape that may be involved in increased erythrophagocytosis. *Cytometry,* 2010, Part B 00B:000–000.

INDEX

D

E

H

I

J

K

L

M

N

U

V

W

X

Y

Z